Lecture Notes in Mathematics

Edited by A. Dold and B. Eckmann

661

P. T. Johnstone R. Paré
R. D. Rosebrugh D. Schumacher
R. J. Wood G. C. Wraith

Indexed Categories and Their Applications

Edited by
P. T. Johnstone and R. Paré

Springer-Verlag
Berlin Heidelberg New York 1978

Editors
Peter T. Johnstone
Department of Pure Mathematics
University of Cambridge
16 Mill Lane
Cambridge CB2 1SB/England

Robert Paré
Department of Mathematics
Dalhousie University
Halifax, N.S. B3H 4H8/Canada

AMS Subject Classifications (1970): 18A05, 18C10, 18D20, 18D30

ISBN 3-540-08914-4 Springer-Verlag Berlin Heidelberg New York
ISBN 0-387-08914-4 Springer-Verlag New York Heidelberg Berlin

Printed in Germany

Printing and binding: Beltz Offsetdruck, Hemsbach/Bergstr.
2141/3140-543210

PREFACE

The recent development of the theory of indexed categories has its origins in the programme, first proposed by F.W. Lawvere, of learning how to develop familiar mathematical notions over an arbitrary base topos in order to remove their dependence on classical set theory. This programme has already led to great advances in our understanding of continuously variable structures (sheaves, bundles, etc.) and their relation to the corresponding constant structures. However, in order to develop category theory itself in this way, it was realized that one needed to develop a notion of "family of objects (or morphisms) indexed by a variable set (= an object of the base topos)", and a suitably flexible technique for handling such indexed families.

This realization was first made by Lawvere himself, who mentioned it in his lectures at Dalhousie University in 1970. In 1972-73 he developed, in unpublished notes written at Perugia, a detailed theory of families for complete categories with small homs.

Independently, J. Penon was developing his theory of locally internal categories (= indexed categories with small homs) from the point of view of enriched categories. Then J. Bénabou and J. Céleyrette, knowing of Penon's work but not of Lawvere's, developed their theory of families using fibrations. Their results were presented by Bénabou in a series of lectures at the Seminaire de Mathématiques Supérieures of the Université de Montréal in the summer of 1974.

In 1973-74, R. Paré and D. Schumacher felt the need for a theory of indexed categories, mainly as a tool for establishing the adjoint functor theorems and using them to prove the existence of free algebras in a topos. Although they were aware of Lawvere's 1970 remarks, their own development of the theory was well advanced when they received his Perugia notes and then, a little later, attended Bénabou's lectures in Montréal.

At about the same time, P. Johnstone and G. Wraith first learned of Penon's work, and realized that it provided the natural setting for their own work on internal algebraic theories and recursion in a topos. Later, Johnstone made some further developments of Penon's theory with this application in mind, although these largely duplicated work of others.

In view of the multiplicity of independent developments of the subject, it has become rather difficult to assign credit for the first introduction of a particular idea. However, both authors of this preface feel that particular credit is due to Lawvere for being the first to recognize the need for the theory, and to Bénabou for being the first to emphasize the importance of the Beck condition, which plays a vital part in the description of indexed limits.

In the belief that the method of indexed categories will come to be seen as one of the most important tools of topos theory, we have attempted in this volume to put together a detailed account of the method with some examples of its applications to problems of universal algebra in a topos. The first paper (by Paré and Schumacher) develops the basic theory, taking as its goal the adjoint functor theorems for indexed categories, and introducing on the way the indexed versions of such important concepts as completeness and well-poweredness. Wood's paper presents a further development of the theory, enabling one to replace the cartesian (finite limit) structure on the base category by a general monoidal structure. The paper of Johnstone and Wraith is a detailed study of how standard notions of universal algebra may be lifted to a general base topos; the language of indexed categories is used freely throughout. It is mainly concerned with finitary algebraic theories, though its final chapter explores the infinitary case; this exploration is carried further in Rosebrugh's paper, which obtains a partial solution to the important (and difficult) problem of showing that categories of algebras have colimits.

Montréal, February 1978

Peter Johnstone
Robert Paré

Authors' addresses

Peter T. Johnstone,
Department of Pure Mathematics,
University of Cambridge,
16 Mill Lane,
Cambridge CB2 1SB, England.

Robert Paré,
Department of Mathematics,
Dalhousie University,
Halifax, N.S. B3H 4H8, Canada.

Robert D. Rosebrugh,
Department of Mathematics,
York University,
4700 Keele Street,
Downsview, Ont. M3J 1P3, Canada.

Dietmar Schumacher,
Department of Mathematics,
Acadia University,
Wolfville, N.S. BOP 1XO, Canada.

Richard J. Wood,
Department of Mathematics,
University of Windsor,
Windsor, Ont. N9B 3P4, Canada.

Gavin C. Wraith,
Mathematics Division,
University of Sussex,
Falmer,
Brighton BN1 9QH, England.

CONTENTS

Abstract Families
and the
Adjoint Functor Theorems
by

Robert Paré and Dietmar Schumacher

0. Introduction:

The research presented in this paper originated from the following considerations. In [Sc], Schumacher had shown that in a topos with natural numbers object, absolutely free algebras with finitely many internal arities (internally projective and pointed objects) existed. In [LS], Le Saffre showed that free algebras on an external finitely presented theory existed. The most efficient way to show the existence of such free algebras, in the category of sets, is by using Freyd's general adjoint functor theorem ([Frl], [ML]). In the case of an elementary topos we cannot apply the adjoint functor theorem directly since most of the categories involved are not complete. However, a topos is "internally complete" and so should many categories constructed from this topos. What was needed was a version of the general adjoint functor theorem which exploited this.

In [Mkl], Mikkelsen had shown that a finitely complete cartesian closed category with subobject classifier is finitely cocomplete. This seemed like a good setting in which to use Freyd's special adjoint functor theorem; finite completeness and cartesian closedness is (some form of) internal completeness, the subobject classifier is an internal cogenerator ($\Omega^{(\ )}$ is faithful [Pa]), and the subobject classifier together with cartesian closedness gives "power set" objects and so the category is internally well-powered.

Although the above results had already been obtained by direct (and often economical) means, the methods were *ad hoc*. The adjoint functor theorems would permit straightforward generalization of the methods used for sets. In any case, the general and special adjoint functor theorems have proved to be useful theorems in ordinary category theory and so should be equally useful in category theory over a topos.

What is needed on a category $\underline{A}$ "constructed from" a topos $\underline{E}$ to permit the statement of the general adjoint functor theorem? First of all, $\underline{A}$ must be $\underline{E}$-complete and to define this we probably should know what the functor category $\underline{A}^{\mathbb{C}}$ is, for any category object $\mathbb{C}$ in $\underline{E}$. In particular, if I is an object of $\underline{E}$ we should know the functor category $\underline{A}^I$ (where I is considered as a discrete category object), i.e. we should know the category of I-indexed families of objects of $\underline{A}$. To talk of $\mathbb{C}$-limits we should also have a diagonal functor $\underline{A} \longrightarrow \underline{A}^{\mathbb{C}}$ and in particular a diagonal functor $\underline{A} \longrightarrow \underline{A}^I$ for any object I. In fact, if the objects of $\underline{A}^I$ are to be functors from I to $\underline{A}$ in any sense, we should have even more: for any morphism $\alpha\colon J \longrightarrow I$ in $\underline{E}$ we should have a "substitution" functor $\alpha^*\colon \underline{A}^I \longrightarrow \underline{A}^J$.

It seems reasonable that if we know $\underline{A}^I$ and α^* for all I and α we will be able to define $\underline{A}^{\mathbb{C}}$ and the diagonal functors. Furthermore, there is an accepted notion of I-indexed family of objects of a topos (an object over I) and of the substitution functor α^* (pulling back along α), and this gives us the corresponding concepts for most categories constructed from that topos. We will therefore *take as fundamental concepts the notions of I-indexed families and substitution functors* α^*.

It turns out that once we have given the $\underline{E}$-indexed families (and the substitution functors) as extra structure on $\underline{A}$, not only can we define the functor categories $\underline{A}^{\mathbb{C}}$, $\underline{E}$-limits and $\underline{E}$-completeness of $\underline{A}$, but also what it means for $\underline{A}$ to have $\underline{E}$-small homs, state the solution set condition, and state and prove the general adjoint functor theorem. We can also say what it means for $\underline{A}$ to be $\underline{E}$-well-powered, to have a generating family, and then state and prove the special adjoint functor theorem. We also get a general theory of $\underline{E}$-smallness conditions (small categories, small homs, well-powered categories, etc.). Finally, we see that it is not necessary that the category of indexing objects be a topos, although for most of the applications we have in mind, it is. As Bénabou has pointed out [Bn2], any category with finite limits will do, and developing the theory in this generality gives insight into the more special case.

The adjoint functor theorems are not an end in themselves but rather a convenient point on which to focus in developing the theory. The main point is the theory of families itself and the fact that it is a useful and suggestive language with which to study topoi and related categories. It is in this spirit that the paper has been written (we certainly could have taken a faster and more direct route to the adjoint functor theorems).

We would also like to say a word on the foundations we have chosen. *A priori* there are three possible approaches to giving the association $I \longmapsto \underline{A}^I$ for an indexed category: as a functor into $\underline{\text{Cat}}$, as a pseudo-functor into $\underline{\text{Cat}}$, or as a fibration.

Although functors are easier to give and to work with, the association $I \longmapsto \underline{E}/I$ is not, in the main example of topoi, a functor into $\underline{\text{Cat}}$. In this example the α^* are obtained by pulling back along α, and this is only functorial up to canonical isomorphism. It is possible to make it functorial by changing the $\underline{E}/I$ to equivalent categories but this is artificial and leads to problems later on (e.g. with adjoints).

The association $I \longmapsto \underline{E}/I$ is a pseudo-functor into $\underline{\text{Cat}}$ (or some "large enough" $\underline{\text{Cat}}$) but pseudo-functors involve canonical isomorphisms and coherence conditions which are very bothersome and seem to add nothing to the comprehension of the situation. In practice these isomorphisms are usually ignored and then inserted afterwards.

These isomorphisms can be eliminated by using fibrations. In this approach, the categories $\underline{A}^I$ are all given simultaneously as one category over the indexing category. The α^* are not given as functors but assumed to exist, satisfying a universal property. This avoids the necessity of choosing pullbacks in order to make the α^* into functors. Also, by working within the 2-category of fibrations, many of the constructions on indexed categories can be performed in an elegant way (see [Bn2,3,4]).

We view the theory of families as a useful tool for studying categories over (constructed from) topoi (and more general categories) and so we want this paper to be as accessible as possible to people working in these fields. The theory with fibrations often has a forbidding formal aspect and for many purposes is less intuitive than with pseudo-functors. For example, important constructions such as the functor categories $\underline{A}^{\mathbb{C}}$ and $\underline{A}^{\mathbf{2}}$ and the category of group objects in $\underline{A}$ can best be understood by transforming (if only mentally) the fibration $\underline{A}$ into a pseudo-functor, performing the construction on this, and then transforming back to a fibration. The construction $\underline{A}^{op}$ is difficult and unintuitive in the fibration setting whereas it is perfectly trivial for pseudo-functors.

Although fibrations eliminate the choice of values for the α^* inside the indexed category, a compromise is made for morphisms between fibrations. These morphisms *are* functors. But for discussing completeness, diagonal functors, Kan extensions, and many other concepts, we want the α^* to be morphisms between indexed categories.

One other thing which should be noted is that if we start with a category object $\mathbb{C}$ in the base category, the association $I \longmapsto [I,\mathbb{C}]$ actually is functorial. Although we can consider it as a pseudo-functor or make it into a fibration, something is lost in the processes. Such indexed categories play a central role in our theory and should be considered as being strictly functorial.

We have tried to make our theory conform as closely as possible to actual mathematical practice. To take into account the preceding remarks, we have adopted the following foundations. All categories are considered as being equipped with a subgroupoid of isomorphisms, called canonical (usually consisting of all identities or all isomorphisms). Morphisms between such categories are functors defined up to canonical isomorphism (see I,0). An indexed category is then given as a "pseudo-functor" for which the α^* are functors defined up to canonical isomorphism.

We would like to acknowledge our indebtedness to Bill Lawvere and Jean Bénabou. Lawvere suggested, in lectures at Dalhousie in 1970, that it would be useful to develop a theory of families and it was this that got us started when we needed such a theory in 1973. His Perugia Notes [La] confirmed that we were on the right track and gave us new perspectives on the theory. Bénabou's lectures [Bn2] at the *Université de Montréal* in 1974 covered almost everything we had done at the time and in several respects his theory was at a more advanced stage. It is difficult to over-estimate the influence of his lectures on these notes. We would also like to thank Luzius Grünenfelder, Robert Rosebrugh, and Richard Wood whose patience and criticism were greatly appreciated when these ideas were presented in detail in the 1974-75 Topos Seminar at Dalhousie.

I. Abstract Families:

0. Preliminaries:

In this paper we shall be using, what in practice are, functors defined up to canonical isomorphism. We also use ordinary functors (i.e. defined up to equality) and combining these with those defined up to isomorphism in various ways gives rise to functors defined up to canonical isomorphism to a certain extent. Thus the basic structure which we consider is a category $\underline{A}$ equipped with a specified class of isomorphisms, called *canonical*. Identities are canonical and composites and inverses of canonical isomorphisms are canonical. There are two extreme cases; all isomorphisms are canonical, or only the identities are canonical. These are the two cases which will be used most often. It will usually be clear what the canonical isomorphisms are from the context and then we will omit mention of them.

We shall not give a technical definition of what we mean by a functor $F: \underline{A} \longrightarrow \underline{B}$ defined up to canonical isomorphism, but take it as being intuitively clear (or it should become so from the examples). What is important is that, once values for F have been chosen for each object of $\underline{A}$, F becomes a functor in a unique way and this functor preserves canonical isomorphisms. If different values for F are chosen, the new functor is canonically isomorphic to the old one (i.e. by a natural transformation whose components are canonical). Two functors which are canonically isomorphic are considered as being "the same functor". So as to keep everything as concrete as possible, all our functors defined up to canonical isomorphism come equipped with a choice of representative functor. In practice such a choice is always available.

Natural transformations between functors defined up to canonical isomorphism are the same thing as natural transformations between the corresponding representative functors.

Throughout this paper, $\underline{S}$ will denote a fixed category, all of whose isomorphisms are canonical. We assume that $\underline{S}$ has finite limits. The category of sets will be denoted $\underline{Set}$.

1. Definitions:

(1.1) An $\underline{S}$-*indexed category* $\underline{A}$ consists of the following data

(i) for every object I of $\underline{S}$ a category $\underline{A}^I$ (with specified canonical isomorphisms),

(ii) for every morphism $\alpha: J \to I$ of $\underline{S}$ a functor (defined up to canonical isomorphism) $\alpha^*: \underline{A}^I \to \underline{A}^J$,

subject to the conditions

(a) $(1_I)^* \cong 1_{\underline{A}^I}$ (canonically)

(b) $(\alpha\beta)^* \cong \beta^*\alpha^*$ (canonically).

These natural isomorphisms are required to satisfy well known coherence conditions (see [Gy2] or [Gk]) which do not play a central role and so will not be given here.

The category $\underline{A}^I$ is called the category of *I-indexed families* of objects of $\underline{A}$. The functor α^* is called the *substitution* functor determined by α .

(1.2) An $\underline{S}$-*indexed functor* $F: \underline{A} \to \underline{B}$ between two indexed categories consists of the following data

(i) for every object I of $\underline{S}$, a functor (defined up to canonical isomorphism)

$$F^I: \underline{A}^I \to \underline{B}^I$$

subject to the condition

(a) for every $\alpha: J \to I$ in $\underline{S}$

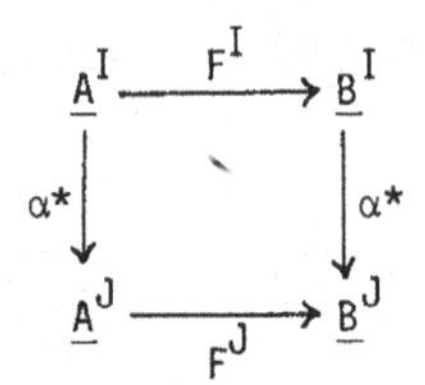

commutes up to canonical isomorphism.

These natural isomorphisms must also satisfy well-known coherence conditions which we do not give here (see [Gy2] or [Gk]).

(1.3) An $\underline{S}$-*indexed natural transformation* $t: F \to G$ between two indexed functors consists of the following data

(i) for every object I of $\underline{S}$, a natural transformation

$$t^I: F^I \to G^I$$

subject to the condition

(a) for every morphism $\alpha: J \to I$ in $\underline{S}$ we have $\alpha^* \cdot t^I = t^J \cdot \alpha^*$.

This last equality must of course be interpreted by using the isomorphisms given in (1.1) and (1.2).

(1.4) If we let $\underline{\text{Cat}}$ denote the (meta-) category of *all* categories, then an indexed category, all of whose canonical isomorphisms are identities, is simply a functor $\underline{A}^{(\ )}: \underline{S}^{op} \to \underline{\text{Cat}}$. An indexed functor between two such indexed categories is a natural transformation $F^{(\ )}: \underline{A}^{(\ )} \to \underline{B}^{(\ )}$. An indexed natural transformation $t: F \to G$ is a *modification* $t^{(\ )}: F^{(\ )} \to G^{(\ )}$ (see [Bn1] or [Gy2]).

(1.5) By working at I, we can compose indexed functors and we can compose indexed natural transformations in the two usual ways, and thus we get the (meta-) 2-category (see [Gy2]) of $\underline{S}$-indexed categories $\underline{S}$-ind. cat . In particular we know what it means for two indexed functors to be adjoint.

(1.6) We have an underlying 2-functor $\underline{S}$-ind. cat $\to$ Cat given by evaluating at 1 , the terminal object of $\underline{S}$. We usually denote the underlying category, functor, and natural transformation by the same symbol as the indexed ones. Thus we identify $\underline{A}^1$ with $\underline{A}$, F^1 with F , and t^1 with t . Also if $\tau: I \to 1$ then we denote τ^* by $\Delta_I: \underline{A} \to \underline{A}^I$.

In practice these identifications do not lead to confusion and reflect more accurately our way of thinking of indexed categories.

2. Examples:

(2.1) Every category $\underline{A}$ is a Set-indexed category in a canonical way. For any set I , $\underline{A}^I$ is taken to be the I-fold product of $\underline{A}$. For $\alpha: J \to I$ we define $\alpha^*: \underline{A}^I \to \underline{A}^J$ as follows: if $\langle A_i | i \in I \rangle$ is a typical object of $\underline{A}^I$ then

$$\alpha^* \langle A_i | i \in I \rangle = \langle A_{\alpha(j)} | j \in J \rangle \quad .$$

A functor $F: \underline{A} \to \underline{B}$ extends uniquely to an indexed functor by defining

$$F^I \langle A_i | i \in I \rangle \ = \ \langle FA_i | i \in I \rangle \quad .$$

Similarly, natural transformations extend uniquely to indexed ones. It is good to keep this example in mind for motivation, since this is the situation we are generalizing.

A category $\underline{A}$ can be indexed by Set in different ways, some of which actually occur in mathematics and are useful (see (2.4) below).

(2.2) The basic example is the $\underline{S}$-indexing of $\underline{S}$ itself. We define $\underline{S}^I$ to be the category $\underline{S}/I$ of objects over I. A typical object of $\underline{S}/I$ is a morphism $p: X \to I$ which should be thought of as the family $\langle p^{-1}(i) | i \in I \rangle$. A morphism from $p: X \to I$ to $q: Y \to I$ in $\underline{S}/I$ is a commutative triangle

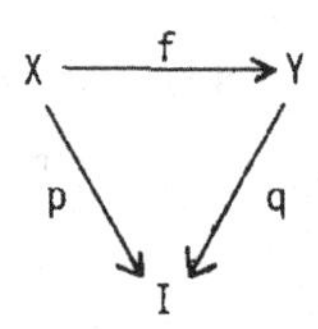

$\underline{S}/I$ comes equipped with the canonical isomorphisms induced from $\underline{S}$, i.e. all isomorphisms are canonical.

If $\alpha: J \to I$ then $\alpha^*: \underline{S}/I \to \underline{S}/J$ is defined by pulling back along α. Since pullbacks are only defined up to isomorphism, it follows that α^* is only defined up to isomorphism. This is the basic example of a functor defined up to canonical isomorphism and most other ones come from this one in some way.

(2.3) The category $Gr(\underline{S})$ of group objects in $\underline{S}$ can be $\underline{S}$-indexed by defining $Gr(\underline{S})^I = Gr(\underline{S}^I)$ and since the α^* preserve products, they also preserve group objects and thus extend to $\alpha^*: Gr(\underline{S})^I \to Gr(\underline{S})^J$. The required properties of $(\ )^*$ follow from the fact that they hold for $\underline{S}$.

More generally, any theory $\underline{T}$ whose models can be interpreted in all of the categories $\underline{S}^I$ and such that all of the $\alpha^*: \underline{S}^I \to \underline{S}^J$ preserve $\underline{T}$-models, will give us an indexed category $\underline{T}\text{-mod}(\underline{S})$ by defining

$$\underline{T}\text{-mod}(\underline{S})^I = \underline{T}\text{-mod}(\underline{S}^I)$$

and then extending the α^* to the categories of models.

For example, if $\underline{S}$ is an elementary topos we could define a topological space object to be a pair (U,X) where U is a subobject of Ω^X closed under finite intersections and arbitrary unions. A morphism $f: X \to Y$ is continuous if there exists a g such that

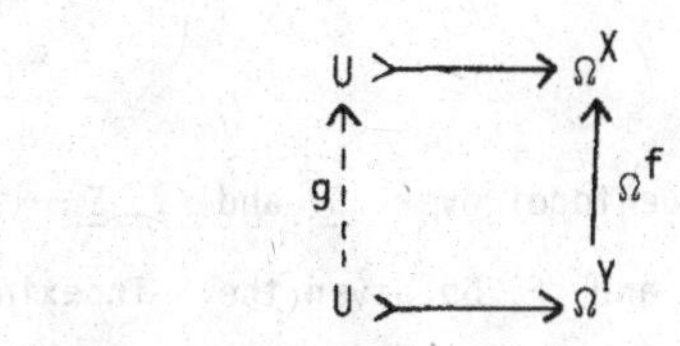

commutes. (For more details see [St].)

If $Top(\underline{S})$ denotes the category of topological space objects in $\underline{S}$ then since $\underline{S}^I = \underline{S}/I$ is a topos for any I $Top(\underline{S}^I)$ makes sense and so we define $Top(\underline{S})^I = Top(\underline{S}^I)$. Since $\alpha^*: \underline{S}^I \to \underline{S}^J$ is a logical morphism, it preserves topological space objects and thus extends to $\alpha^*: Top(\underline{S})^I \to Top(\underline{S})^J$.

(2.4) Let $\underline{S}$ be a topos and $\underline{E}$ a topos defined over $\underline{S}$, i.e. we have a geometric morphism $p: \underline{E} \to \underline{S}$. $\underline{E}$ has a canonical $\underline{S}$-indexing given by $\underline{E}^I = \underline{E}/p^*I$ and where α^* is given by pulling back along $p^*(\alpha)$. When working with $\underline{S}$-topoi this is always the $\underline{S}$-indexing which $\underline{E}$ is given.

When $\underline{S}$ is $\underline{Set}$ and $\underline{E}$ is a $\underline{Set}$-topos, the I-indexed families of objects of $\underline{E}$ in the above sense are ordinary I-indexed families $\langle A_i | i \in I \rangle$ which are bounded in the sense that there exist an object B and monomorphisms $A_i \rightarrowtail B$ for all i. If $\underline{E}$ is not cocomplete (and Barr has given an example to show that it need not be) then there are fewer bounded families than general families. So this gives an example of a category which is naturally indexed by $\underline{Set}$ in a different way than the usual.

Clearly, the above indexing of $\underline{E}$ by $\underline{S}$ makes no use of the fact that $\underline{E}$ and $\underline{S}$ are topoi nor that p is a geometric morphism (only p^* is used). We shall use this indexing in the future when $\underline{E}$ and $\underline{S}$ are simply categories with finite limits, but for smallness questions (see Ch.II) it is useful to have a "geometric morphism" $p\colon \underline{E} \to \underline{S}$.

Let $p\colon \underline{E} \to \underline{S}$ and $q\colon \underline{F} \to \underline{S}$ be topoi over $\underline{S}$ and $f\colon \underline{E} \to \underline{F}$ a geometric morphism over $\underline{S}$. Let $\underline{E}$ and $\underline{F}$ be given the $\underline{S}$-indexing described above.

(2.4.1) Proposition: The inverse image functor $f^*\colon \underline{F} \to \underline{E}$ extends to an indexed functor.

Proof: Define $f^{*I}\colon \underline{F}^I \to \underline{E}^I$ by sending

$$\begin{array}{c} Y \\ \downarrow \\ q^*I \end{array} \quad \text{to} \quad \begin{array}{c} f^*Y \\ \downarrow \\ f^*q^*I \\ \| \\ p^*I \end{array} .$$

That this actually gives an indexed functor follows, by an easy computation, from the fact that f^* preserves pullbacks. □

(2.4.2) Proposition: The direct image functor $f_*\colon \underline{E} \to \underline{F}$ extends to an indexed functor.

Proof: Define $f_*^I\colon \underline{E}^I \to \underline{F}^I$ by sending

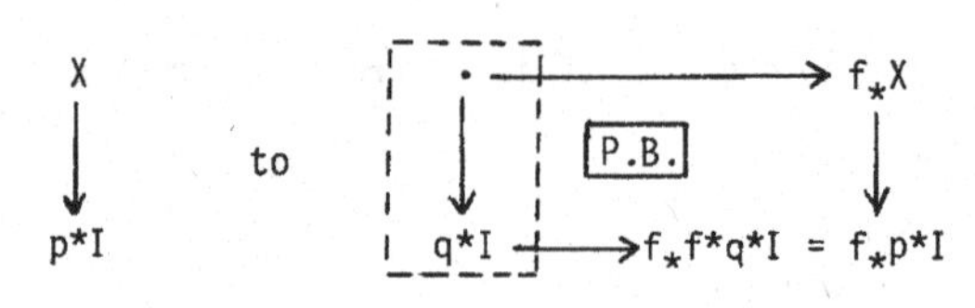

where the bottom morphism is the unit, η , for the adjunction $f^* \dashv f_*$. To show that this makes f_* indexed, consider any $\alpha: J \to I$. The top path around

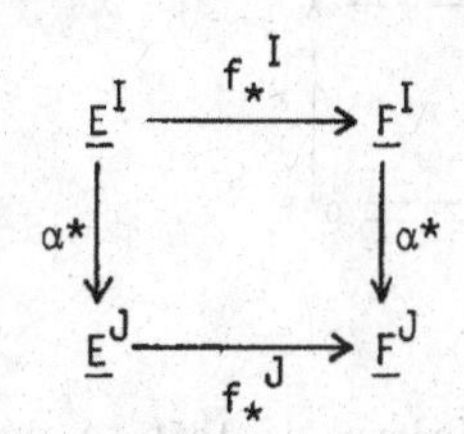

is first applying f_* , then pulling back along ηq^*I and then along $q^*\alpha$, whereas the bottom path is first pulling back along $p^*\alpha$, then applying f_* and then pulling back along ηq^*J . The result now follows from the commutivity of

$$\begin{array}{ccccc}
q^*J & \xrightarrow{\eta q^*J} & f_*f^*q^*J & = & f_*p^*J \\
\downarrow{\scriptstyle q^*\alpha} & & \downarrow{\scriptstyle f_*f^*q^*\alpha} & & \downarrow{\scriptstyle f_*p^*\alpha} \\
q^*I & \xrightarrow[\eta q^*I]{} & f_*f^*q^*I & = & f_*p^*I
\end{array}$$

and the fact that f_* preserves pullbacks. □

(2.4.3) Proposition: f^* is an indexed left adjoint to f_* .

Proof: Let $x: X \to p^*I$ be an object of $\underline{E}^I$ and $y: Y \to q^*I$ an object of $\underline{F}^I$. There is a natural bijection between morphisms $\phi: y \to f_*^I(x)$ in $\underline{F}^I$ and morphisms $\psi: Y \to f_*X$ making the outside square commute in the following diagram:

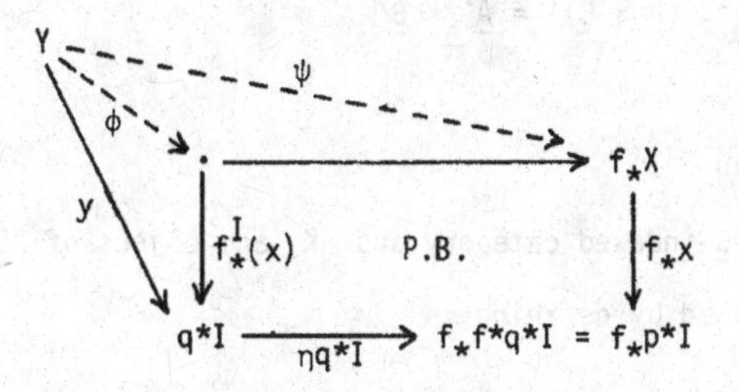

This square, when transposed by the adjointness $f^* \dashv f_*$, becomes

$$\begin{array}{ccc} f^*Y & \xrightarrow{\bar{\psi}} & X \\ {\scriptstyle f^*y}\downarrow & & \downarrow{\scriptstyle x} \\ f^*q^*I & = & p^*I \end{array} .$$

Thus $\bar{\psi}: f^{*I}(y) \to x$ in $\underline{E}^I$.

That this natural bijection is indexed is left to the reader. □

(2.4.4) Corollary: If $p: \underline{E} \to \underline{S}$ is a topos over $\underline{S}$, then p_* and p^* are indexed adjoints. □

(2.5) If $\underline{A}$ is an $\underline{S}$-indexed category we can index $\underline{A}^{op}$ by defining $(\underline{A}^{op})^I = (\underline{A}^I)^{op}$.

(2.6) If $\underline{A}$ is an indexed category we define the indexed category $\underline{A}^{\mathbf{2}}$ by

$$(\underline{A}^{\mathbf{2}})^I = (\underline{A}^I)^{\mathbf{2}}$$

with the obvious α^* .

In fact, if $\underline{X}$ is any category (not indexed) we can index $\underline{A}^{\underline{X}}$ by the formula

$$(\underline{A}^{\underline{X}})^I = (\underline{A}^I)^{\underline{X}} .$$

(2.7) If $\underline{A}$ and $\underline{B}$ are $\underline{S}$-indexed categories then we get an $\underline{S}$-indexed category $\underline{A} \times \underline{B}$ by defining

$$(\underline{A} \times \underline{B})^I = \underline{A}^I \times \underline{B}^I$$

with obvious α^* .

(2.8) If $\underline{A}$ is an $\underline{S}$-indexed category and K any object of $\underline{S}$ then the category $\underline{A}^K$ can be indexed by defining

$$(\underline{A}^K)^I = \underline{A}^{K\times I}$$

and for $\alpha\colon J \to I$ we define α^* for $\underline{A}^K$ to be $(K\times\alpha)^*\colon \underline{A}^{K\times I} \to \underline{A}^{K\times J}$.

With this definition $\underline{A}^1 \cong \underline{A}$ as indexed categories and so this justifies our confusion of $\underline{A}^1$ with $\underline{A}$.

(2.9) If $\kappa\colon K' \to K$ is a morphism of $\underline{S}$ then $\kappa^*\colon \underline{A}^K \to \underline{A}^{K'}$ can be made into an indexed functor by defining $(\kappa^*)^I\colon (\underline{A}^K)^I \to (\underline{A}^{K'})^I$ to be $(\kappa\times I)^*\colon \underline{A}^{K\times I} \to \underline{A}^{K'\times I}$. For $\alpha\colon J \to I$ the diagram

$$\begin{array}{ccc} (\underline{A}^K)^I & \xrightarrow{(\kappa^*)^I} & (\underline{A}^{K'})^I \\ \downarrow{\scriptstyle \alpha^*} & & \downarrow{\scriptstyle \alpha^*} \\ (\underline{A}^K)^J & \xrightarrow[(\kappa^*)^J]{} & (\underline{A}^{K'})^J \end{array}$$

is, by definition of α^* and κ^* ,

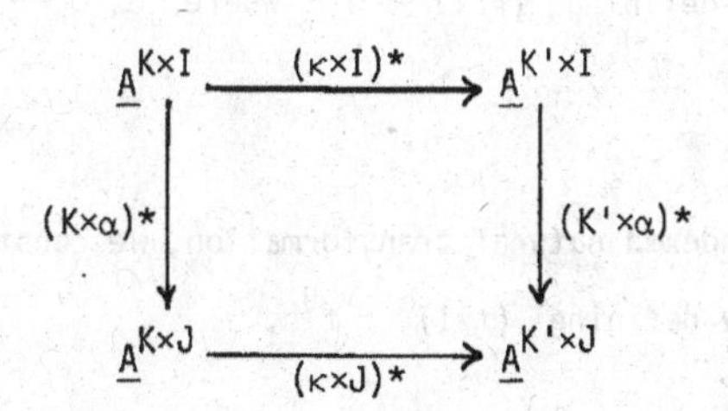

which clearly commutes up to canonical isomorphism.

(2.10) If $F\colon \underline{A} \to \underline{C}$ and $G\colon \underline{B} \to \underline{C}$ are indexed functors, then we have a canonical indexing for the comma category (F,G) given by defining $(F,G)^I$ to be (F^I,G^I) . The usual projection functors $P_{\underline{A}}\colon (F,G) \to \underline{A}$ and $P_{\underline{B}}\colon (F,G) \to \underline{B}$ are indexed and so is the canonical natural transformation

$t: FP_{\underline{A}} \to GP_{\underline{B}}$.

3. Localization:

(3.1) Let $\underline{A}$ be an $\underline{S}$-indexed category and I an object of $\underline{S}$. We construct an $\underline{S}/I$-indexed category $\underline{A}/I$, called the *localization of* $\underline{A}$ *at* I , as follows: if $\alpha: J \to I$ is an object of $\underline{S}/I$ we define $(\underline{A}/I)^{\alpha}$ to be the category $\underline{A}^J$, and if

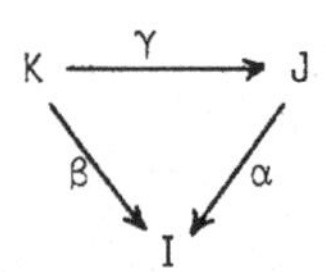

is a morphism in $\underline{S}/I$, then $\gamma^*: (\underline{A}/I)^{\alpha} \to (\underline{A}/I)^{\beta}$ is defined to be $\gamma^*: \underline{A}^J \to \underline{A}^K$.

If $F: \underline{A} \to \underline{B}$ is an $\underline{S}$-indexed functor, we construct an $\underline{S}/I$-indexed functor, $F/I: \underline{A}/I \to \underline{B}/I$ by defining $(F/I)^{\alpha} = F^J$ where $\alpha: J \to I$. F/I is called the *localization of* F *at* I .

If $t: F \to G$ is an $\underline{S}$-indexed natural transformation, we construct the *localization of* t *at* I by defining $(t/I)^{\alpha} = t^J$.

Since the terminal object of $\underline{S}/I$ is $1: I \to I$, we see that $(\underline{A}/I)^1 = \underline{A}^I$, $(F/I)^1 = F^I$, and $(t/I)^1 = t^I$ and so $\underline{A}/I$, F/I , and t/I are like $\underline{A}^I$, F^I, t^I respectively. However, as will be seen in (3.2), $\underline{A}/I$, F/I , and t/I are like $\underline{A}$, F , and t in the $\underline{S}/I$-indexed world. It is this double way of viewing localization at I which makes it useful. Results obtained for $\underline{A}$ at 1 can often be extended to $\underline{A}$ at I by considering $\underline{A}/I$.

When $\underline{A} = \underline{S}$, $\underline{S}/I$ already has a meaning and also has a canonical $\underline{S}/I$ indexing. If $\alpha: J \to I$ is any object of $\underline{S}/I$, then for $\underline{S}/I$ as the localization of $\underline{S}$ we have

$$(\underline{S}/I)^{\alpha} = \underline{S}^{J} = \underline{S}/J$$

and for $\underline{S}/I$ indexed by itself as in (2.2)

$$(\underline{S}/I)^{\alpha} = (\underline{S}/I)/\alpha \cong \underline{S}/J$$

thus the two possible interpretations of $\underline{S}/I$ are canonically isomorphic as $\underline{S}/I$-indexed categories.

(3.2) The 2-category of $\underline{S}$-indexed categories can be given a canonical indexing by $\underline{S}$. For any object I of $\underline{S}$, define

$$(\underline{S}\text{-}\underline{\underline{\mathrm{ind.cat}}})^{I} = \underline{S}/I\text{-}\underline{\underline{\mathrm{ind.cat}}}.$$

If $\alpha: J \to I$ is any morphism in $\underline{S}$ and $\underline{A}$ is any $\underline{S}/I$-indexed category, we define an $\underline{S}/J$-indexed category $\alpha^*\underline{A}$ by

$$(\alpha^*A)^{\beta} = \underline{A}^{\alpha\beta}$$

where $\beta: K \to J$ is any object of $\underline{S}/J$. Then $\alpha^*: \underline{S}/I\text{-}\underline{\underline{\mathrm{ind.cat}}} \to \underline{S}/J\text{-}\underline{\underline{\mathrm{ind.cat}}}$ gives us a 2-functor (see [Gy2]) and makes $\underline{S}$-$\underline{\underline{\mathrm{ind.cat}}}$ into an $\underline{S}$-indexed 2-category (obvious definition).

If $\Delta_I: \underline{S}\text{-}\underline{\underline{\mathrm{ind.cat}}} \to \underline{S}/I\text{-}\underline{\underline{\mathrm{ind.cat}}}$ is the 2-functor corresponding to $I \to 1$, then $\Delta_I(\underline{A}) = \underline{A}/I$. Thus $\underline{A}/I$ can be thought of as the I-indexed family of $\underline{S}$-indexed categories which is constantly $\underline{A}$. It is in this sense that $\underline{A}/I$ is like $\underline{A}$ in the $\underline{S}/I$-world. Similar remarks also hold for F/I and t/I .

(3.3) A property of indexed categories, indexed functors, and indexed natural transformations is said to be *stable under localization* if whenever it holds for some given indexed categories, functors, and natural transformations, it also holds for their localizations. A construction performed on indexed categories, functors, and natural transformations is stable under localization if it commutes with localization.

We shall not attempt a rigorous definition of the terms "property" and "construction" but rather illustrate what we mean through the following examples. These examples will be needed later in the paper.

(3.3.1) The construction $\underline{S} \rightsquigarrow Gr(\underline{S})$ is stable under localization since

$$(Gr(\underline{S})/I)^{\alpha} = Gr(\underline{S})^{J} = Gr(\underline{S}^{J})$$

and

$$Gr(\underline{S}/I)^{\alpha} = Gr((\underline{S}/I)^{\alpha}) \cong Gr(\underline{S}^{J})$$

thus

$$Gr(\underline{S})/I \cong Gr(\underline{S}/I)$$

as $\underline{S}/I$-indexed categories.

More generally, the construction $\underline{S} \rightsquigarrow \underline{T}\text{-mod}(\underline{S})$, as given in (2.3), is stable under localization.

(3.3.2) The property of $\underline{S}$ being a topos is stable under localization. It is well known that if $\underline{S}$ is a topos then $\underline{S}/I$ is also a topos.

(3.3.3) The construction $\underline{A} \rightsquigarrow \underline{A}^{op}$ is stable under localization.

$$[(\underline{A}/I)^{op}]^{\alpha} = [(\underline{A}/I)^{\alpha}]^{op} = (\underline{A}^{J})^{op}$$

and

$$(\underline{A}^{op}/I)^{\alpha} = (\underline{A}^{op})^{J} = (\underline{A}^{J})^{op}$$

thus $(\underline{A}/I)^{op} \cong \underline{A}^{op}/I$ as $\underline{S}/I$-indexed categories.

(3.3.4) The construction $\underline{A} \rightsquigarrow \underline{A}^2$ is stable under localization. The proof is similar to the preceding one.

(3.3.5) The construction $\underline{A},\underline{B} \rightsquigarrow \underline{A}\times\underline{B}$ is stable under localization.

(3.3.6) The construction $F,G \rightsquigarrow (F,G)$ (comma category) is also stable under localization.

We have an indexed category $\mathbb{1}$ which is given by $\mathbb{1}^I = \mathbb{1}$ (the one morphism category) for all I. An indexed functor $\mathbb{1} \rightarrow \underline{A}$ is essentially an object of $\underline{A}$ and so we can talk of objects in the setting described above. If A is an object of $\underline{A}$ then it is easily seen that the localization of A at I is the object $\Delta_I A$ of $\underline{A}^I$. Thus when localizing properties or constructions containing fixed objects, we must apply Δ_I to them.

(3.3.7) The construction $\underline{A} \rightsquigarrow \underline{A}^K$ is stable under localization. Here K is a fixed object of $\underline{S}$ and so stability would mean that

$$\underline{A}^K/I \cong (\underline{A}/I)^{\Delta_I K}$$

as $\underline{S}/I$-indexed categories. This is indeed the case since

$$(\underline{A}^K/I)^{\alpha} = (\underline{A}^K)^J = \underline{A}^{K\times J}$$

and

$$[(\underline{A}/I)^{\Delta_I K}]^{\alpha} = (\underline{A}/I)^{\Delta_I K\times\alpha} \cong \underline{A}^{K\times J}$$

where the last isomorphism follows because $\Delta_I K \times \alpha$ is the product in $\underline{S}/I$, i.e. the pullback

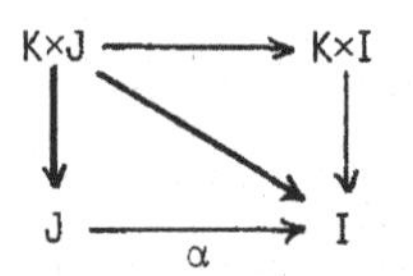

In a similar way we can talk of morphisms of $\underline{A}$ (either as functors $\mathbf{2} \to \underline{A}$, as objects of $\underline{A}^{\mathbf{2}}$, or as indexed natural transformations between $\mathbb{1} \rightrightarrows \underline{A}$) and diagrams in general. We permit quantification over indexed categories, functors, and natural transformations so we can do the same for objects and morphisms. However, care must be taken to distinguish between variables and constants, as a variable object of $\underline{A}$ gets localized to a variable object of $\underline{A}^I$ but a constant object, A , of $\underline{A}$ gets localized to the object $\Delta_I A$ of $\underline{A}^I$.

Thus the property of $m: A \to B$ being a monomorphism is localized to the property of $\Delta_I m: \Delta_I A \to \Delta_I B$ being a monomorphism in $\underline{A}^I$. On the other hand, the property of m being a monomorphism into B is localized to the property of m being a monomorphism into $\Delta_I B$ in $\underline{A}^I$, not just monomorphisms of the form $\Delta_I m$.

The property of being a monomorphism is not in general stable under localization although for $\underline{S}$ it is, as well as for many categories constructed from $\underline{S}$ such as $Gr(\underline{S})$. On the other hand the property of being an epimorphism might not be stable even for $\underline{S}$.

We shall be using the process of localization as follows: if we want to prove something about the indexed category $\underline{A}$, we first prove it about the category $\underline{A}^1$ where our intuition can guide us, and then if everything used was stable under localization, it follows by passing to $\underline{A}/I$ that the result holds for $(\underline{A}/I)^1 = \underline{A}^I$. Thus we adopt as a general methodological principle that everything must be stable under localization. If some properties are not then we pass to the stabilized property.

Thus we always use stable monomorphisms, epimorphisms, finite limits and colimits, i.e. those which are preserved by Δ_I . By localizing, we see that in $\underline{A}^I$ these concepts should be preserved by all α^* .

II. Smallness:

1. Small Categories:

(1.1) A *category object* $\mathbb{C}$ in $\underline{S}$ is a diagram

$$C_2 \overset{\pi_0}{\underset{\pi_1}{\overset{\gamma}{\rightrightarrows\!\!\!\rightarrow}}} C_1 \overset{\partial_0}{\underset{\partial_1}{\overset{\overset{id}{\longleftarrow}}{\rightrightarrows}}} C_0$$

where C_2 is $C_1 \underset{C_0}{\times} C_1$, or more precisely

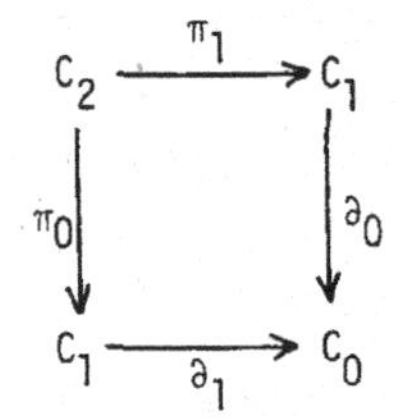

is a pullback diagram, satisfying the following conditions

(a) $\partial_0 \cdot id = 1_{C_0} = \partial_1 \cdot id$

(b) $\gamma \cdot (1_{C_1}, id \cdot \partial_1) = 1_{C_1}$

$\gamma \cdot (id \cdot \partial_0, 1_{C_1}) = 1_{C_1}$

(c) $\gamma \cdot (\gamma \underset{C_0}{\times} C_1) = \gamma \cdot (C_1 \underset{C_0}{\times} \gamma)$.

C_0 is called the object of objects, C_1 the object of morphisms, and C_2 the object of composable pairs of $\mathbb{C}$.

A category object in $\underline{Set}$ is the same thing as a small category.

If $\mathbb{C}$ and $\mathbb{D}$ are two category objects in $\underline{S}$ then an *internal functor* (functorial morphism) $\mathbb{F}: \mathbb{C} \to \mathbb{D}$ is a triple of morphisms (F_2, F_1, F_0) making corresponding squares commute in the diagram

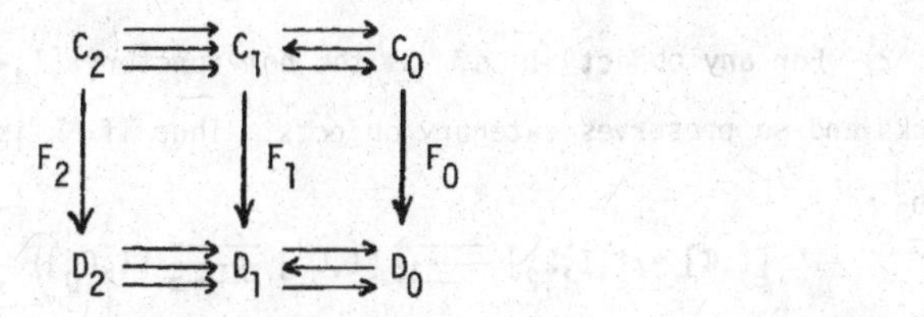

Again, in Set this is the same as a functor between small categories.

If $\mathbb{F}$ and $\mathbb{G}$ are two internal functors $\mathbb{C} \to \mathbb{D}$ then an *internal natural transformation* $t: \mathbb{F} \to \mathbb{G}$ is a morphism $t: C_0 \to D_1$ such that

(a) $\partial_0 t = F_0$
$\partial_1 t = G_0$

(b) $\gamma\cdot(t\partial_0, G_1) = \gamma\cdot(F_1, t\partial_0)$.

In Set this is just an ordinary natural transformation.

Internal functors can be composed and internal natural transformations can be composed in the two usual ways, thus giving us a 2-category $\underline{\underline{Cat}}(\underline{S})$. This can be $\underline{S}$-indexed by defining

$$\underline{\underline{Cat}}(\underline{S})^I = \underline{\underline{Cat}}(\underline{S}^I) .$$

Since $\alpha^*: \underline{S}^I \to \underline{S}^J$ preserves finite limits, it also preserves category objects, functors, and natural transformations and so extends to

$$\alpha^*: \underline{\underline{Cat}}\ (\underline{S})^I \to \underline{\underline{Cat}}(\underline{S})^J .$$

Furthermore, it is apparent that this construction is stable under localization.

More information on category objects can be found in [Di 1,2].

(1.2) For any object I of $\underline{S}$ the hom functor $[I,-]: \underline{S} \to \underline{Set}$ preserves pullbacks and so preserves category objects. Thus if $\mathbb{C}$ is a category object in $\underline{S}$ then

$$[I,\mathbb{C}] = ([I,C_2] \rightrightarrows [I,C_1] \rightleftarrows [I,C_0])$$

is a category object in $\underline{Set}$, i.e. an ordinary small category. For any morphism $\alpha: J \to I$ in $\underline{S}$ we get a diagram in $\underline{Set}$

$$\begin{array}{ccccccc}
[I,\mathbb{C}] & & [I,C_2] & \rightrightarrows & [I,C_1] & \rightleftarrows & [I,C_0] \\
\downarrow{\scriptstyle [\alpha,\mathbb{C}]} & & \downarrow{\scriptstyle [\alpha,C_2]} & & \downarrow{\scriptstyle [\alpha,C_1]} & & \downarrow{\scriptstyle [\alpha,C_0]} \\
[J,\mathbb{C}] & & [J,C_2] & \rightrightarrows & [J,C_1] & \rightleftarrows & [J,C_0]
\end{array}$$

in which the corresponding squares commute, i.e. an ordinary functor from $[I,\mathbb{C}]$ to $[J,\mathbb{C}]$. This clearly makes $[-,\mathbb{C}]$ into a contravariant functor from $\underline{S}$ to $\underline{Cat}$. Thus we get an $\underline{S}$-indexed category which we denote by $[\mathbb{C}]$ and call the *externalization* of $\underline{\mathbb{C}}$. Only identities are considered canonical in $[\mathbb{C}]$, reflecting the fact that the isomorphisms (a) and (b) of (I,1.1) are actually equalities.

If $\mathbb{F}: \mathbb{C} \to \mathbb{D}$ is an internal functor and I any object of $\underline{S}$, we get a diagram in $\underline{Set}$

$$\begin{array}{ccccccc}
[I,\mathbb{C}] & & [I,C_2] & \rightrightarrows & [I,C_1] & \rightleftarrows & [I,C_0] \\
\downarrow{\scriptstyle [I,\mathbb{F}]} & & \downarrow{\scriptstyle [I,F_2]} & & \downarrow{\scriptstyle [I,F_1]} & & \downarrow{\scriptstyle [I,F_0]} \\
[I,\mathbb{D}] & & [I,D_2] & \rightrightarrows & [I,D_1] & \rightleftarrows & [I,D_0]
\end{array}$$

where the corresponding squares commute, i.e. a functor $[I,\mathbb{F}]: [I,\mathbb{C}] \to [I,\mathbb{D}]$.

If we denote this functor by $[\mathbb{F}]^I : [\mathbb{C}]^I \to [\mathbb{D}]^I$ then clearly $[\mathbb{F}]$ is an indexed functor $[\mathbb{C}] \to [\mathbb{D}]$. $[\mathbb{F}]$ is called the externalization of $\mathbb{F}$.

In the same way we see that an internal natural transformation $t: \mathbb{F} \to \mathbb{G}$ can be externalized to give $[t]: [\mathbb{F}] \to [\mathbb{G}]$.

If $\mathbb{C}$ and $\mathbb{D}$ are two category objects in $\underline{S}$ and if $F: [\mathbb{C}] \to [\mathbb{D}]$ is an indexed functor, then since all canonical isomorphisms in $[\mathbb{D}]$ must be identities it follows by the Yoneda lemma that there exists a unique internal functor $\mathbb{F}: \mathbb{C} \to \mathbb{D}$ such that $F = [\mathbb{F}]$. For the same reasons every indexed natural transformation is the externalization of a unique internal one.

In fact we have constructed an embedding

$$\underline{\text{Cat}}(\underline{S}) \to \underline{S}\text{-}\underline{\underline{\text{ind.cat}}}$$

which is essentially the Yoneda embedding.

The reason we introduced canonical isomorphisms was to take into account the fact that indexed categories of the form $[\mathbb{C}]$ are strict (i.e. the canonical isomorphisms are equalities), and that categories such as $\underline{S}$ and $\text{Gr}(\underline{S})$ are not strict. In all the indexed categories introduced before $[\mathbb{C}]$ all isomorphisms were canonical. Categories constructed from other categories, such as $[\mathbb{C}] \times \underline{S}$ have obvious canonical isomorphisms which may be intermediate between "all isomorphisms" and "only identities".

It is usually easier to work with the externalizations of category objects than with the category objects themselves, and in what follows we do so. However, if we want our results to give results about the category objects under consideration, the externalization process should be reversible to some extent.

This is why the categories $[\mathbb{C}]$ must be strict. If we consider these categories as non-strict, the above embedding is not full and $\mathbb{C}$ is not uniquely determined by $[\mathbb{C}]$.

(1.3) Two indexed categories $\underline{A}$ and $\underline{B}$ are said to be *canonically equivalent* (written $\underline{A} \simeq \underline{B}$) if there are indexed functors $F: \underline{A} \to \underline{B}$ and $G: \underline{B} \to \underline{A}$ and canonical natural isomorphisms $FG \cong 1_{\underline{B}}$ and $GF \cong 1_{\underline{A}}$.

An indexed category $\underline{A}$ is called *small* if it is canonically equivalent to $[\mathbb{C}]$ for some category object $\mathbb{C}$. Since categories of the form $[\mathbb{C}]$ are strict, it follows that $\mathbb{C}$ is uniquely determined up to isomorphism by $\underline{A}$.

The functor $\Delta_I: \underline{S} \to \underline{S}/I$ preserves pullbacks and therefore also category objects. Δ_I has a left adjoint $\Sigma_I: \underline{S}/I \to \underline{S}$ defined by

$$\Sigma_I \, (\alpha: J \to I) = J \, .$$

For $\mathbb{C}$ a category object in $\underline{S}$, $[\Delta_I\mathbb{C}]$ is an $\underline{S}/I$-indexed category such that

$$[\Delta_I\mathbb{C}]^\alpha = [\alpha,\Delta_I\mathbb{C}] \cong [\Sigma_I\alpha,\mathbb{C}] = [J,\mathbb{C}] = [\mathbb{C}]^J = ([\mathbb{C}]/I)^\alpha \, .$$

Thus $[\mathbb{C}]/I \cong [\Delta_I\mathbb{C}]$ and so the property of being small is stable under localization. This isomorphism also supports our claim that $\underline{A}/I$ should be thought of as the I-indexed family of $\underline{S}$-indexed categories which is constantly $\underline{A}$

(1.4) If K is an object of $\underline{S}$ we can construct the *discrete* category object determined by K ,

$$K \overset{\longrightarrow}{\underset{\longrightarrow}{\longrightarrow}} K \overset{\longrightarrow}{\underset{\longrightarrow}{\longleftarrow}} K$$

where all the morphisms are identities. We denote this category object by the same symbol K. The externalization $[K]$ has as I-indexed families $[K]^I$, the discrete category with set of objects $[I,K]$.

In general, we define an indexed category $\underline{A}$ to be *discrete* if it is canonically equivalent to a category $\underline{B}$ for which $\underline{B}^I$ is discrete for every I.

(1.5) Now that internal categories have been externalized we can consider indexed functors from small categories to arbitrary indexed categories, and by a Yoneda lemma argument get an internal description of these.

Let $\mathbb{C}$ be a category object in $\underline{S}$ and $\underline{A}$ an arbitrary $\underline{S}$-indexed category. We define an *internal functor* from $\mathbb{C}$ to $\underline{A}$ to be a pair (X,ξ) where X is an object of $\underline{A}^{C_0}$ (the C_0-indexed family of objects of $\underline{A}$ which consists of the values of the functor on the objects) and $\xi\colon \partial_0^*X \to \partial_1^*X$ is a morphism of $\underline{A}^{C_1}$ (the C_1-indexed family of morphisms which has as members the values of the functor on the morphisms) subject to the conditions

(a) $id^*(\xi) = 1_X$ (preservation of identities),

(b) $\pi_1^*(\xi)\cdot\pi_0^*(\xi) = \gamma^*(\xi)$ (preservation of composition).

The above equalities must be interpreted by inserting the canonical isomorphisms between the appropriate objects.

If (X,ξ) and (Y,θ) are two internal functors from $\mathbb{C}$ to $\underline{A}$ then we define an *internal natural transformation* $t\colon (X,\xi) \to (Y,\theta)$ to be a morphism $t\colon X \to Y$ in $\underline{A}^{C_0}$ such that

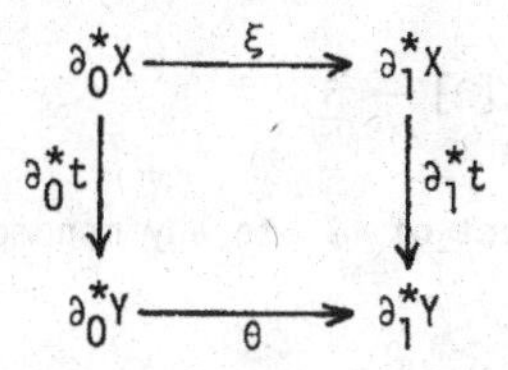

commutes.

Internal natural transformations are easily composed.

(1.5.1) <u>Proposition</u>: The category of indexed functors from $[\mathbb{C}]$ to $\underline{A}$ is equivalent to the category of internal functors from $\mathbb{C}$ to $\underline{A}$. □

If we take I to be a discrete category object, then the previous proposition tells us that the category of indexed functors from $[I]$ to $\underline{A}$ is equivalent to $\underline{A}^I$. Thus a functor $[I] \to \underline{A}$ is essentially the same as an I-indexed family of objects of $\underline{A}$.

(1.6) We can also consider indexed functors from an arbitrary indexed category $\underline{A}$ to a small category $[\mathbb{C}]$. Although there is no internal description of these (because the Yoneda lemma does not work in that direction) there is one important case which should be mentioned.

Suppose that $\underline{S}$ is a topos and Ω its subobject classifier. Ω is a partially ordered object and so is a category object. $[\Omega]^I$ is the partially ordered set of subobjects of I , and the substitution functors are defined by inverse image.

We have an indexed functor

$$\sigma\colon \underline{S} \to [\Omega]$$

defined at I by sending $p\colon X \to I$ to the image of p . σ is the *support* functor.

We also have an indexed functor

$$\text{inc}\colon [\Omega] \to \underline{S}$$

defined at I by sending a subobject of I to any monomorphism which represents it (defined up to canonical isomorphism). We call inc the *inclusion* of $[\Omega]$ into $\underline{S}$.

σ is left adjoint to inc as indexed functors. This gives us an example of a small category which is a reflective subcategory of a large one. We shall return to this example in section (4.4).

2. General Smallness Conditions:

(2.0) Now that we have defined what small indexed categories are, we can discuss general smallness conditions. If we wish to consider the smallness of the class of diagrams in $\underline{A}$ satisfying some property (e.g. morphisms $A \to B$, subobjects of A, etc.), we construct the indexed category (often discrete) of such diagrams and study its smallness. If $P(\underline{A})$ is such a category and is small we say that $\underline{A}$ has a *small class of P's*. If $P(\underline{A}) \simeq [\mathbb{C}]$ then C_0 is called the *object of P's* of $\underline{A}$. The object X of $P(\underline{A})^{C_0}$ corresponding to $1_{C_0} \in [\mathbb{C}]^{C_0}$ is called the *generic family of P's* of $\underline{A}$. X has the following universal property: for any J and any $Y \in P(\underline{A})^J$ there exists a unique $\alpha: J \to C_0$ such that $\alpha^*X \cong Y$ (canonically). Thus we have a bijection between morphisms $J \to C_0$ and canonical isomorphism classes of objects of $P(\underline{A})^J$. We express this by saying that there is a bijection

$$\cong \frac{J \to C_0}{\text{J-indexed families of P's}}$$

which is natural in J.

We now give several examples to illustrate these rather vague concepts.

(2.1) Let $Ob(\underline{A})$ be the indexed category defined by letting $Ob(\underline{A})^I$ be the category whose objects are the same as those of $\underline{A}^I$ and whose morphisms are the canonical isomorphisms of $\underline{A}^I$. We think of $Ob(\underline{A})$ as being the category of objects of $\underline{A}$. $Ob(\underline{A})$ is discrete. It is necessary to include the canonical isomorphisms for otherwise $Ob(\underline{A})$ would not be an indexed category.

If $Ob(\underline{A})$ is small we say that $\underline{A}$ has a *small class of objects* and if $Ob(\underline{A}) \simeq [I]$ we say that I is the object of objects of $\underline{A}$. As mentioned in (2.0), we have a *generic family of objects* X in $\underline{A}^I$ with the property that for any J and any A in $\underline{A}^J$ there exists a unique $\alpha: J \to I$ such that $\alpha^*X \cong A$ (canonically). Thus we have a bijection

$$\cong \frac{J \longrightarrow I}{\text{Canonical isomorphism classes of objects of } \underline{A}^J}$$

If $\underline{A}$ is a small category, then it has a small class of objects. Indeed, if $\underline{A} \simeq [\mathbb{C}]$ then $Ob(\underline{A}) \simeq [C_0]$ and so the object of objects is C_0 . The generic family of objects corresponds to $1_{C_0} \in [C_0, \mathbb{C}] = [\mathbb{C}]^{C_0}$.

An important construction, in this connection, is that of the full subcategory determined by a family. Let $\underline{A}$ be an $\underline{S}$-indexed category and A an I-indexed family of objects of $\underline{A}$, i.e. $A \in \underline{A}^I$. We define an indexed category $\underline{\mathrm{Full}}(A)$ as follows: the objects of $\underline{\mathrm{Full}}(A)^J$ are in bijection with morphisms $\phi: J \to I$ (denote the object corresponding to ϕ by $[\phi]$), and the morphisms from $[\phi]$ to $[\psi]$ are the morphisms from ϕ^*A to ψ^*A in $\underline{A}^J$. In $\underline{\mathrm{Full}}(A)^J$ only identities are canonical. For $\beta: K \to J$, $\beta^*: \underline{\mathrm{Full}}(A)^J \longrightarrow \underline{\mathrm{Full}}(A)^K$ is given by $\beta^*[\phi] = [\phi\beta]$ on the objects and by the β^* for $\underline{A}$ on the morphisms.

There is a canonical indexed functor $F: \underline{\mathrm{Full}}(A) \to \underline{A}$ defined at J by

$$F^J[\phi] = \phi^*A$$

$$F^J(a) = a$$

which is clearly full and faithful.

$\underline{\text{Full}}(A)$ has a small class of objects since $\text{Ob}(\underline{\text{Full}}(A)) \cong [I]$. The generic family of objects is A.

If $\underline{A}$ has a small class of objects and X is the generic family of objects, then by choosing for each A a canonical isomorphism $\alpha^*X \cong A$, we see that $\underline{A}$ is canonically equivalent to $\underline{\text{Full}}(X)$.

(2.2) Let $\text{Mor}(\underline{A})$ be the indexed category defined by letting $\text{Mor}(\underline{A})^I$ have as objects the class of morphisms of $\underline{A}^I$ and as morphisms the obvious commutative squares made up of two morphisms and two canonical isomorphisms. $\text{Mor}(\underline{A})$ is discrete. If $\text{Mor}(\underline{A})$ is small we say that $\underline{A}$ has a *small class of morphisms*. If $\text{Mor}(\underline{A}) \simeq [I]$ then I is called the *object of morphisms* of $\underline{A}$. $\underline{A}$ has a *generic family of morphisms* $x\colon X_1 \to X_2$ in $\underline{A}^I$ which, is intuitively the family of all morphisms of $\underline{A}$ indexed by the object of all morphisms.

Any small category $\underline{A}$ has a small class of morphisms. If $\underline{A} \simeq [\mathbb{C}]$, then $\text{Mor}(\underline{A}) \simeq [C_1]$.

(2.2.1) Proposition: If $\underline{A}$ has a small class of objects and a small class of morphisms, then it is small. □

(2.3) Let A and B be two objects of $\underline{A}$ and define the indexed category $H(A,B)$ by letting $H(A,B)^I$ be the discrete category of morphisms from $\Delta_I A$ to $\Delta_I B$ in $\underline{A}^I$. We define $\alpha^*\colon H(A,B)^I \to H(A,B)^J$ as follows: $f\colon \Delta_I A \to \Delta_I B$ gets sent to

$$\Delta_J A \cong \alpha^*\Delta_I A \xrightarrow{\alpha^*(f)} \alpha^*\Delta_I B \cong \Delta_J B .$$

Even though $\underline{A}$ may not be strict, $H(A,B)$ is because of the coherence conditions which the above isomorphisms satisfy.

If $H(A,B)$ is small we say that $\underline{A}$ has a *small class of morphisms from A to B*. We denote the object which represents $H(A,B)$ by $Hom(A,B)$ (or $Hom_{\underline{A}}(A,B)$). Thus $Hom(A,B)$ is an object of $\underline{S}$ with the property that there is a natural bijection

$$\cong \frac{I \to Hom(A,B)}{\Delta_I A \to \Delta_I B} \quad .$$

This is a reasonable condition, since both top and bottom can be interpreted as I-indexed families of morphisms from A to B.

We have a *generic family of morphisms* from A to B indexed by $Hom(A,B)$, i.e. a morphism $x\colon \Delta_{Hom(A,B)}A \longrightarrow \Delta_{Hom(A,B)}B$ with the usual universal property (see (2.0)). We should think of x as the family of all morphisms from A to B.

This example will be studied in greater detail in §3.

(2.4) The property of being a monomorphism is not stable under localization since the α^* do not necessarily preserve monos. However, in the case where the base category is $\underline{Set}$ the α^* do preserve them. It appears that the appropriate generalization of monomorphism to the S-indexed case is that of *stable monomorphism*. A monomorphism in $\underline{A}$ is called *stable* if it is preserved by $\Delta_I\colon \underline{A} \to \underline{A}^I$ for every I. Localizing this, we say that a monomorphism in $\underline{A}^I$ is stable if it is preserved by $\alpha^*\colon \underline{A}^I \to \underline{A}^J$ for every α. *Stable epimorphisms* are defined similarly.

We are interested in subobjects and stable subobjects. We define a *subobject* of A to be an equivalence class of monomorphisms $A_0 \rightarrowtail A$, where $A_0 \rightarrowtail A$ is equivalent to $A_1 \rightarrowtail A$ if there exists a *canonical* isomorphism

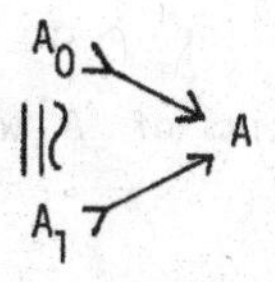

Our definition of subobject differs from the accepted one in that we require our isomorphism to be canonical. This should cause no problem since in practice we are mainly interested in subobjects in large categories (built out of Set in some way) and in these categories all isomorphisms are canonical.

A *stable subobject* of A is a subobject for which any (and therefore every) representative monomorphism is stable. We define *stable quotients* similarly.

Let A be an object of $\underline{A}$ and construct the discrete indexed category $\underline{\text{Mon}}(A)$ by letting $\underline{\text{Mon}}(A)^I$ have as objects all stable monomorphisms $\cdot \rightarrowtail \Delta_I A$ in $\underline{A}^I$. The only morphisms in $\underline{\text{Mon}}(A)^I$ are the obvious commutative triangles consisting of two monos and one canonical isomorphism

$$\begin{array}{ccc} A_0 & \searrow & \\ \wr\Vert & & \Delta_I A \\ A_1 & \nearrow & \end{array} .$$

If $\underline{\text{Mon}}(A)$ is small we say that A has a *small class of (stable) subobjects*. We denote the object which represents the strict category equivalent to $\underline{\text{Mon}}(A)$ by $\text{Sub}(A)$. Thus we have a natural bijection

$$\cong \frac{I \longrightarrow \text{Sub}(A)}{\text{stable subobjects of } \Delta_I A} .$$

A has a generic family of subobjects $A_0 \overset{\sigma}{\rightarrowtail} \Delta_{\text{Sub}(A)} A$ which has the property that for every I and every subobject $A_1 \overset{m}{\rightarrowtail} \Delta_I A$ there exists a unique $\phi\colon I \longrightarrow \text{Sub}(A)$ such that $\phi^*(\sigma) = m$ (equal as subobjects, i.e. canonically isomorphic).

Similarly we can say what it means for A to have a small class of stable quotients. The object of stable quotients of A will be denoted $\text{Quo}(A)$.

Remark: It might have been more convenient to define $\underline{\text{Mon}}(A)$ as above on objects but to take as morphisms all inclusions. The smallness of this $\underline{\text{Mon}}(A)$ implies the other and under mild conditions the converse is also true (see section (4.3)).

(2.5) Let $U\colon \underline{B} \longrightarrow \underline{A}$ be an indexed functor and A an object of $\underline{A}$. Define an indexed category $\underline{U}^{-1}(A)$ by letting $\underline{U}^{-1}(A)^I$ consist of all objects $B \in \underline{B}^I$ such that $U^I(B) \cong \Delta_I A$ (canonically) and all morphisms $b\colon B \longrightarrow B'$ such that

$$\begin{array}{ccc} U^I B & \xrightarrow{U^I b} & U^I B' \\ \| \wr & & \| \wr \\ \Delta_I A & = & \Delta_I A \end{array}$$

commutes.

If $\underline{U}^{-1}(A)$ is small, we say that the *fibre of U over A is small.* Rosebrugh [Rs1] and Penon [Pn2] have studied functors with small fibres (i.e. functors, all of whose fibres, including the ones at I , are small).

3. Small Homs:

(3.0) In this paragraph we introduce the important concept of a category having small homs (locally small categories in Bénabou's vocabulary). We first introduce the notion of having small homs at 1 which is more intuitive and then localize this definition to get the concept of small homs. All our results are first proved at 1 and then localized. Having small homs at 1 is a straightforward generalization of a $\underline{Set}$ based category having only a set of morphisms between any pair of objects. Having small homs is the "correct" generalization because it takes families of objects into account.

(3.1) $\underline{A}$ is said to have *small homs at 1* if for every A , B in $\underline{A}$, there is a small class of morphisms from A to B , i.e. there is an object $\mathrm{Hom}(A,B)$ in $\underline{S}$ such that there is a bijection

$$\frac{\Delta_I A \longrightarrow \Delta_I B \text{ in } \underline{A}^I}{I \longrightarrow \mathrm{Hom}(A,B) \text{ in } \underline{S}}$$

which is natural in I .

(3.2) $\underline{S}$ can be considered as a monoidal category with the monoidal structure given by cartesian product. If $\underline{A}$ has small homs at 1 , then $\underline{A}^1$ is enriched over $\underline{S}$ in the following way. For any A , B , C in $\underline{A}^1$ and any I in $\underline{S}$

$$[I,\mathrm{Hom}(A,B)\times\mathrm{Hom}(B,C)] \cong [I,\mathrm{Hom}(A,B)]\times[I,\mathrm{Hom}(B,C)]$$

$$\cong [\Delta_I A,\Delta_I B] \times [\Delta_I B,\Delta_I C] \longrightarrow [\Delta_I A,\Delta_I C]$$

$$\cong [I,\mathrm{Hom}(A,C)]$$

and since all these morphisms are natural in I , the Yoneda lemma gives us a, morphism

$$\mathrm{Hom}(A,B) \times \mathrm{Hom}(B,C) \longrightarrow \mathrm{Hom}(A,C) .$$

We get $id_A\colon 1 \longrightarrow \mathrm{Hom}(A,A)$ corresponding to $1_A\colon A \longrightarrow A$ in $\underline{A}^1$.

That these morphisms satisfy the associative and unitary laws follows easily from the Yoneda lemma and the fact that $\underline{A}^I$ is a category for all I .

It follows that the hom functor

$$[-,-]: \underline{A}^{1\ op} \times \underline{A}^{1} \longrightarrow \underline{Set}$$

lifts to a functor (not indexed)

$$\text{Hom}: \underline{A}^{1\ op} \times \underline{A}^{1} \longrightarrow \underline{S} .$$

If $a: A' \to A$ and $b: B \to B'$ are morphisms in $\underline{A}^1$ then for each I we get

$$[I,\text{Hom}(A,B)] \cong [\Delta_I A, \Delta_I B] \xrightarrow{[\Delta_I a, \Delta_I b]} [\Delta_I A', \Delta_I B'] \cong [I,\text{Hom}(A',B')]$$

which is natural in I and so gives a morphism

$$\text{Hom}(A,B) \longrightarrow \text{Hom}(A',B')$$

which we denote $\text{Hom}(a,b)$.

If $\underline{A}$ and $\underline{B}$ are indexed categories with small homs at 1 , and if $F: \underline{A} \to \underline{B}$ is an indexed functor, then $F^1: \underline{A}^1 \to \underline{B}^1$ is strong. We use the usual Yoneda lemma argument to get the strength of F^1 :

$$[I,\text{Hom}(A,A')] \cong [\Delta_I A, \Delta_I A'] \longrightarrow [F^I \Delta_I A, F^I \Delta_I A']$$

$$\cong [\Delta_I F^1 A, \Delta_I F^1 A'] \cong [I,\text{Hom}(F^1 A, F^1 A')]$$

which is natural in I at each stage and so we get a morphism

$$\text{Hom}(A,A') \longrightarrow \text{Hom}(F^1 A, F^1 A') .$$

That this satisfies the required properties to make F^1 strong follows by similar arguments.

It is also easily checked that indexed natural transformations between functors as above are strong at 1 .

Because we are dealing with indexed categories, the situation is slightly better than simply having a category enriched over $\underline{S}$. In the indexed case $\mathrm{Hom}(A,B)$ is determined up to isomorphism by the fact that $[I,\mathrm{Hom}(A,B)] \cong [\Delta_I A,\Delta_I B]$ and so having small homs at 1 is a property rather than extra structure. Then, as we saw above, indexed functors and indexed natural transformations are automatically strong.

(3.3) If $\underline{A}$ is an ordinary category canonically indexed by $\underline{\mathrm{Set}}$, then to say that $\underline{A}$ has small homs at 1 means simply that the morphisms between any two objects form a set (not a proper class).

(3.4) $\underline{S}$ has a canonical $\underline{S}$-indexing as mentioned in (I,2.2). $\underline{S}$ has small homs at 1 if and only if it is cartesian closed, as can easily be seen from the definition.

<u>Proposition</u>: If $\underline{S}$ is cartesian closed, then the $\underline{S}$-indexed category $\underline{S}^I$ has small homs at 1 .

<u>Proof</u>: Let $p\colon X \to I$ and $q\colon Y \to I$ be two objects of $\underline{S}^I$ (= $\underline{S}/I$) . Define $\mathrm{Hom}_{\underline{S}^I}(p,q)$ to be the pullback

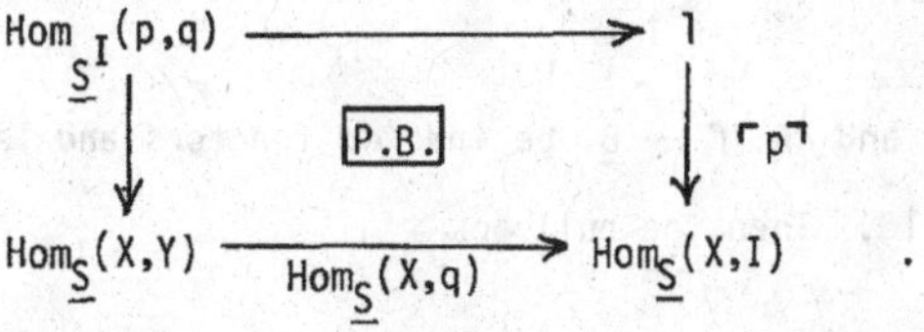

$$\begin{array}{ccc} \mathrm{Hom}_{\underline{S}^I}(p,q) & \longrightarrow & 1 \\ \downarrow & \boxed{\text{P.B.}} & \downarrow \ulcorner p \urcorner \\ \mathrm{Hom}_{\underline{S}}(X,Y) & \xrightarrow[\mathrm{Hom}_{\underline{S}}(X,q)]{} & \mathrm{Hom}_{\underline{S}}(X,I) \end{array} .$$

We have the following sequence of natural bijections

$$\cong \frac{J \to \mathrm{Hom}_{\underline{S}^I}(p,q)}{J \xrightarrow{\ulcorner\phi\urcorner} \mathrm{Hom}_{\underline{S}}(X,Y) \quad \text{s.t.} \quad \mathrm{Hom}_{\underline{S}}(X,q)\cdot\ulcorner\phi\urcorner = \ulcorner p\urcorner}$$

$$\cong \frac{}{\Delta_J X \xrightarrow{\phi} \Delta_J Y \quad \text{s.t.} \quad \Delta_J q\cdot\phi = \Delta_J P}$$

$$\cong \frac{}{J\times X \xrightarrow{\phi} J\times Y \quad \text{s.t.} \quad J\times q\cdot\phi = J\times p}$$

$$\cong \quad \Delta_J(p) \to \Delta_J(q) \quad \text{in} \quad (\underline{S}^I)^J$$

which shows that the $\mathrm{Hom}_{\underline{S}^I}$ which we have defined has the correct property. □

(3.5) If $\underline{A}$ has small homs at 1 then $\underline{A}^{op}$ and $\underline{A}^2$ also have. The case of $\underline{A}^{op}$ is obvious. For $a\colon A_0 \to A_1$ and $b\colon B_0 \to B_1$ two objects of $\underline{A}^2$, $\mathrm{Hom}_{\underline{A}^2}(a,b)$ is defined to be the pullback

$$\begin{array}{ccc} \mathrm{Hom}_{\underline{A}^2}(a,b) & \longrightarrow & \mathrm{Hom}_{\underline{A}}(A_0,B_0) \\ \downarrow & \boxed{\text{P.B.}} & \downarrow \mathrm{Hom}_{\underline{A}}(A_0,b) \\ \mathrm{Hom}_{\underline{A}}(A_1,B_1) & \xrightarrow[\mathrm{Hom}_{\underline{A}}(a,B_1)]{} & \mathrm{Hom}_{\underline{A}}(A_0,B_1) \end{array} .$$

If $\underline{A}$ and $\underline{B}$ have small homs at 1 then clearly so does $\underline{A}\times\underline{B}$.

Let $F\colon \underline{A} \to \underline{B}$ and $G\colon \underline{C} \to \underline{B}$ be indexed functors and let $\underline{A}$, $\underline{B}$, $\underline{C}$ have small homs at 1. Then the pullback

$$\begin{array}{ccc} \underline{D} & \longrightarrow & \underline{A} \\ \downarrow & & \downarrow F \\ \underline{C} & \xrightarrow[G]{} & \underline{B} \end{array}$$

also has small homs at 1. Indeed, if (A,C) and (A',C') are two objects of $\underline{D}$ then $\mathrm{Hom}_{\underline{D}}((A,C),(A',C'))$ is defined to be the pullback

$$\begin{array}{ccc} \mathrm{Hom}_{\underline{D}}((A,C),(A',C')) & \longrightarrow & \mathrm{Hom}_{\underline{A}}(A,A') \\ \downarrow & \boxed{\text{P.B.}} & \downarrow st_F \\ \mathrm{Hom}_{\underline{C}}(C,C') & \xrightarrow[st_G]{} & \mathrm{Hom}_{\underline{B}}(GC,GC') = \mathrm{Hom}_{\underline{B}}(FA,FA') \end{array}$$

where st_F and st_G are the strengths of F and G respectively. It is easily checked that $\mathrm{Hom}_{\underline{D}}$ has the required universal property.

Clearly the category $\mathbb{1}$ has small homs at 1 and so we conclude that any finite limit of categories with small homs at 1 also has this property.

It follows that for F and G as above, the comma category (F,G) also has small homs at 1.

(3.6) If $\underline{A}$ is a small category then it has small homs at 1. If $\underline{A} \simeq [\mathbb{C}]$ and $x,y\colon 1 \to C_0$ are the morphisms corresponding to A, B in $\underline{A}^1$ then $\mathrm{Hom}(A,B)$ is defined by the pullback

$$\begin{array}{ccc} \mathrm{Hom}(A,B) & \longrightarrow & C_1 \\ \downarrow & \boxed{\text{P.B.}} & \downarrow (\partial_0,\partial_1) \\ 1 & \xrightarrow[(x,y)]{} & C_0 \times C_0 \end{array} .$$

(3.7) Most categories constructed, using "finite limit type constructions", from categories with small homs at 1 will have small homs at 1 as long as the I-indexed families of the constructed category depend only on the I-indexed families of the given categories. For example, if $\underline{X}$ is a (non-indexed) finite category and $\underline{A}$ has small homs at 1 then so has $\underline{A}^{\underline{X}}$ (see I,2.6). Also, if $\underline{S}$ has small homs at 1 then so has $Gr(\underline{S})$, the category of group objects in $\underline{S}$.

If $\underline{A}$ has small homs at 1 it *does not* follow that $\underline{A}^I$ has. For example, let $\underline{S}$ be the category of at most denumerable sets and let $\underline{A}$ be the monoid $\mathbb{Z}/(2)$. This is a small category so has small homs at 1 . If I is an infinite set in $\underline{S}$ then $\underline{A}^I$ does not have small homs at 1 .

(3.8) It is desirable that any property concerned with indexed categories be stable under localization (i.e. that the property should hold not only for single objects but also for families of objects) and having small homs at 1 is not. This is reflected in the fact that the hom functor $\mathrm{Hom}: \underline{A}^{op} \times \underline{A} \to \underline{S}$ is *not* indexed. This leads us to localize the "small homs at 1" property and make the following definition. $\underline{A}$ is said to have *small homs* (or in Bénabou's notation, to be *locally small)* if for every I the $\underline{S}/I$-indexed category $\underline{A}/I$ has small homs at 1 . In more concrete terms this means that for every I and every A , B in $\underline{A}^I$ there is an object $\mathrm{Hom}^I(A,B): X \to I$ in $\underline{S}/I$ with the property that for every $\alpha: J \to I$ there is a bijection

$$\cong \frac{\alpha \to \mathrm{Hom}^I(A,B) \quad \text{in } \underline{S}/I}{\alpha^*A \to \alpha^*B \quad \text{in } \underline{A}^J}$$

which is natural in α .

By what was said above for categories with small homs at 1 we conclude that $\underline{A}^I$ (which is $(\underline{A}/I)^1$) is enriched over $\underline{S}/I$. It follows that Hom^I extends to a functor

$$\mathrm{Hom}^I: \underline{A}^{I\ op} \times \underline{A}^I \to \underline{S}/I \ .$$

In fact we get more than this. We get an indexed hom functor

$$\text{Hom}: \underline{A}^{op} \times \underline{A} \to \underline{S}$$

as indicated below.

For any $\alpha: J \to I$ and any $\beta: K \to J$ we have the following natural bijections

$$\cong \frac{\beta \to \text{Hom}^J(\alpha^*A,\alpha^*B) \quad \text{in } \underline{S}/J}{\beta^*(\alpha^*A) \to \beta^*(\alpha^*B) \quad \text{in } \underline{A}^K}$$
$$\cong \frac{}{(\alpha\beta)^*A \to (\alpha\beta)^*B \quad \text{in } \underline{A}^K}$$
$$\cong \frac{}{\alpha\beta \to \text{Hom}^I(A,B) \quad \text{in } \underline{S}/I}$$
$$\cong \frac{}{\beta \to \alpha^*\text{Hom}^I(A,B) \quad \text{in } S/J} .$$

Thus $\alpha^*\text{Hom}^I(A,B) \cong \text{Hom}^J(\alpha^*A,\alpha^*B)$ and so Hom is indexed.

Conversely if $\underline{A}^I$ is enriched over $\underline{S}^I$ for every I and if $\text{Hom}^I: \underline{A}^{I\ op} \times \underline{A}^I \longrightarrow \underline{S}^I$ gives us an indexed functor then $\underline{A}$ has small homs. Indeed, for any $\alpha: J \to I$ we have the following sequence of natural bijections

$$\cong \frac{\alpha \to \text{Hom}^I(A,B) \quad \text{in } \underline{S}/I}{1_J \to \alpha^*\text{Hom}^I(A,B) \quad \text{in } \underline{S}/J}$$
$$\cong \frac{}{1_J \longrightarrow \text{Hom}^J(\alpha^*A,\alpha^*B) \quad \text{in } \underline{S}/J}$$
$$\cong \frac{}{\alpha^*A \to \alpha^*B \quad \text{in } \underline{A}^J} .$$

Thus $\text{Hom}^I(A,B)$ has the required property.

Furthermore, the above sequence of bijections tells us $\text{Hom}^I(A,B)$ is uniquely determined up to isomorphism by the requirements that $\underline{A}^I$ be enriched over $\underline{S}^I$ for every I and that the hom functor be indexed.

It follows from (3.2) that if $\underline{A}$ and $\underline{B}$ have small homs and $F: \underline{A} \to \underline{B}$ is an indexed functor, then F^I is $\underline{S}/I$-strong for every I .

(3.9) All of the examples for small homs at 1 give corresponding examples for categories with small homs.

If $\underline{A}$ is a category canonically indexed by $\underline{\text{Set}}$ then $\underline{A}$ has small homs if and only if it has small homs at 1 .

$\underline{S}$ itself has small homs if and only if $\underline{S}/I$ is cartesian closed for every I . This holds, for example, if $\underline{S}$ is an elementary topos.

If $\underline{E}$ is an $\underline{S}$-topos with geometric morphism $p: \underline{E} \to \underline{S}$ then $\underline{E}$ is an $\underline{S}$-indexed category (I,2.4). $\underline{E}$ has small homs. Indeed, define $\text{Hom}(A,B) = p_*(B^A)$. Then we have the following sequence of natural bijections

$$\cong \frac{I \to p_*(B^A) \text{ in } \underline{S}}{p^*I \to B^A \text{ in } \underline{E}}$$

$$\cong \frac{p^*I \to B^A \text{ in } \underline{E}}{A \times p^*I \to B \text{ in } \underline{E}}$$

$$\cong \frac{A \times p^*I \to B \text{ in } \underline{E}}{\begin{array}{c} A \times p^*I \longrightarrow B \times p^*I \text{ in } \underline{E}/p^*I \\ \searrow \quad \swarrow \\ p^*I \end{array}}$$

$$\cong \frac{\begin{array}{c} A \times p^*I \longrightarrow B \times p^*I \text{ in } \underline{E}/p^*I \\ \searrow \quad \swarrow \\ p^*I \end{array}}{\Delta_I A \to \Delta_I B \text{ in } \underline{E}^I} \quad .$$

This shows that $\underline{E}$ has small homs at 1 .

Now being a topos over $\underline{S}$ is stable under localization and so the result follows by localization.

Any small category has small homs since small categories have small homs at 1 and the property of being a small category is stable under localization.

If $\underline{A}$ has small homs then so have $\underline{A}^{op}$ and $\underline{A}^{\mathbf{2}}$. A finite limit of categories with small homs has small homs and comma categories built from functors between categories with small homs again have small homs. All of these examples are true because the corresponding examples for small homs at 1 are true and each construction is stable under localization.

Similarly, if $\underline{A}$ has small homs and $\underline{X}$ is a (non-indexed) finite category, then $\underline{A}^{\underline{X}}$ also has small homs. If $\underline{S}$ has small homs then so does $Gr(\underline{S})$.

(3.10) If $\underline{A}$ has small homs then it does not follow that $\underline{A}^I$ has. However, we have the following results.

Proposition: If $\underline{S}$ is cartesian closed and $\underline{A}$ has small homs, then $\underline{A}^I$ has small homs at 1.

Proof: Given A and B in $\underline{A}^I$ we have $Hom^I(A,B)$ in $\underline{S}^I$ by small homs. Since $\underline{S}$ is cartesian closed, $\underline{S}^I$ has small homs at 1 by proposition (3.4). Define

$$Hom_{\underline{A}^I}(A,B) = Hom_{\underline{S}^I}(1,Hom^I(A,B))$$

where 1 is the terminal object of $\underline{S}^I$. For any J in $\underline{S}$ we have the following sequence of natural bijections

$$\begin{array}{rc} & J \to Hom_{\underline{A}^I}(A,B) \text{ in } \underline{S} \\ \cong & \overline{J \to Hom_{\underline{S}^I}(1,Hom^I(A,B)) \text{ in } \underline{S}} \\ \cong & \overline{\Delta_J(1) \to \Delta_J Hom^I(A,B) \text{ in } (\underline{S}^I)^J} \\ \cong & \overline{1 \to p_I^* Hom^I(A,B) \text{ in } \underline{S}^{I\times J}} \\ \cong & \overline{1 \to Hom^{I\times J}(p_I^*A,p_I^*B) \text{ in } \underline{S}^{I\times J}} \\ \cong & \overline{p_I^*A \to p_I^*B \text{ in } \underline{A}^{I\times J}} \\ \cong & \overline{\Delta_J A \to \Delta_J B \text{ in } (\underline{A}^I)^J} \end{array}$$

where $p_I\colon I\times J \to I$ is the projection. This shows that $\mathrm{Hom}_{\underline{A}^I}(A,B)$ has the right property. □

Corollary: If $\underline{S}$ has small homs and $\underline{A}$ has small homs then $\underline{A}^I$ has small homs for every I .

Proof: $\underline{S}$ has small homs implies that $\underline{S}$ is cartesian closed. Thus by the above proposition, $\underline{A}^I$ has small homs at 1 . Since the hypotheses are stable under localization, it follows that $\underline{A}^I$ has small homs. □

(3.11) The following theorem is an important generalization of Bénabou's construction of the full subcategory of a topos, generated by an internal family. The theorem was also known to Bénabou [Bn2].

(3.11.1) Theorem: If $\underline{A}$ has small homs and a small class of objects, then $\underline{A}$ is small.

Proof: Let $\mathrm{Ob}(\underline{A}) \simeq [C_0]$ with generic family of objects $A_0 \in \underline{A}^{C_0}$. Let $p_1, p_2\colon C_0 \times C_0 \to C_0$ be the first and second projections and let

$$\mathrm{Hom}^{C_0\times C_0}(p_1^*A_0, p_2^*A_0)\colon \; C_1 \to C_0 \times C_0$$

be the object of $\underline{S}^{C_0\times C_0}$ given by small homs. We have the following sequence of natural bijections

$$\cong \frac{I \to C_1 \text{ in } \underline{S}}{\phi: I \to C_0\times C_0 \text{ in } \underline{S} \text{ and } \phi \to \mathrm{Hom}^{C_0\times C_0}(p_1^*A_0, p_2^*A_0) \text{ in } \underline{S}/C_0\times C_0}$$

$$\cong \frac{\phi: I \to C_0\times C_0 \text{ in } \underline{S} \text{ and } \phi^*p_1^*A_0 \to \phi^*p_2^*A_0 \text{ in } \underline{A}^I}{}$$

$$\cong \frac{\phi: I \to C_0\times C_0 \text{ in } \underline{S} \text{ and } (p_1\phi)^*A_0 \to (p_2\phi)^*A_0 \text{ in } \underline{A}^I}{}$$

$$\cong \frac{\phi_1,\phi_2: I \to C_0 \text{ in } \underline{S} \text{ and } \phi_1^*A_0 \to \phi_2^*A_0 \text{ in } \underline{A}^I}{}$$

$$\cong \frac{A_1, A_2 \text{ in } \underline{A}^I \text{ (determined up to iso) and } A_1 \to A_2 \text{ in } \underline{A}^I}{\text{Morphisms of } \underline{A}^I}$$

which shows that $\mathrm{Mor}(\underline{A}) \cong [C_1]$, i.e. $\underline{A}$ has a small class of morphisms. It follows by proposition (2.2.1) that $\underline{A}$ is small. □

(3.11.2) <u>Corollary</u>: Let $\underline{A}$ have small homs and take an object A in $\underline{A}^I$ Then <u>Full</u>(A) is small.

<u>Proof</u>: From paragraph (2.1) we know that <u>Full</u>(A) has a small class of objects. <u>Full</u>(A) has the same homs as $\underline{A}$ and so has small homs. The result now follows from theorem (3.11.1). □

4. <u>Well-powered Categories</u>:

(4.1) In this paragraph we introduce the notion of an indexed category being well-powered. As with small homs, we first introduce the notion of being well-powered at 1 which is more intuitive, and then localize to get the definition of well-powered categories.

We say that $\underline{A}$ *is well-powered at 1* if every object A of $\underline{A}^1$ has a small class of (stable) subobjects (see (2.4)). $\underline{A}$ *is cowell-powered at 1* if every object A of $\underline{A}^1$ has a small class of (stable) quotient objects. $\underline{A}$ is cowell-powered at 1 if and only if $\underline{A}^{op}$ is well-powered at 1 .

To say that $\underline{A}$ is well-powered at 1 means that for every A in $\underline{A}^1$

there is an object Sub(A) in $\underline{S}$ such that there is a natural bijection

$$\cong \frac{I \longrightarrow \text{Sub}(A) \quad \text{in } \underline{S}}{\text{stable subobjects of } \Delta_I A \quad \text{in } \underline{A}^I} \quad .$$

It follows by taking the identity on top that we get a generic family of stable subobjects $A_0 \overset{\sigma_A}{\rightarrowtail} \Delta_{\text{Sub}(A)}A$ in $\underline{A}^{\text{Sub}(A)}$ with the property that for any I and any stable subobject $A' \rightarrowtail \Delta_I A$ there exists a unique $\alpha: I \longrightarrow \text{Sub}(A)$ such that $\alpha^* A_0 \cong A'$ (canonically as subobjects).

Dually, to say that $\underline{A}$ is cowell-powered at 1 means that for every A in $\underline{A}^1$ there is an object Quo(A) in $\underline{S}$ such that there is a natural bijection

$$\cong \frac{I \longrightarrow \text{Quo}(A) \quad \text{in } \underline{S}}{\text{stable quotients of } \Delta_I A \quad \text{in } \underline{A}^I}$$

As for Sub(A) there is a generic family of quotient objects $\Delta_{\text{Quo}(A)}A \overset{q_A}{\twoheadrightarrow} A_0$ with a similar universal property.

(4.2) If $\underline{A}$ is an ordinary category canonically indexed by $\underline{\text{Set}}$ then $\underline{A}$ is well-powered at 1 in the above sense if and only if it is well-powered in the usual sense, i.e. every object of $\underline{A}$ has only a set (not a proper class) of subobjects.

(4.2.1) Proposition: $\underline{S}$ considered as an $\underline{S}$-indexed category is well-powered at 1 if and only if it is an elementary topos.

Proof: If $\underline{S}$ is an elementary topos and A any object then we have the following sequence of natural bijections

$$\cong \frac{I \to \Omega^A \text{ in } \underline{S}}{\dfrac{A\times I \to \Omega \text{ in } \underline{S}}{\dfrac{\cdot \rightarrowtail A\times I \text{ in } \underline{S}}{\cdot \rightarrowtail \Delta_I A \text{ in } \underline{S}^I}}}$$

and since all subobjects are stable in $\underline{S}$ it follows that $\Omega^A = \mathrm{Sub}(A)$. Conversely, if $\underline{S}$ is well-powered at 1 then we let $\Omega = \mathrm{Sub}(1)$. Then the bijections

$$\begin{array}{c} I \to \mathrm{Sub}(A) \text{ in } \underline{S} \\ \hline \cdot \rightarrowtail \Delta_I A \text{ in } \underline{S}^I \\ \hline \cdot \rightarrowtail A\times I \text{ in } \underline{S} \\ \hline \cdot \rightarrowtail \Delta_{A\times I} 1 \text{ in } \underline{S}^{A\times I} \\ \hline A\times I \to \mathrm{Sub}(1) \text{ in } \underline{S} \end{array}$$

tell us that $\mathrm{Sub}(A) \cong \Omega^A$. It is well known that a category with finite limits and Ω^A for every A is an elementary topos (see [Pa]). □

(4.2.2) Proposition: If $\underline{S}$ is a topos, then the $\underline{S}$-indexed category $\underline{S}^I$ is well-powered at 1 .

Proof: If $p: X \to I$ is an arbitrary object of $\underline{S}^I$ then we have the following sequence of natural bijections:

$$\begin{array}{c} \text{Stable subobjects of } \Delta_J(p) \text{ in } (\underline{S}^I)^J \\ \hline \text{Stable subobjects of } p\times J: X\times J \to I\times J \text{ in } \underline{S}^{I\times J} \\ \hline \text{Stable subobjects of } X\times J \text{ in } \underline{S} \\ \hline X\times J \to \Omega \text{ in } \underline{S} \\ \hline J \to \Omega^X \text{ in } \underline{S} \end{array}$$

Therefore $\mathrm{Sub}(p) \cong \Omega^X$. □

(4.2.3) Proposition: If $\underline{S}$ is a topos, then $\underline{S}$ is cowell-powered at 1.

Proof: In a topos every epimorphism is the coequalizer of its kernel pair and is stable under localization (i.e. stable under pullback). Thus there is a bijection between stable quotients of X and kernel pairs into X and so we get $\mathrm{Quo}(X)$ as the subobject of $\Omega^{X\times X}$ consisting of the kernel pairs.

Consider the following sequence of natural transformations (not necessarily bijections) going from top to bottom

$$\begin{array}{cc}
 & I \to \Omega^{X\times X} \quad \text{in } \underline{S} \\
\cong & \overline{} \\
 & X_0 \rightarrowtail \Delta_I(X\times X) \quad \text{in } \underline{S}/I \\
\cong & \overline{} \\
 & X_0 \rightarrowtail \Delta_I X \times \Delta_I X \quad \text{in } \underline{S}/I \\
\downarrow & \overline{} \\
 & X_0 \rightrightarrows \Delta_I X \quad \text{in } \underline{S}/I \\
\downarrow & \overline{} \\
 & \Delta_I X \twoheadrightarrow (\mathrm{coeq}) \quad \text{in } \underline{S}/I \\
\downarrow & \overline{} \\
 & (\ker p) \rightrightarrows \Delta_I X \quad \text{in } \underline{S}/I \\
\downarrow & \overline{} \\
 & \cdot \rightarrowtail \Delta_I X \times \Delta_I X \quad \text{in } \underline{S}/I \\
\cong & \overline{} \\
 & \cdot \rightarrowtail \Delta_I(X\times X) \quad \text{in } \underline{S}/I \\
\cong & \overline{} \\
 & I \to \Omega^{X\times X} \quad \text{in } \underline{S}
\end{array}$$

where the fourth and fifth steps are natural because coequalizers and kernel pairs are stable. The Yoneda lemma gives us a morphism

$$\phi\colon \Omega^{X\times X} \to \Omega^{X\times X}$$

which represents the composite natural transformation. Let $\mathrm{Quo}(X)$ be the equalizer of ϕ with the identity. Then we have the bijections

$$\begin{array}{cc}
 & I \to \mathrm{Quo}(X) \\
\cong & \overline{} \\
 & \psi\colon I \to \Omega^{X\times X} \quad \text{such that} \quad \phi\psi = \psi \\
\cong & \overline{} \\
 & X_0 \rightrightarrows \Delta_I X \quad \text{which are kernel pairs} \\
\cong & \overline{} \\
 & \Delta_I X \twoheadrightarrow \cdot \quad \text{in } \underline{S}^I
\end{array}$$

which indicate that $\mathrm{Quo}(X)$ has the right universal property. □

(4.2.4) Proposition: If $\underline{S}$ is an elementary topos then the $\underline{S}$-indexed category $\underline{S}^I$ is cowell-powered at 1.

Proof: Since $\underline{S}$ is a topos, $\underline{S}/I$ is also a topos and so by Proposition (4.2.3) $\underline{S}/I$ is cowell-powered at 1 as an $\underline{S}/I$-indexed category. For $p: X \to I$, let $Quo^I(p)$ be the object in $\underline{S}/I$ of quotient objects of p. Let $Quo_I(p)$ be the object of $\underline{S}$ given by $Hom_{\underline{S}^I}(1, Quo^I(p))$ which exists since $\underline{S}^I$ has small homs at 1 by proposition (3.4). The remainder of the proof is similar to that of proposition (3.10). □

It is not known whether $\underline{S}$ must be an elementary topos in order to be cowell-powered at 1.

Although small categories are not necessarily well-powered at 1 there are several important cases where they are. If a small category $\underline{A}$ has stable kernel pairs or stable image factorizations (i.e. every morphism factors through a unique smallest stable subobject), then it is easily seen that $\underline{A}$ is well-powered at 1.

A less obvious fact is that if $\underline{S}$ has small homs, then every small category is well-powered. Before proving this, we need the following lemma, which says that for every morphism f of $\underline{A}$ there is an object of $\underline{S}$ which "measures the extent to which f is monic".

(4.2.5) Lemma: Let $\underline{A}$ be a small category and assume that $\underline{S}$ has small homs at 1. For every morphism $f: A \to A'$ of $\underline{A}$, there is a subobject of 1 in $\underline{S}$, $M(f)$, with the property that for any I in $\underline{S}$ $\Delta_I(f)$ is a stable monomorphism in $\underline{A}^I$ if and only if there is a (necessarily unique) morphism $I \to M(f)$.

Proof: The discrete indexed category, which at I consists of pairs of morphisms $\cdot \overset{x}{\underset{y}{\rightrightarrows}} \Delta_I A$ such that $\Delta_I f \cdot x = \Delta_I f \cdot y$, is clearly small. Let P be the object of $\underline{S}$ which represents it. Similarly, let Q be the object which represents the discrete indexed category consisting of morphisms $\cdot \xrightarrow{x} \Delta_I A$.

We have a monomorphism $\delta\colon Q \rightarrowtail P$ which sends $\cdot \xrightarrow{x} \Delta_I A$ to $\cdot \overset{x}{\underset{x}{\rightrightarrows}} \Delta_I A$.

Since $\underline{S}$ has small homs at 1, so has $\underline{S}^P$ (proposition (3.4)). Define $M(f) = \mathrm{Hom}_P(1_P, \delta)$. We have the following sequence of natural bijections:

$$\cong \frac{I \to M(f) \text{ in } \underline{S}}{I \to \mathrm{Hom}_P(1_P,\delta) \text{ in } \underline{S}}$$

$$\cong \frac{}{\Delta_I(1_P) \to \Delta_I(\delta) \text{ in } (\underline{S}^P)^I}$$

$$\cong \frac{}{\begin{array}{ccc} P\times I & \xrightarrow{(\phi,\psi)} & Q\times I \\ & {\scriptstyle 1_{P\times I}}\searrow \quad \swarrow{\scriptstyle \delta\times I} & \\ & P\times I & \end{array} \text{ in } \underline{S}/P\times I}$$

$$\cong \frac{}{\begin{array}{ccc} P\times I & \xrightarrow{\phi} & Q \\ & {\scriptstyle p_1}\searrow \quad \swarrow{\scriptstyle \delta} & \\ & P & \end{array} \text{ in } \underline{S}/P}$$

$$\cong \frac{}{\begin{cases} 1 & \text{if } p_1 \text{ factors through } \delta \\ 0 & \text{otherwise} \end{cases}}$$

Now p_1 factors through δ if and only if for every $J \xrightarrow{\alpha} I$ and $J \xrightarrow{\theta} P$, θ factors through δ. But $\theta\colon J \to P$ corresponds to a pair of morphisms $\cdot \overset{x}{\underset{y}{\rightrightarrows}} \Delta_J A$ such that $\Delta_J f \cdot x = \Delta_J f \cdot y$, and θ factors through δ if and only if $x = y$. It now follows that p_1 factors through δ if and only if for every $\alpha\colon J \to I$, $\Delta_J f$ is monic, i.e. if and only if $\Delta_I f$ is a stable monomorphism in $\underline{A}^I$. □

(4.2.6) Proposition: If $\underline{S}$ has small homs, then every small category is well-powered at 1 .

Proof: Let $\underline{A}$ be a small category and A an object of $\underline{A}$. Let Q be the object of all morphisms into A as in the previous lemma. We have a natural bijection

$$\cong \frac{\phi: I \to Q \text{ in } \underline{S}}{X \to \Delta_I A \text{ in } \underline{A}^I} \quad .$$

There is a generic family of morphisms $f: X \to \Delta_Q A$ and the above bijection is given by $\phi \mapsto \phi^*(f)$.

$\underline{A}/Q$ is a small $\underline{S}/Q$ category (see (1.3)) and $\underline{S}/Q$ has small homs at 1 . Define Sub(A) to be the domain of $M^Q(f)$ (which is the object of $\underline{S}/Q$ given by the previous lemma). We have the following sequence of natural bijections:

$$\begin{array}{l} \cong \overline{\qquad I \to \text{Sub}(A) \text{ in } \underline{S} \qquad} \\ \cong \overline{\quad I \xrightarrow{\phi} Q \text{ and } \phi \to M^Q(f) \text{ in } \underline{S}/Q \quad} \\ \cong \overline{I \xrightarrow{\phi} Q \text{ such that } \Delta_\phi(f) \text{ stable mono in } (\underline{A}/Q)^\phi} \\ \cong \overline{\quad I \xrightarrow{\phi} Q \text{ such that } \phi^*(f) \text{ stable mono in } \underline{A}^I \quad} \\ \qquad \text{stable monos} \rightarrowtail \Delta_I A \text{ in } \underline{A}^I \quad . \end{array}$$ □

(4.3) If for every I , $\underline{A}^I$ has stable pullbacks (i.e. pullbacks preserved by every α^*) then Sub(-) can be made into a functor $\underline{A}^{1\ op} \to \underline{S}$ as follows. For $f: A \to B$ any morphism in $\underline{A}^1$ we get a sequence of natural transformations

$$\cong \frac{I \to \mathrm{Sub}(B) \quad \text{in } \underline{S}}{\cong \frac{\bullet \rightarrowtail \Delta_I B \quad \text{in } \underline{A}^I}{\begin{array}{c} \bullet \rightarrowtail \Delta_I B \\ \uparrow \boxed{P.B.} \uparrow \Delta_I f \quad \text{in } \underline{A}^I \\ \bullet \rightarrowtail \Delta_I A \end{array}}}$$

$$\downarrow \overline{I \to \mathrm{Sub}(A) \quad \text{in } \underline{S}}$$

where the third step is natural in I since pullbacks are stable, and thus by the Yoneda lemma we get a morphism

$$\mathrm{Sub}(f)\colon \quad \mathrm{Sub}(B) \to \mathrm{Sub}(A)$$

called the inverse image morphism. Usual arguments show that this makes Sub into a functor $\underline{A}^{1\ \mathrm{op}} \to \underline{S}$. However, this functor is not indexed since it is only defined at 1 .

In the presence of stable pullbacks we have stable intersections which can be internalized to give a morphism

$$\cap\colon \mathrm{Sub}(A) \times \mathrm{Sub}(A) \to \mathrm{Sub}(A)$$

as follows. For any I we get the following sequence of natural transformations

$$\cong \frac{I \to \mathrm{Sub}(A) \times \mathrm{Sub}(A)}{\cong \frac{I \rightrightarrows \mathrm{Sub}(A)}{\begin{array}{c} A_1 \rightarrowtail \\ \qquad \Delta_I A \\ A_2 \rightarrowtail \end{array}}}$$

$$\downarrow \overline{\cong \frac{A_1 \cap A_2 \rightarrowtail \Delta_I A}{I \to \mathrm{Sub}(A)}}$$

which induces the morphism $\cap\colon \mathrm{Sub}(A) \times \mathrm{Sub}(A) \to \mathrm{Sub}(A)$. In fact, as is easily seen, we get a natural transformation

$$\cap : \mathrm{Sub}(-) \times \mathrm{Sub}(-) \longrightarrow \mathrm{Sub}(-) .$$

The intersection endows Sub(A) with a canonical order $\leq$ defined as the following equalizer

$$\circledS \overset{\leq}{\rightarrowtail} \mathrm{Sub}(A) \times \mathrm{Sub}(A) \underset{\cap}{\overset{p_1}{\rightrightarrows}} \mathrm{Sub}(A) .$$

We shall see later (Corollary (III,4.4)) that (co-) completeness of $\underline{A}$ (in the indexed sense) implies that $(\mathrm{Sub}(A),\leq)$ is a (co-) complete ordered object.

By following the identity through the above sequence of natural transformations we see that $\cap$ could alternately have been defined as the morphism from $\mathrm{Sub}(A) \times \mathrm{Sub}(A)$ to Sub(A) corresponding, by the universal property of Sub(A), to the intersection of $p_1^*\sigma_A$ with $p_2^*\sigma_A$ ($p_i : \mathrm{Sub}(A) \times \mathrm{Sub}(A) \to \mathrm{Sub}(A)$ is the ith projection and $\sigma_A : A_0 \rightarrowtail \Delta_{\mathrm{Sub}(A)}A$ is the generic family of subobjects).

For any object A of an indexed category $\underline{A}$, we get an indexed poset which we denote by $\underline{\mathrm{Sub}}(A)$ and which is defined by letting $\underline{\mathrm{Sub}}(A)^I$ be the poset of stable monomorphisms $\rightarrowtail \Delta_I A$ in $\underline{A}^I$. If $\underline{A}$ is well-powered at 1 then $\underline{\mathrm{Sub}}(A)$ has a small class of objects but it may not be small. However, if $\underline{A}$ has stable intersections then $\underline{\mathrm{Sub}}(A)$ *is* small. Indeed if $\circledS$ is the order relation as defined above then we have the following sequence of natural bijections:

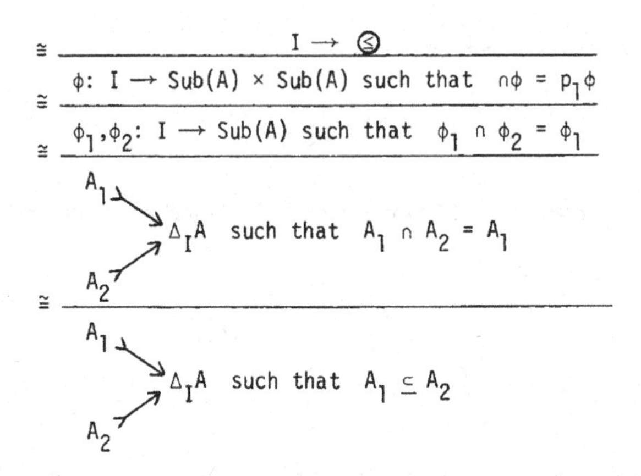

It follows that Ⓢ is the object of morphisms of a category which is canonically equivalent to $\underline{\mathrm{Sub}}(A)$.

For any A in $\underline{A}$ we have the comma category $\underline{A}/A$ which is indexed by $(\underline{A}/A)^I = \underline{A}^I/\Delta_I A$ (see (I.2.10) with $G = \ulcorner A \urcorner : \mathbb{1} \longrightarrow \underline{A}$). Clearly $\underline{\mathrm{Sub}}(A)$ is the full subcategory of $\underline{A}/A$ determined by the stable subobjects of 1_A .

<u>Proposition</u>: If $\underline{A}$ is well-powered at 1 and has small homs, then $\underline{\mathrm{Sub}}(A)$ is small.

<u>Proof</u>: If $\underline{A}$ has small homs, then by (3.9) so has $\underline{A}/A$, and $\underline{\mathrm{Sub}}(A)$ being a full subcategory of $\underline{A}/A$ must also have small homs. Well-poweredness implies that $\underline{\mathrm{Sub}}(A)$ has a small class of objects and the result now follows from (3.11.3). □

(4.4) If $\underline{A}$ has stable image factorizations (i.e. for every I and every morphism $f: A \to A'$ of $\underline{A}^I$, there is a unique smallest stable subobject $\mathrm{Im}(f) \rightarrowtail A'$ through which f factors, and for every $\alpha: J \to I$, $\alpha^*\mathrm{Im}(f) = \mathrm{Im}(\alpha^* f)$) then we get an indexed functor

$$\exists_f\colon \underline{Sub}(A) \to \underline{Sub}(B)$$

for every $f\colon A \to B$ as follows. For any I we let $\exists_f^I$ take $A_0 \rightarrowtail \Delta_I A$ to its image under $\Delta_I f$. This is easily checked to be an indexed functor. If $\underline{A}$ is well-powered at 1 then this induces a morphism $\exists_f\colon Sub(A) \to Sub(B)$ in $\underline{S}$. It can easily be seen that $\exists_f$ is functorial in f thus giving us a functor $\underline{A}^1 \to \underline{S}$.

If $\underline{A}$ has stable factorizations we also get an indexed functor

$$Im\colon \underline{A}/A \to \underline{Sub}(A)$$

defined at I by taking $a\colon A' \to \Delta_I A$ to any mono $\cdot \rightarrowtail \Delta_I A$ which determines the image of a. This is an indexed functor since the factorizations were assumed to be stable. Im is an indexed left adjoint to the inclusion functor

$$\underline{Sub}(A) \rightarrowtail \underline{A}/A.$$

If F is an indexed functor $\underline{A} \to \underline{B}$, $\underline{A}$ and $\underline{B}$ are well-powered at 1 and $\underline{B}$ has stable image factorizations, then the sequence of indexed functors

$$\underline{Sub}(A) \rightarrowtail \underline{A}/A \xrightarrow{\overline{F}} \underline{B}/FA \xrightarrow{Im} \underline{Sub}(FA)$$

gives a morphism $Sub(A) \to Sub(FA)$ in $\underline{S}$; it is the morphism which corresponds to the image of $F(\sigma_A)$ under the bijection which characterizes $Sub(FA)$. This example illustrates how external operations, such as F, can be internalized to give morphisms in $\underline{S}$. The morphisms $Sub(f)$ and $\exists_f$ can also be obtained by similar methods.

(4.5) We say that $\underline{A}$ is <u>well-powered</u> if $\underline{A}/I$ is well-powered at 1 as an $\underline{S}/I$-indexed category, for every I. Thus well-poweredness is stable under localization, and in fact is the stabilization of the property of being well-powered at 1 . More concretely, A is well-powered if for every I and every A in A^I there is an object $Sub^I(A): X \to I$ in S/I such that for every $\alpha: J \to I$ there is a natural bijection

$$\cong \frac{\alpha \to Sub^I(A) \text{ in } S/I}{\cdot \rightarrowtail \alpha^*A \text{ in } A^J}$$

where the bottom line represents stable subobjects of α^*A .

Dually, A is cowell-powered if A/I is cowell-powered at 1 for every I . Thus for every I and every A in A^I there is an object $Quo^I(A): X \to I$ in S/I such that for every α there is a natural bijection

$$\frac{\alpha \longrightarrow Quo^I(A) \text{ in } S/I}{\alpha^*A \twoheadrightarrow \cdot \text{ in } A^J} .$$

(4.6) If A is an ordinary category canonically indexed by Set then it is well-powered (resp. cowell-powered) in the sense defined above if and only if it is in the usual sense.

If S is an elementary topos then S and S^I are well-powered and cowell-powered. We have already seen that this is true at 1 and the property of being a topos is stable under localization and the construction of S^I from S is also stable under localization and so the result follows by localizing.

(4.6.1) Proposition: If E is an S-topos, then as an S-indexed category (I,2.4) E is well-powered and cowell-powered.

Proof: Let $p: E \to S$ be the structural geometric morphism. For any A in E , define $Sub(A) = p_*(\Omega^A)$. Then for any I in S we have the natural bijections

$$\cong \frac{I \to p_*(\Omega^A) \text{ in } \underline{S}}{p^*I \to \Omega^A \text{ in } \underline{E}} \cong \frac{}{\cdot \rightarrowtail \Delta_{p^*I}A \text{ in } \underline{E}/p^*I} \cong \frac{}{\cdot \rightarrowtail \Delta_I A \text{ in } \underline{E}^I}$$

which shows that $\underline{E}$ is well-powered at 1 . That $\underline{E}$ is well-powered follows by localization.

The cowell-poweredness is similar. □

(4.6.2) Proposition: If $\underline{S}$ has small homs, then every small category is well-powered and cowell-powered.

Proof: The result follows immediately from proposition (4.2.6) by localization and dualization. □

Definition: A category $\underline{A}$ has *boundable families* if for every I and every A in $\underline{A}^I$ there exists B in $\underline{A}^1$ and a stable monomorphism $A \rightarrowtail \Delta_I B$.

(4.6.3) Proposition: Let $\underline{A}$ be a category with stable intersections (of families) and boundable families. If $\underline{A}$ is well-powered at 1 then $\underline{A}$ is well-powered.

Proof: Let A be any object of $\underline{A}^I$ and B the bounding object of $\underline{A}^1$, i.e. $A \rightarrowtail \Delta_I B$. The following sequence of natural transformations

$$\cong \frac{J \to I \times \mathrm{Sub}(B)}{\alpha\colon J \to I \text{ and } X \rightarrowtail \Delta_J B} \downarrow \frac{}{X \cap \alpha^*A \rightarrowtail \Delta_J B} \downarrow \frac{}{J \to \mathrm{Sub}(B)}$$

induces a morphism

$$(\ )\cap A\colon I \times \mathrm{Sub}(B) \to \mathrm{Sub}(B)$$

in $\underline{S}$. Define $\mathrm{Sub}^I(A)$ by means of the following equalizer diagram

$$X \xrightarrow{(\mathrm{Sub}^I(A),\xi)} I \times \mathrm{Sub}(B) \underset{p_2}{\overset{(\ )\cap A}{\rightrightarrows}} \mathrm{Sub}(B) \ .$$

Now, for any $\alpha\colon J \to I$ we have the following sequence of natural bijections

$$\cong\ \frac{\alpha \to \mathrm{Sub}^I(A) \quad \text{in } \underline{S/I}}{J \xrightarrow{\phi} X \quad \text{such that} \quad \mathrm{Sub}^I(A)\cdot\phi = \alpha \quad \text{in } \underline{S}}$$

$$\cong\ \frac{}{J \xrightarrow{(\alpha,\psi)} I \times \mathrm{Sub}(B) \quad \text{such that} \quad (\ )\cap A\cdot(\alpha,\psi) = p_2\cdot(\alpha,\psi) \quad \text{in } \underline{S}}$$

$$\cong\ \frac{}{X \rightarrowtail \Delta_J B \quad \text{such that} \quad X \cap \alpha^*A = X \quad \text{in } \underline{A}^J}$$

$$\cong\ \frac{}{X \rightarrowtail \Delta_J B \quad \text{such that} \quad X \subseteq \alpha^*A \quad \text{in } \underline{A}^J}$$

$$\cong\ \frac{}{X \rightarrowtail \alpha^*A \quad \text{in } \underline{A}^J}$$

Thus $\mathrm{Sub}^I(A)$ has the required universal property. □

This proof is probably best understood as follows. First, $\mathrm{Sub}^I(\Delta_I B)$ always exists and is given by $\Delta_I \mathrm{Sub}(B)$. Then, since stable intersections exist, we internalize the operation of intersecting with A to get a morphism $\mathrm{Sub}^I(\Delta_I B) \to \mathrm{Sub}^I(\Delta_I B)$. Finally, this allows us to define $\mathrm{Sub}^I(A)$ to consist of those subobjects of $\Delta_I B$ which are equal to their intersection with A , i.e. contained in A .

In the case where $\underline{A} = \underline{S}$, well-powered at 1 means that $\underline{S}$ is a topos and well-powered means that for every I , $\underline{S/I}$ is a topos. So proposition (4.6.2) gives a different proof of the fact that being a topos is stable under localization.

(4.7) Assume that $\underline{A}$ is well-powered. For any $K \xrightarrow{\beta} J \xrightarrow{\alpha} I$ in $\underline{S}$

and any A in $\underline{A}^I$ we have the following natural bijections

$$\begin{array}{l} \cong \dfrac{\beta \longrightarrow \mathrm{Sub}^J(\alpha^*A) \quad \text{in } \underline{S}/J}{\cdot \rightarrowtail \beta^*\alpha^*A \quad \text{in } \underline{A}^K} \\ \cong \dfrac{}{\cdot \rightarrowtail (\alpha\beta)^* \quad \text{in } \underline{A}^K} \\ \cong \dfrac{}{\alpha\beta \longrightarrow \mathrm{Sub}^I(A) \quad \text{in } \underline{S}/I} \\ \cong \dfrac{}{\beta \longrightarrow \alpha^*\mathrm{Sub}^I(A) \quad \text{in } \underline{S}/J} \end{array} .$$

Thus $\mathrm{Sub}^J(\alpha^*A) \cong \alpha^*\mathrm{Sub}^I(A)$ and so, when $\underline{A}$ has stable pullbacks, we get an indexed functor

$$\mathrm{Sub}\colon \underline{A}^{op} \longrightarrow \underline{S} .$$

(If $\underline{A}$ does not have stable pullbacks, we still get an indexed functor $\mathrm{Sub}\colon \ \mathrm{Ob}(\underline{A})^{op} \longrightarrow \underline{S}$ (see (2.1)).)

(4.8) If $\underline{A}$ is well-powered it does not follow that $\underline{A}^I$ is well-powered. However we do have the following result analogous to corollary (3.10).

Proposition: If $\underline{A}$ is well-powered and $\underline{S}$ has small homs, then $\underline{A}^I$ is well-powered for every I.

Proof: For any A in $\underline{A}^I$ we have $\mathrm{Sub}^I(A)$ in $\underline{S}^I$ by well-poweredness. Consider the object of $\underline{S}$ given by $\mathrm{Hom}_{\underline{S}^I}(1,\mathrm{Sub}^I(A))$ where 1 is the terminal object of $\underline{S}^I$. For any J in $\underline{S}$ we have the following sequence of natural bijections

$$\begin{array}{c} J \to \mathrm{Hom}_{\underline{S}^I}(1, \mathrm{Sub}^I(A)) \quad \text{in } \underline{S} \\ \hline \cong \quad \Delta_J(1) \to \Delta_J \mathrm{Sub}^I(A) \quad \text{in } (\underline{S}^I)^J \\ \hline \cong \quad 1 \to p_1^* \mathrm{Sub}^I(A) \quad \text{in } \underline{S}^{I\times J} \\ \hline \cong \quad 1 \to \mathrm{Sub}^{I\times J}(p_1^* A) \quad \text{in } \underline{S}^{I\times J} \\ \hline \cong \quad \cdot \rightarrowtail p_1^* A \quad \text{in } \underline{A}^{I\times J} \\ \hline \cong \quad \cdot \rightarrowtail \Delta_J A \quad \text{in } (\underline{A}^I)^J \end{array}$$

thus $\mathrm{Hom}_{\underline{S}^I}(1, \mathrm{Sub}^I(A))$ gives us $\mathrm{Sub}(A)$ for $\underline{A}^I$ as an $\underline{S}$-indexed category. We have shown that $\underline{A}^I$ is well-powered at 1 and now the result follows by localizing. □

III. Limits:

1. Functor Categories:

(1.1) Let $\underline{A}$ and $\underline{B}$ be indexed categories. We define an I-indexed family of indexed functors from $\underline{B}$ to $\underline{A}$ to be an indexed functor $\underline{B} \to \underline{A}^I$. If for each I the category of indexed functors $\underline{B} \to \underline{A}^I$ with indexed natural transformations is legitimate (i.e. there is only a set of indexed natural transformations between any two indexed functors) then this gives us the I-indexed families of an indexed category which we denote $\underline{A}^{\underline{B}}$. Thus $(\underline{A}^{\underline{B}})^I = (\underline{A}^I)^{\underline{B}}$ by definition. For $\alpha: J \to I$, $\alpha^*: \underline{A}^I \to \underline{A}^J$ is an indexed functor so given any indexed functor $G: \underline{B} \to \underline{A}^I$ we get another one by composition $\alpha^* \cdot G: \underline{B} \to \underline{A}^J$. This defines the substitution functor $\alpha^*: (\underline{A}^{\underline{B}})^I \to (\underline{A}^{\underline{B}})^J$. Thus $\alpha^*(G) = \alpha^* \cdot G$.

$\underline{A}^{\underline{B}}$ is clearly 2-functorial in $\underline{A}$ and $\underline{B}$, covariant in $\underline{A}$ and contravariant on functors and covariant on natural transformations in $\underline{B}$. In particular, corresponding to the unique functor $\underline{B} \to \mathbb{1}$ we get an indexed functor $\underline{A}^{\mathbb{1}} \to \underline{A}^{\underline{B}}$ and we denote $\underline{A} \cong \underline{A}^{\mathbb{1}} \to \underline{A}^{\underline{B}}$ by $\Delta_{\underline{B}}$. For any I and any A in $\underline{A}^I$, $\Delta^I_{\underline{B}}(A): \underline{B} \to \underline{A}^I$ is given at J by

$$\Delta^I_{\underline{B}}(A)^J(B) = \Delta_J A$$

where $\Delta_J: \underline{A}^I \to (\underline{A}^I)^J$.

(1.1.1) Proposition: There is an isomorphism of categories between the category of $\underline{S}$-indexed functors $\underline{C} \to \underline{A}^{\underline{B}}$ and the category of $\underline{S}$-indexed functors $\underline{B} \times \underline{C} \to \underline{A}$.

Proof (Sketch): Let $\Phi: \underline{C} \to \underline{A}^{\underline{B}}$ be an indexed functor. For any C in $\underline{C}^I$, $\Phi^I(C)$ is an indexed functor $\underline{B} \to \underline{A}^I$ whose I^{th} component is a functor $\Phi^I(C)^I: \underline{B}^I \to (\underline{A}^I)^I = \underline{A}^{I \times I}$. Let $\Delta: I \to I \times I$ be the diagonal morphism. We define an indexed functor $\check{\Phi}: \underline{B} \times \underline{C} \to \underline{A}$ at I by the formula $\check{\Phi}^I(B,C) = \Delta^* \Phi^I(C)^I(B)$.

It is routine to check that $\check{\Phi}^I$ is a functor and that this gives us an indexed functor $\check{\Phi}$.

Conversely, given $\Psi\colon \underline{B}\times\underline{C} \longrightarrow \underline{A}$ we define an indexed functor $\hat{\Psi}\colon \underline{C} \longrightarrow \underline{A}^{\underline{B}}$ as follows. For any C in $\underline{C}^I$, $\hat{\Psi}^I(C)$ should be an indexed functor $\underline{B} \longrightarrow \underline{A}^I$. Let B be an object in $\underline{B}^J$ and let $\pi_1\colon I\times J \longrightarrow I$ and $\pi_2\colon I\times J \longrightarrow J$ be projections. Then $\hat{\Psi}^I(C)^J(B)$ is defined to be $\Psi^{I\times J}(\pi_2^*B,\pi_1^*C)$ which is an object in $\underline{A}^{I\times J} = (\underline{A}^I)^J$. The details of the verification that $\hat{\Psi}^I(C)$ is an indexed functor $\underline{B}\longrightarrow\underline{A}^I$, $\hat{\Psi}$ is an indexed functor $\underline{C} \longrightarrow \underline{A}^{\underline{B}}$, and $(\hat{\ })$ and $(\check{\ })$ give the desired isomorphism are left to the reader. □

It is easily seen that $\Delta_{\underline{B}}\colon \underline{A} \longrightarrow \underline{A}^{\underline{B}}$ as defined above, corresponds to the projection $\underline{B}\times\underline{A} \longrightarrow \underline{A}$ under the isomorphism just described.

(1.1.2) <u>Corollary</u>: $(\underline{A}^{\underline{B}})^{\underline{C}} \cong \underline{A}^{\underline{B}\times\underline{C}}$ as indexed categories. □

The above proposition shows that our definition of the indexed functor category, and in particular of I-indexed families of functors, is the correct one. However, in (I,3.2) we claim that the $\underline{S}$/I-indexed category $\underline{A}/I$ should be thought of as the I-indexed family of $\underline{S}$-indexed categories which is constantly $\underline{A}$, and this should give an alternate definition of I-indexed family of functors. The next proposition says that these two definitions are equivalent.

(1.1.3) <u>Proposition</u>: The category of $\underline{S}$-indexed functors $\underline{B} \longrightarrow \underline{A}^I$ is isomorphic to the category of $\underline{S}$/I-indexed functors $\underline{B}/I \longrightarrow \underline{A}/I$.

<u>Proof</u> (Sketch): Given an $\underline{S}$-indexed functor $\Phi\colon \underline{B} \longrightarrow \underline{A}^I$, define an $\underline{S}$/I-indexed functor $\check{\Phi}\colon \underline{B}/I \longrightarrow \underline{A}/I$ by letting $\check{\Phi}^\alpha$ be the composite

$$(\underline{B}/I)^\alpha = \underline{B}^J \xrightarrow{\Phi^J} (\underline{A}^I)^J = \underline{A}^{I\times J} \xrightarrow{(\alpha,J)^*} \underline{A}^J = (\underline{A}/I)^\alpha$$

for $\alpha: J \to I$.

Conversely given an $\underline{S}/I$-indexed functor $\Psi: \underline{B}/I \to \underline{A}/I$, define an $\underline{S}$-indexed functor $\hat{\Psi}: \underline{B} \to \underline{A}^I$ by letting $\hat{\Psi}^J$ be the composite

$$\underline{B}^J \xrightarrow{\pi_2^*} \underline{B}^{I\times J} = (\underline{B}/I)^{\pi_1} \xrightarrow{\Psi^{\pi_1}} (\underline{A}/I)^{\pi_1} = \underline{A}^{I\times J} = (\underline{A}^I)^J$$

where $\pi_1: I\times J \to I$ and $\pi_2: I\times J \to J$ are the projections.

It is easily checked that this establishes the desired isomorphism. □

(1.1.4) Corollary: The construction $\underline{A}^{\underline{B}}$ is stable under localization, i.e. for every I we have an $\underline{S}/I$-indexed isomorphism $(\underline{A}^{\underline{B}})/I \cong (\underline{A}/I)^{(\underline{B}/I)}$.

Proof: For any $\alpha: J \to I$ we have the following sequence of natural isomorphisms of categories

$$\cong \frac{\underline{S}/I\text{-indexed functors } \ \underline{B}/I \to (\underline{A}/I)^\alpha}{(\underline{S}/I)/\alpha\text{-indexed functors } \ (\underline{B}/I)/\alpha \to (\underline{A}/I)/\alpha}$$

$$\cong \frac{\underline{S}/J\text{-indexed functors } \ \underline{B}/J \to \underline{A}/J}{\underline{S}\text{-indexed functors } \ \underline{B} \to \underline{A}^J}$$

with the third $\cong$ preceding the last line. □

(1.2) Let $\underline{A}$ be an indexed category with small homs. Then we have an indexed functor $\mathrm{Hom}: \underline{A}^{op} \times \underline{A} \to \underline{S}$ and by (1.1.1) we get an indexed functor $Y: \underline{A} \to \underline{S}^{\underline{A}^{op}}$ called the Yoneda functor (assuming all the categories involved are legitimate). For $A \in \underline{A}$, $Y(A): \underline{A}^{op} \to \underline{S}$ is defined at I by

$$Y(A)^I(B) = \mathrm{Hom}^I(B, \Delta_I A) .$$

Whether the categories used to define $\underline{S}^{\underline{A}^{op}}$ are legitimate or not, $Y(A)$ makes sense.

Proposition: For any indexed functor $\Phi: \underline{A}^{op} \to \underline{S}$ and any object A of

$\underline{A}$, the class of indexed natural transformations from $Y(A)$ to Φ is small and is represented by $\Phi(A)$.

Proof: According to (II,2.3) we are required to demonstrate a natural bijection

$$\cong \frac{I \longrightarrow \Phi(A) \quad \text{in } \underline{S}}{\Delta_I \cdot Y(A) \longrightarrow \Delta_I \cdot \Phi \quad \text{in } (\underline{S}^I)^{\underline{A}^{op}}} \quad .$$

Since $\Delta_I \cdot Y(A)$ and $\Delta_I \cdot \Phi$ are functors into $\underline{S}^I$, an $\underline{S}$-indexed natural transformation $\Delta_I \cdot Y(A) \longrightarrow \Delta_I \cdot \Phi$ is the same as an $\underline{S}$-indexed natural transformation $I \times \mathrm{Hom}(-,A) \longrightarrow \Phi$.

Given a natural transformation $t: I \times \mathrm{Hom}(-,A) \to \Phi$ we get a morphism

$$I \cong I \times 1 \xrightarrow{I \times \ulcorner 1_A \urcorner} I \times \mathrm{Hom}(A,A) \xrightarrow{t(A)} \Phi(A) \ .$$

Now, we have a canonical natural transformation

$$\Phi(A) \times \mathrm{Hom}(-,A) \longrightarrow \Phi$$

which corresponds to the strength of Φ . If we are given a morphism $I \longrightarrow \Phi(A)$ then we get an indexed natural transformation

$$I \times \mathrm{Hom}(-,A) \longrightarrow \Phi(A) \times \mathrm{Hom}(-,A) \longrightarrow \Phi \ .$$

The details showing that these two processes are inverse to each other, being computational and straightforward, are omitted. □

Corollary: $Y: \underline{A} \longrightarrow \underline{S}^{\underline{A}^{op}}$ is full and faithful.

Proof: For A and B any two objects of $\underline{A}$, the preceding proposition says that I-indexed families of natural transformations from $Y(A)$ to $Y(B)$ are in natural bijection with morphisms

$$I \longrightarrow Y(B)(A) = \mathrm{Hom}(A,B)$$

i.e. I-indexed families of morphisms from A to B in $\underline{A}$. One has only to check, now, that this bijection is induced by Y. □

(1.3) The case we are mainly interested in is when $\underline{B}$ is $[\mathbb{C}]$. As was seen in (II,1.5.1), the category of indexed functors $[\mathbb{C}] \longrightarrow \underline{A}^I$ is equivalent to the category of internal functors $\mathbb{C} \longrightarrow \underline{A}^I$. This category of internal functors is clearly a legitimate category and so $\underline{A}^{[\mathbb{C}]}$ is always defined. We denote the category of internal functors $\mathbb{C} \longrightarrow \underline{A}$ by $\underline{A}^{\mathbb{C}}$. This is an indexed category with $(\underline{A}^{\mathbb{C}})^I = (\underline{A}^I)^{\mathbb{C}}$ and $\underline{A}^{\mathbb{C}} \simeq \underline{A}^{[\mathbb{C}]}$ as indexed categories.

If we let I denote the discrete category object with object of objects I in $\underline{S}$, then $\underline{A}^{[I]} \simeq \underline{A}^I$ where $\underline{A}^I$ represents the indexed category of I-indexed families of $\underline{A}$ as defined in (I,2.8). Thus the two possible interpretations of $\underline{A}^I$ agree.

It follows that an object of $(\underline{A}^{\underline{B}})^I$ may be viewed as a functor $\underline{B}\times[I] \longrightarrow \underline{A}$. In particular $(\underline{A}^{\mathbb{C}})^I \cong \underline{A}^{\mathbb{C}\times I}$.

(1.4) Proposition: If $\underline{A}^{C_0}$ and $\underline{A}^{C_1}$ have small homs at 1 then so does $\underline{A}^{\mathbb{C}}$.

Proof: If (X,ξ) and (Y,θ) are internal functors from $\mathbb{C}$ to $\underline{A}$, then the equalizer of

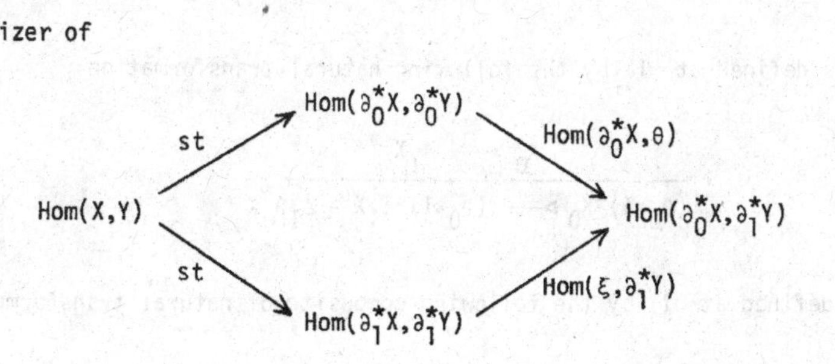

is easily seen to have the universal property of $\mathrm{Hom}((X,\xi),(Y,\theta))$. □

Remark: All that is needed in the preceding proposition is that $\underline{A}^{C_1}$ have small homs for then it follows that $\underline{A}^{C_0}$ has also. If A,A' are in $\underline{A}^{C_0}$ then $\mathrm{Hom}(A,A')$ may be computed as the equalizer of

$$\mathrm{Hom}(\partial_0^* A,\partial_0^* A') \underset{st_{(id\cdot\partial_0)^*}}{\overset{1}{\rightrightarrows}} \mathrm{Hom}(\partial_0^* A,\partial_0^* A') \ .$$

Corollary: If $\underline{S}$ and $\underline{A}$ have small homs then so does $\underline{A}^{\mathbb{C}}$.

Proof: This follows immediately from the preceding proposition, corollary (II,3.10), and localization. □

(1.5) Proposition: Assume that $\underline{A}^I$ has stable pullbacks for every I in $\underline{S}$. If $\underline{A}^{C_0}$ and $\underline{A}^{C_1}$ are well-powered at 1 then $\underline{A}^{\mathbb{C}}$ is well-powered at 1 .

Proof: The forgetful functor $\underline{A}^{\mathbb{C}} \longrightarrow \underline{A}^{C_0}$ creates stable pullbacks and therefore it preserves and reflects monomorphisms (all of which are necessarily stable).

For any internal functor $(X,\xi)\colon \mathbb{C} \longrightarrow \underline{A}$, define $\mathrm{Sub}(X,\xi)$ to be the equalizer of

$$\mathrm{Sub}(X) \underset{\psi}{\overset{\phi}{\rightrightarrows}} \mathrm{Sub}(\partial_0^* X)$$

where ϕ is defined at I by the following natural transformation

$$\downarrow \frac{X_0 \rightarrowtail \Delta_I X}{(\partial_0\times I)^* X_0 \rightarrowtail (\partial_0\times I)^*\Delta_I X = \Delta_I \partial_0^* X}$$

and ψ is defined at I by the following composite of natural transformations

$$\begin{array}{c} X_0 \rightarrowtail \Delta_I X \\ \hline (\partial_0\times I)^*X_0 \rightarrowtail (\partial_0\times I)^*\Delta_I X \text{ and } (\partial_1\times I)^*X_0 \rightarrowtail (\partial_1\times I)^*\Delta_I X \\ \hline (\partial_0\times I)^*X_0 \rightarrowtail \Delta_I\partial_0^* X \text{ and } (\partial_1\times I)^*X_0 \rightarrowtail \Delta_I\partial_1^* X \\ \uparrow \qquad \text{P.B.} \qquad \uparrow \Delta_I\xi \\ X_1 \rightarrowtail \Delta_I\partial_0^* X \\ \hline (\partial_0\times I)^*X_0 \cap X_1 \rightarrowtail \Delta_I\partial_0^* X \end{array}$$

It is routine to check that $Sub(X,\xi)$ has the right universal property. □

Remark: It is not necessary to require that $\underline{A}^{C_0}$ be well-powered at 1 since it follows from the fact that $\underline{A}^{C_1}$ is well-powered at 1. For A in $\underline{A}^{C_0}$ we can define $Sub(A)$ to be the equalizer of

$$Sub(\partial_0^*A) \overset{1}{\underset{\alpha}{\rightrightarrows}} Sub(\partial_0^*A)$$

where α is defined at I by the following natural transformation

$$\begin{array}{c} A_0 \rightarrowtail \Delta_I\partial_0^*A \\ \hline (id\cdot\partial_0\times I)^*A_0 \rightarrowtail (id\cdot\partial_0\times I)^*\Delta_I\partial_0^*A \cong \Delta_I\partial_0^*A \end{array}.$$

Corollary: Assume that $\underline{A}^I$ has stable pullbacks for every I and that $\underline{S}$ has small homs. If $\underline{A}$ is well-powered then $\underline{A}^{\mathbb{C}}$ is also.

Proof: This follows immediately from the previous proposition and proposition (II,4.8), by localization. □

(1.6) We end this section by stating the following proposition whose straightforward proof is left to the reader.

Proposition: Assume that $\underline{S}$ is cartesian closed, i.e. has small homs.

If $\underline{A}$ is small, then so is $\underline{A}^{\mathbb{C}}$. □

2. Adjoints:

(2.1) In ordinary category theory, adjoints are defined in either of the two following ways:

(a) $U: \underline{B} \to \underline{A}$ has a left adjoint $F: \underline{A} \to \underline{B}$ if there are natural transformations $\varepsilon: FU \to 1_{\underline{B}}$ and $\eta: 1_{\underline{A}} \to UF$ such that $\varepsilon F \cdot F\eta = 1_F$ and $U\varepsilon \cdot \eta U = 1_U$ (or if there is a bijection $[A,UB] \cong [FA,B]$, natural in A and B),

(b) $U: \underline{B} \to \underline{A}$ has a left adjoint if for every A in $\underline{A}$ a certain universal problem has a solution (or the functor $[A,U-]: \underline{B} \to \underline{Set}$ is representable).

The difference is that in (a) the functor F is given explicitly whereas in (b) it is not. If U has a left adjoint as in (a) then it also has one as in (b). If U has a left adjoint as in (b), then choosing a representative from each isomorphism class of solutions to the universal problem, we get a functor F which is left adjoint to U as defined in (a). Thus definitions (a) and (b) are not quite equivalent; going from (b) to (a) requires a choice. In the indexed situation, this is a more important consideration and we get two (related) concepts of adjoint.

The following definition corresponds to (a) and is available in any 2-category. It is the one to which we have been referring since the beginning of the paper.

Definition: Let $U: \underline{B} \to \underline{A}$ be an indexed functor. We say that U has a *canonical left adjoint* if there are an indexed functor $F: \underline{A} \to \underline{B}$ and indexed natural transformations $\varepsilon: FU \to 1_{\underline{B}}$ and $\eta: 1_{\underline{A}} \to UF$ such that $\varepsilon F \cdot F\eta = 1_F$

and $U\varepsilon \cdot \eta U = 1_U$.

Proposition: $U: \underline{B} \to \underline{A}$ has a canonical left adjoint if and only if for each I in $\underline{S}$, $U^I: \underline{B}^I \to \underline{A}^I$ has a left adjoint F^I and for every $\alpha: J \to I$ the canonical morphism $F^J\alpha^* \to \alpha^*F^I$ is a canonical isomorphism.

Proof: If U has a canonical left adjoint F then F^I is left adjoint to U^I for all I . If we denote by ϕ_F the isomorphism $\alpha^*F^J \to F^I\alpha^*$ (similarly for ϕ_U , etc.), the following diagram shows that ϕ_F is equal to the canonical (induced by adjointness and ϕ_U) morphism in the statement of the proposition (which is the path around the top):

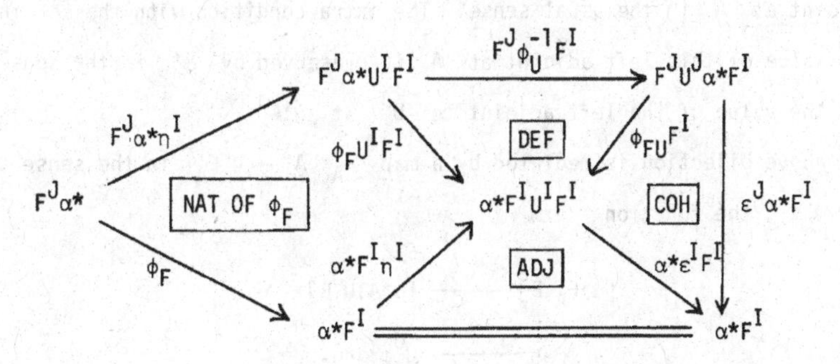

where the "Coh" triangle commutes by the "well-known" coherence conditions mentioned in (I,1.2).

The proof of the converse is similar. □

(2.2) The concept of adjoint which is the more useful to us is the one which corresponds to definition (b). In this case we shall say that U has a left adjoint (not canonical). We first introduce the preliminary notions of left adjoint at 1 and left adjoint at a single object, which we find useful in practice.

(2.2.1) Definition: Let $U: \underline{B} \to \underline{A}$ be an indexed functor and A an object of $\underline{A}^I$ for some I. We say that *U has a left adjoint at A* if there is an object F_A in $\underline{B}^I$ such that for every $\alpha: J \to I$ and every B in $\underline{B}^J$ we have a bijection

$$\cong \frac{\alpha^*F_A \to B \text{ in } \underline{B}^J}{\alpha^*A \to U^J B \text{ in } \underline{A}^J}$$

which is natural in B and α.

By taking $\alpha = 1_I$ we see that the above condition implies that U^I has a left adjoint at A in the usual sense. The extra condition with the α insures that the value of this left adjoint at A is preserved by α^* in the sense that α^*F_A is the value of the left adjoint to U^J at α^*A.

The above bijection is mediated by a map $\eta_A: A \to U^I F_A$ in the sense that for every α, the function

$$[\alpha^*F_A, B] \longrightarrow [\alpha^*A, U^J B]$$
$$\phi \longmapsto U^J\phi \cdot \alpha^*\eta_A$$

is a bijection.

(2.2.2) Proposition: U has a left adjoint at A in $\underline{A}^1$ if and only if the indexed comma category (A,U) has a stable initial object.

Proof: Straightforward computation. □

(2.2.3) Definition: We say that *U has a left adjoint at 1* if U has a left adjoint at A for every A in $\underline{A}^1$.

Remark: If $\underline{A}$ has a terminal object then there could be some confusion between the two previous definitions. It will usually be clear from the context which we mean but if confusion is possible we will specify whether we mean "at the object 1" or "defined on the category $\underline{A}^1$ ".

If U has a left adjoint at 1, then in particular the functor $U^1: \underline{B}^1 \to \underline{A}^1$ has an ordinary left adjoint which we denote by F^1 .

(2.2.4) Definition: We say that *U has a left adjoint* if it has a left adjoint at every A in $\underline{A}^I$ for every I .

This is the concept which will be of most use to us. Having a left adjoint is a concept which is stable under localization. In fact it is the localization of the concept of having a left adjoint at 1 in the sense that U has a left adjoint if and only if U/I has a left adjoint at 1 for every I . In practice it is easier to show that a functor has a left adjoint at 1 and then conclude by localization that it has a left adjoint everywhere.

(2.2.5) Proposition: An indexed functor U has a left adjoint if and only if each functor $U^I: \underline{B}^I \to \underline{A}^I$ has an ordinary left adjoint F^I and for each $\alpha: J \to I$ the canonical morphism $F^J\alpha^* \to \alpha^*F^I$ is an isomorphism (not necessarily canonical in our technical sense).

Proof: Assume that U has a left adjoint. It is clear from the definition that each U^I has a left adjoint F^I . For any $\alpha: J \to I$ we have the following natural bijections

$$\cong \frac{\alpha^*F^IA \to B}{\cong \frac{\alpha^*A \to U^JB}{F^J\alpha^*A \to B}}$$

which induce an isomorphism $\alpha^*F^IA \xrightarrow{\sim} F^J\alpha^*A$. For the same reasons as in proposition (2.1), this is the canonical morphism $\alpha^*F^I \to F^J\alpha^*$.

Conversely if all of the F^I exist and the canonical morphisms $F^J\alpha^* \to \alpha^*F^I$ are isomorphisms, then we have the following natural bijections

$$\cong \frac{\alpha^*F^I A \to B}{\cong \frac{F^J\alpha^* A \to B}{\alpha^* A \to U^J B}} \quad .$$ □

The above proposition shows that although left adjoints do not in general give us indexed functors, we do get something quite close.

(2.2.6) <u>Corollary</u>: If in $\underline{B}$ all isomorphisms are canonical, then every left adjoint is a canonical left adjoint. □

(2.3) For an indexed category $\underline{X}$, let $\tilde{\underline{X}}$ be the same category but in which all isomorphisms are canonical. The identity on $\underline{X}$ is a functor $H_{\underline{X}}: \underline{X} \to \tilde{\underline{X}}$ which has an inverse if and only if all isomorphisms of $\underline{X}$ are canonical. Clearly $(\tilde{\ })$ is a 2-functor. It is clear from propositions (2.1) and (2.2.5) that $U: \underline{B} \to \underline{A}$ has a left adjoint iff $\tilde{U}: \tilde{\underline{B}} \to \tilde{\underline{A}}$ has a canonical left adjoint. Thus, although left adjoints are not indexed functors unless they are canonical, they are indexed functors $\tilde{\underline{A}} \to \tilde{\underline{B}}$ and so can be composed with indexed functors, right adjoints, or left adjoints.

(2.3.1) <u>Proposition</u>: Let $\Phi: \tilde{\underline{X}} \to \tilde{\underline{A}}$ be an indexed functor where $\underline{X}$ has a small class of objects. Then there exists an indexed functor $\Psi: \underline{X} \to \underline{A}$ such that

$$\begin{array}{ccc} \underline{X} & \xrightarrow{\Psi} & \underline{A} \\ {\scriptstyle H_{\underline{X}}}\downarrow & \cong & \downarrow{\scriptstyle H_{\underline{A}}} \\ \tilde{\underline{X}} & \xrightarrow[\Phi]{} & \tilde{\underline{A}} \end{array} \quad .$$

<u>Proof</u>: Let $X_0 \in \underline{X}^{I_0}$ be the generic family of objects (II,2.1). For any I and any X in $\underline{X}^I$ there exists a unique $\xi\colon I \to I_0$ such that $\xi^*(X_0) \cong X$ (canonically in $\underline{X}$). Define $\Psi^I(X) = \xi^*\Phi^{I_0}(X_0)$ and let $t(X)\colon H_{\underline{A}}^I\Psi^I(X) \to \Phi^I H_{\underline{X}}^I(X)$ be

$$H_{\underline{A}}^I\Psi^I(X) = \Psi^I(X) = \xi^*\Phi^{I_0}X_0 \cong \Phi^I\xi^*X_0 \cong \Phi^I X = \Phi^I H_{\underline{X}}^I(X) .$$

For a fixed I, Ψ^I can be made into a functor in a unique way such that $tX\colon \Psi^I X \xrightarrow{\cong} \Phi^I X$ is natural. If $\alpha\colon J \to I$, $(\xi\alpha)^*X_0 \cong \alpha^*\xi^*X_0 \cong \alpha^*X$ and so

$$\Psi^J(\alpha^*X) = (\xi\alpha)^*\Phi^{I_0}X_0 \cong \alpha^*\xi^*\Phi^{I_0}X_0 = \alpha^*\Phi^I X$$

canonically. □

(2.3.2) <u>Corollary</u>: If $U\colon \underline{B} \to \underline{A}$ has a left adjoint and $\underline{A}$ has a small class of objects, then the left adjoint can be chosen to be canonical. □

Almost all of the adjoints which we meet in practice satisfy the hypotheses of either corollary (2.2.6) or corollary (2.3.2), and so from a practical point of view canonical left adjoints are the same as left adjoints.

(2.3.3) <u>Proposition</u>: Let $U\colon \underline{B} \to \underline{A}$ have a left adjoint F and let $\mathbb{C}$ be a category object in $\underline{S}$. Then the functor $U^{\mathbb{C}}\colon \underline{B}^{\mathbb{C}} \to \underline{A}^{\mathbb{C}}$ has a left adjoint $F^{\mathbb{C}}$.

<u>Proof</u>: Let Φ be in $\underline{A}^{\mathbb{C}}$ and define $F^{\mathbb{C}}(\Phi)$ to be the Ψ such that $H_{\underline{B}}\Psi \cong F\tilde{\Phi}H_{[\mathbb{C}]}$ given by proposition (2.3.1). Then for any J and Θ in $(\underline{B}^{\mathbb{C}})^J$ we have the following sequence of natural bijections

$$\begin{array}{c}
\Delta_J\Psi \longrightarrow \Theta \quad \text{in} \quad (\underline{B}^{\mathbb{C}})^J \\
\hline
\Psi p_{\mathbb{C}} \longrightarrow \Theta \quad \text{in} \quad \underline{B}^{\mathbb{C}\times J} \\
\hline
\text{compatible families} \quad \langle \Psi^I p_{\mathbb{C}}^I \longrightarrow \Theta^I \rangle_I \\
\hline
\text{compatible families} \quad \langle F^I \Phi^I p_{\mathbb{C}}^I \longrightarrow \Theta^I \rangle_I \\
\hline
\text{compatible families} \quad \langle \Phi^I p_{\mathbb{C}}^I \longrightarrow U^I \Theta^I \rangle_I \\
\hline
\Phi p_{\mathbb{C}} \longrightarrow U\Theta \quad \text{in} \quad \underline{A}^{\mathbb{C}\times J} \\
\hline
\Delta_J\Phi \longrightarrow (U^{\mathbb{C}})^J(\Theta) \quad \text{in} \quad (\underline{A}^{\mathbb{C}})^J
\end{array}$$

(each horizontal line denoting a bijection $\cong$)

where $p_{\mathbb{C}}\colon \mathbb{C}\times J \longrightarrow \mathbb{C}$ is the projection.

It follows that $F^{\mathbb{C}}$ gives us a left adjoint for $U^{\mathbb{C}}$ at 1 . The result follows by localization. □

(2.4) We end this section with a few examples of adjoints.

(2.4.1) The unique functor $\underline{A} \longrightarrow \mathbb{1}$ has a left adjoint if and only if each $\underline{A}^I$ has an initial object and the functors α^* preserve these initial objects. In this case we say simply that $\underline{A}$ has an initial object.

(2.4.2) $\underline{A}$ has stable binary coproducts, in the sense that each $\underline{A}^I$ has binary coproducts which are preserved by the α^* , if and only if the indexed functor $\Delta\colon \underline{A} \longrightarrow \underline{A}\times\underline{A}$ has a left adjoint.

(2.4.3) More generally, let $\underline{X}$ be any category (not indexed). Then we have an indexed category $\underline{A}^{\underline{X}}$ (I,2.6). If the indexed diagonal $\Delta\colon \underline{A} \longrightarrow \underline{A}^{\underline{X}}$ has a left adjoint, then each of the categories $\underline{A}^I$ has $\underline{X}$-colimits which are preserved by the α^* . In this case we say simply that $\underline{A}$ has $\underline{X}$-colimits.

(2.4.4) If $\underline{A}$ has stable image factorizations (in each $\underline{A}^I$) and A is any object of $\underline{A}$, then $\underline{\mathrm{Sub}}(A) \rightarrowtail \underline{A}/A$ (see II,4.3) has a left adjoint.

Of course, everything said for left adjoints can be dualized for right adjoints.

3. Small Limits:

(3.1) Definition: Let $\underline{A}$ be an indexed category and $\mathbb{C}$ a category object in $\underline{S}$. We say that $\underline{A}$ *has $\mathbb{C}$-limits at 1* if the indexed functor $\Delta_{\mathbb{C}}: \underline{A} \to \underline{A}^{\mathbb{C}}$ has a right adjoint at 1, denoted $\varprojlim_{\mathbb{C}}$ or simply $\varprojlim$. We say that $\underline{A}$ *has $\mathbb{C}$-limits* if $\Delta_{\mathbb{C}}$ has a right adjoint.

Since the construction $\underline{A}^{\mathbb{C}}$ is stable under localization and since having a right adjoint is the localization of having a right adjoint at 1, we conclude that having $\mathbb{C}$-limits is the localization of having $\mathbb{C}$-limits at 1. Thus having $\mathbb{C}$-limits is stable under localization. As before, the important concept here is that of having $\mathbb{C}$-limits but that of having $\mathbb{C}$-limits at 1 is more intuitive and easier to work with. The process of localization can then be used in passing from $\mathbb{C}$-limits at 1 to $\mathbb{C}$-limits everywhere.

The following proposition is an immediate consequence of proposition (2.2.5) and the definition of $\mathbb{C}$-limits.

Proposition: $\underline{A}$ has $\mathbb{C}$-limits if and only if for every I the functor

$\underline{A}^I \xrightarrow{\Delta_{\mathbb{C}}^I} (\underline{A}^{\mathbb{C}})^I \cong (\underline{A}^I)^{\mathbb{C}}$ has a right adjoint $\varprojlim{}_{\mathbb{C}}^I$ and for every $\alpha: J \to I$ the canonical morphism $\alpha^* \cdot \varprojlim{}_{\mathbb{C}}^I \to \varprojlim{}_{\mathbb{C}}^J \cdot \alpha^*$ is an isomorphism. □

It follows from this proposition that if $\underline{A}$ has $\mathbb{C}$-limits then so does $\underline{A}^I$ for any I.

(3.2) For any functor $\Phi: [\mathbb{C}] \to \underline{A}$ we can talk of its limit without requiring all $\mathbb{C}$-limits to exist by requiring $\Delta_{\mathbb{C}}$ to have a right adjoint at Φ. We give an internal description of this.

Let (X,ξ) be the internal functor corresponding to Φ. Thus $X \in \underline{A}^{C_0}$ and $\xi: \partial_0^* X \to \partial_1^* X$ in $\underline{A}^{C_1}$ satisfying the conditions of (II,1.5). A *cone* for (X,ξ) is an object A of $\underline{A}^1$ and a morphism $\mu: \Delta_{C_0} A \to X$ such that

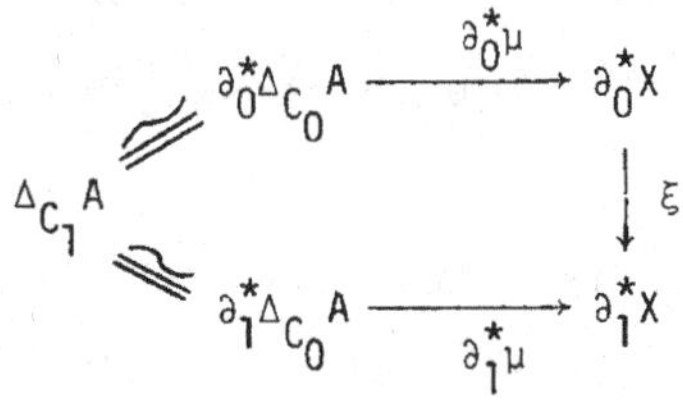

$$\begin{array}{ccccc} & & \partial_0^*\Delta_{C_0} A & \xrightarrow{\partial_0^*\mu} & \partial_0^* X \\ & \nearrow\!\!\!\!\!= & & & \downarrow \xi \\ \Delta_{C_1} A & & & & \\ & \searrow\!\!\!\!\!= & & & \\ & & \partial_1^*\Delta_{C_0} A & \xrightarrow[\partial_1^*\mu]{} & \partial_1^* X \end{array}$$

commutes.

Then $\varprojlim(X,\xi)$ is a universal such cone which is also preserved by the functors Δ_I. This means that $p_1^*\mu: \Delta_{C_0\times I} A \to p_1^* X$, which is a cone from $\Delta_I A$ to $\Delta_I(X,\xi) = (p_1^* X, q_1^*\xi)$ in $\underline{A}^I$, is also universal ($p_1: C_0\times I \to C_0$ and $q_1: C_1\times I \to C_1$ are the projections).

As in ordinary category theory, it is sometimes useful to consider limits of large diagrams. We now externalize the preceding description to study this case.

Let $\underline{X}$ be any indexed category and $\Gamma: \underline{X} \to \underline{A}$ any indexed functor. An *indexed cone* $\mu: A \to \Gamma$ consists of an object A in $\underline{A}^1$ together with an indexed natural transformation $\mu: \Delta_{\underline{X}}(A) \to \Gamma$, i.e. for each I an ordinary cone $\mu^I: \Delta_I A \to \Gamma^I$ such that for each $\alpha: J \to I$, $\alpha^*\cdot\mu^I = \mu^J\cdot\alpha^*$. If $\mu: A \to \Gamma$ is a universal such cone we denote A by $\varprojlim^1 \Gamma$. If, furthermore, Δ_I preserves $\varprojlim^1 \Gamma$ in the sense that $\Delta_I \varprojlim^1 \Gamma \xrightarrow{\sim} \varprojlim^I \Delta_I \Gamma$ then we say that $\varprojlim^1 \Gamma$ is the (indexed) limit of Γ. (Here $\varprojlim^I$ is the same as $\varprojlim^1$ taken in $\underline{A}^I$.)

Colimits are treated dually, i.e. as left adjoints to Δ's .

(3.3) If I is a discrete category then I-limits are called *I-products* and $\varprojlim_I$ is denoted Π_I . I-colimits are called *I-coproducts* and $\varinjlim_I$ is denoted $\amalg_I$ (or sometimes Σ_I) .

The universal properties of Π_I and $\amalg_I$ are easier to state than for general limits and colimits. Let A be an object of $\underline{A}^I$. Then $\Pi_I A$ is an object of $\underline{A}^1$ equipped with a projection morphism $p: \Delta_I \Pi_I A \to A$ such that for every A' in $\underline{A}^1$ and every morphism $q: \Delta_I A' \to A$ there exists a unique $a: A' \to \Pi_I A$ such that $p \cdot \Delta_I a = q$, and for every J , $\Delta_J: \underline{A} \to \underline{A}^J$ preserves $\Pi_I A$ (as before). The case of $\amalg_I$ is dual.

(3.4) Recall (see [Db] or [Kl]) that if $\underline{V}$ is a symmetric monoidal closed category and $\underline{A}$ is a $\underline{V}$-category, $\underline{A}$ is said to be *cotensored* if for every A in $\underline{A}$ and V in $\underline{V}$ there is an object A^V in $\underline{A}$ such that there is a $\underline{V}$-natural isomorphism

$$[-,A^V]_{\underline{A}} \cong [V,[-,A]_{\underline{A}}]_{\underline{V}} .$$

$\underline{A}$ is *tensored* if $\underline{A}^{op}$ is cotensored.

If $\underline{S}$ and $\underline{A}$ have small homs, then $\underline{S}$ is cartesian closed and so is symmetric monoidal closed, and $\underline{A}$ is an $\underline{S}$-category.

Proposition: If $\underline{A}$ has internal products, then $\underline{A}$ is cotensored. Dually, if $\underline{A}$ has internal coproducts, then $\underline{A}$ is tensored.

Proof: Define the cotensor of A by I to be $A^I = \Pi_I \Delta_I A$. We must show that $\mathrm{Hom}(A',A)^I \cong \mathrm{Hom}(A',A^I)$, $\underline{S}$-naturally in A' .

For any J in $\underline{S}$ we have the following sequence of natural bijections

$$
\begin{array}{cc}
 & J \longrightarrow \mathrm{Hom}(A',A^I) \quad \text{in} \quad \underline{S} \\
\cong & \overline{\Delta_J A' \longrightarrow \Delta_J(A^I) \quad \text{in} \quad \underline{A}^J} \\
\cong & \overline{\Delta_J A' \longrightarrow \Delta_J \Pi_I \Delta_I A \quad \text{in} \quad \underline{A}^J} \\
\cong & \overline{\Delta_J A' \longrightarrow \Pi_I^J \Delta_J(\Delta_I A) \quad \text{in} \quad \underline{A}^J} \\
\cong & \overline{\Delta_I(\Delta_J A') \longrightarrow \Delta_J(\Delta_I A) \quad \text{in} \quad (\underline{A}^J)^I \cong (\underline{A}^I)^J} \\
\cong & \overline{\Delta_{I\times J} A' \longrightarrow \Delta_{I\times J} A \quad \text{in} \quad \underline{A}^{I\times J}} \\
\cong & \overline{I\times J \longrightarrow \mathrm{Hom}(A',A) \quad \text{in} \quad \underline{S}} \\
\cong & \overline{J \longrightarrow \mathrm{Hom}(A',A)^I \quad \text{in} \quad \underline{S}}
\end{array}
$$

where the third bijection refers to the diagram

$$
\begin{array}{ccc}
\underline{A}^I & \xrightarrow{\Pi_I} & \underline{A} \\
{\scriptstyle \Delta_J}\downarrow & \cong & \downarrow{\scriptstyle \Delta_J} \\
(\underline{A}^I)^J & \xrightarrow[\Pi_I^J]{} & \underline{A}^J
\end{array} .
$$

It is easily seen, by inspection, that each of the above bijections is natural in A' . Thus we get a natural isomorphism

$$\mathrm{Hom}(A',A)^I \cong \mathrm{Hom}(A',A^I) .$$

This is an isomorphism between indexed functors $\underline{A}^{op} \longrightarrow \underline{S}$, and by localization it is an indexed isomorphism. It is therefore automatically an $\underline{S}$-natural isomorphism (II,3.2).

The tensored part follows by duality. □

<u>Remark</u>: The above proposition also holds if we only have the hypotheses at 1 , i.e. $\underline{S}$ and $\underline{A}$ have small homs at 1 and $\underline{A}$ has internal products at 1 .

The only difference is that it is slightly harder to show that the natural isomorphism is $\underline{S}$-natural. In fact, even if $\underline{S}$ and $\underline{A}$ do not have small homs at 1, A^I has all the properties of the cotensor (see [Rn]).

(3.5) $\underline{S}$ has all I-coproducts. Define $\amalg_I: \underline{S}^I \to \underline{S}$ by $\amalg_I(X \to I) = X$. It is easily checked that $\amalg_I$ is left adjoint at 1 to Δ_I. The result now follows by localization.

(3.5.1) Proposition: $\underline{S}$ has I-products for all I if and only if $\underline{S}$ is cartesian closed.

Proof: We want to construct $\Pi_I: \underline{S}^I \to \underline{S}$ right adjoint to Δ_I, i.e. for every J we want $\Pi_I^J: (\underline{S}^I)^J \to \underline{S}^J$ right adjoint to Δ_I^J. (We cannot use localization here because cartesian closedness is not stable.)

Let $p: X \to I \times J$ be an object of $(\underline{S}^I)^J$. Then $\Pi_I^J(p)$ is defined to be the pullback

$$\begin{array}{ccc} \Pi_I^J(p) & \longrightarrow & X^I \\ \downarrow & \boxed{\text{P.B.}} & \downarrow \\ J & \longrightarrow & (I \times J)^I \end{array}$$

where the morphism $J \to (I \times J)^I$ is the transpose of $1_{I \times J}: I \times J \to I \times J$.

For any $\beta: K \to J$ we have the following bijections

$$\cong \frac{\beta \to \Pi_I^J(p) \text{ in } \underline{S}^J}{\begin{array}{ccc} K & \longrightarrow & X^I \\ \beta \downarrow & /\!/\!/ & \downarrow p^I \\ J & \longrightarrow & (I\times J)^I \end{array} \text{ in } \underline{S}}$$

$$\cong \frac{\begin{array}{ccc} I\times K & \longrightarrow & X \\ I\times\beta \downarrow & /\!/\!/ & \downarrow p \\ I\times J & = & I\times J \end{array} \text{ in } \underline{S}}{\Delta_I^J(\beta) \to p \text{ in } (\underline{S}^I)^J}$$

and we see that $\Delta_I^J \dashv \Pi_I^J$. It is an easy computation to see that the Π_I's are preserved by the α^*'s .

Conversely if $\underline{S}$ has all small products then it is cartesian closed. We define $J^I = \Pi_I \Delta_I J$. The natural bijections

$$\cong \frac{K \to J^I \text{ in } \underline{S}}{K \to \Pi_I \Delta_I J \text{ in } \underline{S}}$$
$$\cong \frac{\qquad}{\Delta_I K \to \Delta_I J \text{ in } \underline{S}^I}$$
$$\cong \frac{\begin{array}{ccc} I\times K & \longrightarrow & I\times J \\ & \searrow \quad \swarrow & \\ & I & \end{array} \text{ in } \underline{S}}{I\times K \to J \text{ in } \underline{S}}$$

show that J^I has the required property. □

(3.5.2) <u>Proposition</u>: If $\underline{S}$ has I-products then $Gr(\underline{S})$ has also.

<u>Proof</u>: The functor $\Pi_I : \underline{S}^I \to \underline{S}$ preserves group objects (since it has a left adjoint) and so extends to a functor $\Pi_I : Gr(\underline{S}^I) \to Gr(\underline{S})$. But $Gr(\underline{S}^I) = Gr(\underline{S})^I$ and it is easily seen that Π_I is a right adjoint at 1 to the diagonal. The result now follows by localization.

(3.5.3) <u>Remark</u>: It is clear that proposition (3.5.2) is true, not only for $Gr(\underline{S})$ but for any category of structures defined by inverse limits in $\underline{S}$, such as $Ring(\underline{S})$, $R\text{-modules}(\underline{S})$ (where R is a ring in $\underline{S}$), $Cat(\underline{S})$, etc.

We shall see in (3.6) that these categories have all $\mathbb{C}$-limits for any $\mathbb{C}$. The question of coproducts is more difficult. In fact they do not exist in general (e.g. groups in finite sets do not have finite coproducts).

(3.5.4) <u>Proposition</u>: Let $G: \underline{A} \to \underline{B}$ and $H: \underline{C} \to \underline{B}$ be indexed functors. If $\underline{A}$ and $\underline{C}$ have I-products and H preserves I-products then the comma category (G,H) has I-products.

<u>Proof</u>: Let $(A, G^I A \xrightarrow{b} H^I C, C)$ be an object of $(G,H)^I = (G^I, H^I)$. We have the following natural bijections

$$\cong \frac{G^1\Pi_I A \to H^1\Pi_I C \cong \Pi_I H^I C \text{ in } \underline{B}}{\Delta_I G^1 \Pi_I A \to H^I C \text{ in } \underline{B}^I}$$
$$\cong \frac{}{G^I \Delta_I \Pi_I A \to H^I C \text{ in } \underline{B}^I}$$

and so we let $\bar{b}: G^1\Pi_I A \to H^1\Pi_I C$ correspond to

$$G^I\Delta_I\Pi_I A \xrightarrow{G^I(p)} G^I A \xrightarrow{b} H^I C$$

where $p: \Delta_I\Pi_I A \to A$ is the projection morphism.

For any (A',b',C') in $(G,H)^1$ morphisms $(A',b',C') \to (\Pi_I A, \bar{b}, \Pi_I C)$ are pairs of morphisms (u,v) with $u: A' \to \Pi_I A$ and $v: C' \to \Pi_I C$ such that

$$\begin{array}{ccccc} G^1 A' & \xrightarrow{G^1 u} & G^1\Pi_I A & \cong & \Pi_I G^I A \\ {\scriptstyle b'}\downarrow & & \downarrow{\scriptstyle \bar{b}} & & \downarrow \\ H^1 C' & \xrightarrow[H^1 v]{} & H^1\Pi_I C & \cong & \Pi_I H^I C \end{array}$$

commutes. This commutivity, when transformed by the adjointness $\Delta_I \dashv \Pi_I$, becomes the commutivity of ★ below (the remaining diagrams commuting for obvious reasons)

$$\begin{array}{ccccc}
G^I\Delta_I A' & \xrightarrow{G^I\Delta_I u} & G^I\Delta_I\Pi_I A & \xrightarrow{G^I(p)} & G^I A \\
\wr\| & & \wr\| & & \\
\Delta_I G^1 A' & \xrightarrow{\Delta_I G^1 u} & \Delta_I G^1\Pi_I A & & \\
\Delta_I b' \downarrow & & & \bigstar & \downarrow b \\
\Delta_I H^1 C' & \xrightarrow[\Delta_I H^1 v]{} & \Delta_I H^1\Pi_I C & & \\
\wr\| & & \wr\| & & \\
H^I\Delta_I C' & \xrightarrow[H^I\Delta_I v]{} & H^I\Delta_I\Pi_I C & \xrightarrow[H^I(p')]{} & H^I C
\end{array}$$

But morphisms $u: A' \to \Pi_I A$ correspond bijectively to morphisms $\bar{u}: \Delta_I A' \to A$ and the correspondence is given by $\bar{u} = p \cdot \Delta_I u$ and similarly for v . Thus the above diagram says exactly that $(\bar{u},\bar{v})$ is a morphism $\Delta_I(A',b',C') \to (A,b,C)$.

Thus $(\Pi_I A,\bar{b},\Pi_I C)$ has the correct universal property. That Δ_J preserves it for every J follows from the fact that Δ_J preserves Π_I in $\underline{A}$ and $\underline{C}$ and that Δ_J commutes with G and H . Thus (G,H) has Π_I at 1 . The result now follows by localization. □

(3.6) As the following proposition shows, the usual way of computing limits as equalizers of products also works in our setting.

Proposition: If $\underline{A}$ has equalizers (as in (2.4.3)), C_0-products and C_1-products then $\underline{A}$ has $\mathbb{C}$-limits.

Proof: Let (X,ξ) be an internal functor from $\mathbb{C}$ to $\underline{A}$. Define E by the following equalizer

$$E \rightarrowtail \Pi_{C_0} X \overset{u}{\underset{v}{\rightrightarrows}} \Pi_{C_1} \partial_1^* X$$

where u and v are defined by the following bijections

$$\cong \frac{\partial_1^*\Delta_{C_0}\Pi_{C_0} X \xrightarrow{\partial_1^*(p)} \partial_1^* X}{\Delta_{C_1}\Pi_{C_0} X \longrightarrow \partial_1^* X} \qquad \cong \frac{\partial_0^*\Delta_{C_0}\Pi_{C_0} X \xrightarrow{\partial_0^*(p)} \partial_0^* X \xrightarrow{\xi} \partial_1^* X}{\Delta_{C_1}\Pi_{C_0} X \longrightarrow \partial_1^* X}$$

$$\cong \frac{\Delta_{C_1}\Pi_{C_0} X \longrightarrow \partial_1^* X}{\Pi_{C_0} X \xrightarrow{u} \Pi_{C_1}\partial_1^* X} \qquad \cong \frac{\Delta_{C_1}\Pi_{C_0} X \longrightarrow \partial_1^* X}{\Pi_{C_0} X \xrightarrow{v} \Pi_{C_1}\partial_1^* A}$$

where $p: \Delta_{C_0}\Pi_{C_0} X \rightarrow X$ is the projection morphism (i.e. the counit of $\Delta_{C_0} \dashv \Pi_{C_0}$).

For any object A of $\underline{A}^I$, morphisms $A \longrightarrow E$ correspond bijectively to morphisms $\phi: A \rightarrow \Pi_{C_0} X$ such that $u\phi = v\phi$. This condition, transformed by the adjointness $\Delta_{C_1} \dashv \Pi_{C_1}$, becomes the commutivity of the outer rectangle in the following diagram (the square on the right is not required to commute)

$$\begin{array}{ccccc}
\partial_0^*\Delta_{C_0} A & \xrightarrow{\partial_0^*\Delta_{C_0}\phi} & \partial_0^*\Delta_{C_0}\Pi_{C_0} X & \xrightarrow{\partial_0^*(p)} & \partial_0^* X \\
\cong & & \cong & & \\
\Delta_{C_1} A & \xrightarrow{\Delta_{C_1}\phi} & \Delta_{C_1}\Pi_{C_0} X & & \downarrow \xi \\
\cong & & \cong & & \\
\partial_1^*\Delta_{C_0} A & \xrightarrow[\partial_1^*\Delta_{C_0}\phi]{} & \partial_1^*\Delta_{C_0}\Pi_{C_0} X & \xrightarrow[\partial_1^*(p)]{} & \partial_1^* X .
\end{array}$$

But morphisms $\phi: A \rightarrow \Pi_{C_0} X$ correspond bijectively to morphisms $\psi: \Delta_{C_0} A \rightarrow X$ and the correspondence is given by $\psi = p \cdot \Delta_{C_0}\phi$. Thus the above rectangle becomes

$$\begin{array}{ccc} \partial_0^* \Delta_{C_1} A & \xrightarrow{\partial_0^* \psi} & \partial_0^* X \\ \wr\| & & \downarrow \xi \\ \partial_1^* \Delta_{C_1} A & \xrightarrow[\partial_1^* \psi]{} & \partial_1^* X \end{array}$$

i.e. ψ is an internal natural transformation $\Delta_{\mathbb{C}} A \longrightarrow (X,\xi)$. Thus $E = \varprojlim^1_{\mathbb{C}}(X,\xi)$.

The same construction can be performed in any $\underline{A}^I$ so we have all $\varprojlim^I_{\mathbb{C}}$. It is clear from the construction that the $\varprojlim_{\mathbb{C}}$ are stable. □

It follows from this proposition that if $\underline{S}$ is cartesian closed, then $\underline{S}$ has $\mathbb{C}$-limits for all $\mathbb{C}$. Also if $\underline{S}$ has stable coequalizers, then $\underline{S}$ has $\mathbb{C}$-colimits for all $\mathbb{C}$. Thus if $\underline{S}$ is an elementary topos it has all $\mathbb{C}$-limits and $\mathbb{C}$-colimits.

If $\underline{S}$ is cartesian closed then all categories of structures defined by finite limits, such as $Gr(\underline{S})$, $Ring(\underline{S})$, $Cat(\underline{S})$, etc., have $\mathbb{C}$-limits for any $\mathbb{C}$.

(3.7) We have already referred to functors preserving certain internal limits. We now make these concepts more precise.

(3.7.1) <u>Definition</u>: Let $\Phi: \underline{X} \longrightarrow \underline{B}$ be an indexed functor and let the indexed cone $\mu: B \longrightarrow \Phi$ (3.2) be the limit of Φ . An indexed functor $U: \underline{B} \longrightarrow \underline{A}$ is said to *preserve* $\varprojlim \Phi$ if $U\mu: UB \longrightarrow U\Phi$ is a limit cone for $U\Phi$.

When $\varprojlim U\Phi$ exists the above definition is equivalent to requiring the comparison morphism $U \varprojlim \Phi \longrightarrow \varprojlim U\Phi$ to be an isomorphism (not necessarily canonical).

(3.7.2) <u>Definitions</u>: We say that $U: \underline{B} \longrightarrow \underline{A}$ *preserves* $\underline{X}$*-limits at 1* if U preserves $\varprojlim \Phi$ for every indexed $\Phi\colon \underline{X} \to \underline{B}$. We say that U *preserves* $\underline{X}$*-limits* if for every I and every $\Phi\colon \underline{X} \to \underline{B}^I$, U^I preserves $\varprojlim \Phi$.

Thus preserving $\underline{X}$-limits is the localization of preserving $\underline{X}$-limits at 1 . It says simply that U not only preserves $\underline{X}$-limits individually but also families of $\underline{X}$-limits.

If $\underline{D}$ is an ordinary category, we can construct an indexed category which we also denote by $\underline{D}$ by letting $\underline{D}^I = \underline{D}$ and the substitution functors $\alpha^* = 1_{\underline{D}}$. An indexed functor $\Phi\colon \underline{D} \to \underline{B}$ is the same as an ordinary functor $\underline{D} \to \underline{B}^1$. $\underline{D}$-limits in the indexed sense are ordinary $\underline{D}$-limits in $\underline{B}^1$ which are stable under substitution.

We shall mainly be interested in two special cases of definitions (3.7.1) and (3.7.2): when $\underline{X} = \underline{D}$ some finite category, and when $\underline{X} = [\mathbb{C}]$.

(3.7.3) For any $\alpha\colon J \to I$, the indexed functor $\alpha^*\colon \underline{A}^I \to \underline{A}^J$ preserves all limits and colimits which exist in $\underline{A}^I$. This follows immediately from the definition of limits and colimits.

(3.7.4) <u>Proposition</u>: If $U\colon \underline{B} \to \underline{A}$ has a left adjoint then U preserves all limits which exist in $\underline{B}$.

<u>Proof</u>: Let $\Phi\colon \underline{X} \to \underline{B}$ be an indexed functor and $\mu\colon B \to \Phi$ a limit cone. We wish to see that $U\mu\colon UB \to U\Phi$ is a limit cone in $\underline{A}$. For any cone $\nu\colon A \to U\Phi$ we get by adjointness a cone $\bar{\nu}\colon FA \to \Phi$ which at I is defined by

$$\cong \frac{\nu^I\colon \Delta_I A \to U^I \Phi^I}{\bar{\nu}^I\colon \Delta_I FA \to \Phi^I} .$$

Thus there exists a unique $b\colon FA \to B$ such that $\mu \cdot b = \bar{\nu}$. Transforming this through the adjointness $F \dashv U$ we get a unique $\bar{b}: A \to UB$ such that $U\mu \cdot \bar{b} = \nu$.

That Δ_I preserves $\varprojlim^1 U\Phi$ follows from the fact that U^I also has a left adjoint and so preserves $\varprojlim^I \Delta_I \Phi$ and that Δ_I preserves $\varprojlim \Phi$ in $\underline{B}$. Thus $\Delta_I \varprojlim^1 U\Phi \cong \Delta_I U^1 \varprojlim^1 \Phi \cong U^I \Delta_I \varprojlim^1 \Phi \cong U^I \varprojlim^I \Delta_I \Phi \cong \varprojlim^I U^I \Delta_I \Phi \cong \varprojlim^I \Delta_I U\Phi$. This shows that U preserves all limits which exist at 1. The result now follows by localization. □

If U only has a left adjoint at 1 we cannot conclude that U preserves those limits which exist at 1. From the above proof we see that U of the limit cone will have the universal property relative to cones with vertex in $\underline{A}^1$ but there is no way of knowing, in general, that the Δ_I preserve this universality. However, since the limit is determined up to isomorphism by the universal property, if we know that U of the given diagram has a limit, then U will preserve the limit under discussion.

(3.7.5) <u>Proposition</u>: Let $U: \underline{B} \to \underline{A}$ have a left adjoint at 1. If $\Phi: \underline{X} \to \underline{B}$ is an indexed functor such that $\varprojlim \Phi$ and $\varprojlim U\Phi$ exist, then $U \varprojlim \Phi \cong \varprojlim U\Phi$. □

(3.7.6) <u>Proposition</u>: If $\underline{B}$ has small homs, then for any B in $\underline{B}$, $\mathrm{Hom}(B,-): \underline{B} \to \underline{S}$ preserves any limits which exist in $\underline{B}$.

<u>Proof (Sketch)</u>: Let $\Phi: \underline{X} \to \underline{B}$ have a limit in $\underline{B}$ and let I be any object of $\underline{S}$. We have the following natural bijections:

$$\cong \frac{I \longrightarrow \mathrm{Hom}(B, \varprojlim \Phi) \text{ in } \underline{S}}{\Delta_I B \longrightarrow \Delta_I \varprojlim \Phi \cong \varprojlim \Delta_I \Phi \text{ in } \underline{B}^I}$$

$$\cong \frac{}{\text{indexed cones } \Delta_I B \longrightarrow \Delta_I \Phi \text{ in } \underline{B}^I}$$

$$\cong \frac{}{\text{compatible families } \langle \Delta_{I\times J} B \longrightarrow p_2^* \Phi^J(X) \rangle_{J\in \underline{S},\ X\in \underline{X}^J}}$$

$$\cong \frac{}{\text{compatible families } \langle \Delta_J I \longrightarrow \mathrm{Hom}^J(\Delta_J B, \Phi^J(X)) \rangle_{J\in \underline{S},\ X\in \underline{X}^J}}$$

$$\cong \frac{}{\text{indexed cones } I \longrightarrow \mathrm{Hom}(B, \Phi(-)) \text{ in } \underline{S}}$$

where $p_2 \colon I\times J \longrightarrow J$ is the projection.

It follows that $\mathrm{Hom}(B, \varprojlim \Phi)$ has the universal property in $\underline{S}$ to make it $\varprojlim \mathrm{Hom}(B, \Phi(-))$.

Now the same argument in $\underline{B}^I$ shows that $\mathrm{Hom}^I(\Delta_I B, \varprojlim^I \Delta_I \Phi)$ has the universal property of $\varprojlim^I \mathrm{Hom}^I(\Delta_I B, \Delta_I \Phi(-))$ in $\underline{S}^I$. But $\mathrm{Hom}^I(\Delta_I B, \varprojlim^I \Delta_I \Phi)$ $\cong \mathrm{Hom}^I(\Delta_I B, \Delta_I \varprojlim \Phi) \cong \Delta_I \mathrm{Hom}(B, \varprojlim \Phi)$. Thus our candidate for $\varprojlim \mathrm{Hom}(B, \Phi(-))$ is preserved by Δ_I and so it is $\varprojlim \mathrm{Hom}(B, \Phi(-))$.

Thus $\mathrm{Hom}(B,-)$ preserves those limits which exist at 1 . The result now follows by localization. □

4. Completeness:

(4.1) Before introducing the notion of completeness we introduce the preliminary notion of completeness at 1 , the localization of which will be completeness. As with previous notions, the important concept here is completeness but we find completeness at 1 more intuitive and easier to work with.

(4.1.1) Definition: We say that $\underline{A}$ is *complete at 1* if $\underline{A}$ has finite (stable) limits at 1 and for every category object $\mathbb{C}$ in $\underline{S}$, $\underline{A}$ has all $\mathbb{C}$-limits at 1 .

(4.1.2) <u>Definition</u>: We say that $\underline{A}$ is *complete* if $\underline{A}/I$ is complete at 1 for every I .

It is clear from the definitions that completeness is the localization of completeness at 1 , and consequently is stable under localization.

(4.1.3) <u>Proposition</u>: If $\underline{A}$ is complete then $\underline{A}$ has $\mathbb{C}$-limits for all $\mathbb{C}$.

<u>Proof</u>: By definition $\underline{A}$ has $\mathbb{C}$-limits at 1 . Being complete is stable under localization and so $\underline{A}$ has $\mathbb{C}$-limits. □

However having $\mathbb{C}$-limits for all $\mathbb{C}$ is not stable under localization. Intuitively, being complete at 1 means that individual limits of diagrams exist, having all $\underline{S}$-limits as in proposition (4.1.3) means that families of limits of the same type (domain category) exist, and being complete means that families of limits of variable type exist. Clearly, we often use this stronger form of completeness in ordinary category theory (e.g. if F is a sheaf, the stalk at x is given by $F_x = \varinjlim_{x \in U} F(U)$; this is clearly a family of colimits of variable type).

The following proposition shows that our definition of completeness is the same as Bénabou's.

(4.1.4) <u>Proposition</u>: $\underline{A}$ is complete if and only if

(i) $\underline{A}^I$ has stable finite limits for every I ,

(ii) for every $\alpha: J \longrightarrow I$ the functor $\alpha^*: \underline{A}^I \longrightarrow \underline{A}^J$ has a right adjoint Π_α ,

(iii) (<u>Beck condition</u>) for every pullback diagram

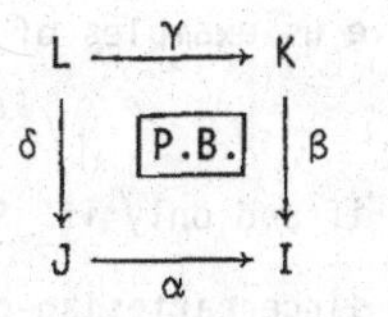

the comparison morphism $\beta^*\Pi_\alpha \to \Pi_\gamma\delta^*$ is an isomorphism (not necessarily canonical)

$$\begin{array}{ccc} \underline{A}^L & \xrightarrow{\Pi_\gamma} & \underline{A}^K \\ {\scriptstyle \delta^*}\uparrow & \cong & \uparrow{\scriptstyle \beta^*} \\ \underline{A}^J & \xrightarrow[\Pi_\alpha]{} & \underline{A}^I \end{array} .$$

<u>Proof</u>: The finite limit conditions are the same in both statements. If $\underline{A}$ is complete, then for $\alpha: J \to I$ an object of $\underline{S}/I$ we have

$$\Pi_\alpha: (\underline{A}/I)^\alpha \to \underline{A}/I$$

right adjoint to Δ_α. At 1 this becomes $\Pi_\alpha: \underline{A}^J \to \underline{A}^I$ right adjoint to α^*. For any $\beta: K \to I$ in $\underline{S}/I$, Δ_β must preserve Π_α, i.e.

$$\begin{array}{ccc} (\underline{A}/I)^\alpha & \xrightarrow{\Pi_\alpha} & \underline{A}/I \\ {\scriptstyle \Delta_\beta}\downarrow & \cong & \downarrow{\scriptstyle \Delta_\beta} \\ (\underline{A}/I)^{\alpha\times\beta} & \xrightarrow[\Pi_\alpha^\beta]{} & (\underline{A}/I)^\beta \end{array}$$

which at 1 gives condition (iii) (note that $\alpha\times\beta$ in $\underline{S}/I$ is the given pullback).

Conversely, taking $I = 1$ in (ii) we get $\Pi_J: \underline{A}^J \to \underline{A}$ right adjoint to Δ_J and (iii) says that Δ_K preserves Π_J. Thus $\underline{A}$ has small products at 1. Thus by proposition (3.6) $\underline{A}$ has all small limits at 1. Now conditions (i), (ii), (iii) are easily seen to be stable under localization. Thus $\underline{A}$ is complete by localization. □

(4.2) The results of (3.5) give us examples of complete categories.

(4.2.1) $\underline{S}$ is complete at 1 if and only if $\underline{S}$ is cartesian closed. This does not imply that $\underline{S}$ is complete since cartesian closedness is not stable under localization. $\underline{S}$ is complete if and only if $\underline{S}/I$ is cartesian closed for each I. This is equivalent to saying that each α^* has a right adjoint Π_α (the Beck condition (condition (iii) of proposition (4.1.4)) is always satisfied).

For example if $\underline{S}$ is an elementary topos then it is complete.

(4.2.2) $\underline{S}$ is cocomplete if and only if $\underline{S}$ has stable finite colimits. For $\alpha: J \longrightarrow I$, $\amalg_\alpha: \underline{S}^J \longrightarrow \underline{S}^I$ is given by

$$\amalg_\alpha(X \xrightarrow{p} J) = X \xrightarrow{\alpha p} I .$$

It is easily seen that this is left adjoint to α^* and that the Beck condition is satisfied.

(4.2.3) If $\underline{S}$ is complete, categories of "structures defined by limits" in $\underline{S}$ will also be complete. Examples of such categories are $\mathrm{Gr}(\underline{S})$, $\mathrm{Ring}(\underline{S})$, $\mathrm{Cat}(\underline{S})$, etc.

(4.2.4) From proposition (4.1.4) it is easily seen that if $\underline{A}$ is complete, then so is $\underline{A}^I$ for any I. Similarly, using propositions (2.3.3) and (4.1.4) we see that if $\underline{A}$ is complete then so is $\underline{A}^{\mathbb{C}}$ for any category object $\mathbb{C}$ in $\underline{S}$. These results can, of course, be dualized for cocompleteness.

(4.2.5) If $\underline{E}$ is an $\underline{S}$-topos (with structure morphism $p: \underline{E} \longrightarrow \underline{S}$) then, as pointed out in (I, 2.4), $\underline{E}$ is an $\underline{S}$-indexed category with $\underline{E}^I = \underline{E}/p^*I$. As an $\underline{E}$-indexed category $\underline{E}$ is complete and cocomplete and so has Π_ϕ and Π_ϕ

satisfying the Beck condition for every morphism ϕ of $\underline{E}$. For $\alpha: J \to I$ in $\underline{S}$, $\alpha^*: \underline{E}^I \to \underline{E}^J$ is given by $(p^*(\alpha))^*$ and so defining $\Pi_\alpha = \Pi_{p^*(\alpha)}$ and $\amalg_\alpha = \amalg_{p^*(\alpha)}$ we get right and left adjoints to α^*. Since p^* preserves pullbacks, Π_α and $\amalg_\alpha$ satisfy the Beck condition. $\underline{E}$ clearly has stable finite limits and colimits and so we conclude that $\underline{E}$ (as an $\underline{S}$-indexed category) is complete and cocomplete.

(4.3) Definition: If $\underline{B}$ is complete at 1 and $U: \underline{B} \to \underline{A}$, we say that U is *continuous at 1* if U preserves all finite limits at 1 and preserves all $\mathbb{C}$-limits at 1. If $\underline{B}$ is complete we say that U is *continuous* if U/I is continuous at 1 for every I.

Clearly, being continuous is the localization of being continuous at 1. Continuity means simply that U not only preserves small limits but also families of small limits.

(4.3.1) Proposition: If $\underline{A}$ and $\underline{B}$ are complete (at 1), then $U: \underline{B} \to \underline{A}$ is continuous (at 1) if and only if U preserves finite limits (at 1) and U preserves Π_α (resp. Π_I) for all α (resp. all I). □

(4.3.2) Proposition: Let $G: \underline{A} \to \underline{B}$ and $H: \underline{C} \to \underline{B}$ be indexed functors. If $\underline{A}$ and $\underline{C}$ are complete (at 1) and H is continuous (at 1) then the comma category (G,H) is complete (at 1).

Proof: This follows immediately from proposition (3.5.4) (and localization).

(4.3.3) Corollary: If $\underline{B}$ is complete (at 1), $U: \underline{B} \to \underline{A}$ continuous (at 1) and A any object of $\underline{A}$, then the comma category (A,U) is complete (at 1). □

(4.3.4) Proposition: If $\underline{B}$ is complete (at 1) and U has a left adjoint, then U is continuous (at 1). If $\underline{B}$ and $\underline{A}$ are complete at 1 and U has a left adjoint at 1 then U is continuous at 1.

Proof: Immediate consequence of propositions (3.7.4) and (3.7.5). □

(4.3.5) Proposition: Suppose that $\underline{B}$ has small products at 1 and that $U: \underline{B} \to \underline{A}$ preserves them. If $F: \underline{A}^1 \to \underline{B}^1$ is an ordinary left adjoint to the (unindexed) functor U^1 then F is a left adjoint to U at 1.

Proof: For any B in $\underline{B}^I$ we have the following natural bijections

$$\cong \frac{\Delta_I FA \to B \text{ in } \underline{B}^I}{FA \to \Pi_I B \text{ in } \underline{B}}$$
$$\cong \frac{}{A \to U^1 \Pi_I B \text{ in } \underline{A}}$$
$$\cong \frac{}{A \to \Pi_I U^I B \text{ in } \underline{A}}$$
$$\cong \frac{}{\Delta_I A \to U^I B \text{ in } \underline{A}^I} \quad .$$

□

(4.3.6) Corollary: If $\underline{B}$ has all small products and U preserves them, then U has a left adjoint if and only if each unindexed functor $U^I: \underline{B}^I \to \underline{A}^I$ has an ordinary left adjoint. □

(4.4) Recall the following facts about reflective subcategories (see Barr [Br 2]). Let $U: \underline{B} \rightarrowtail \underline{A}$ and $U': \underline{B}' \rightarrowtail \underline{A}'$ be full embeddings with left adjoints F and F' respectively. Assume that

$$\begin{array}{ccc} \underline{B} & \xrightarrow{\bar{\Phi}} & \underline{B}' \\ {\scriptstyle U}\downarrow & & \downarrow{\scriptstyle U'} \\ \underline{A} & \xrightarrow[\Phi]{} & \underline{A}' \end{array}$$

commutes up to isomorphism.

If Φ has a left adjoint Ψ then $\overline{\Phi}$ has a left adjoint $\overline{\Psi} = F\Psi U'$.

If, furthermore, the Φ's commute with the F's in the sense that the morphism corresponding to the given isomorphism under the bijections

$$\begin{array}{c} \cong \dfrac{\Phi U \longrightarrow U'\overline{\Phi}}{} \\ \cong \dfrac{F'\Phi U \longrightarrow \overline{\Phi}}{F'\Phi \longrightarrow \overline{\Phi}F} \end{array}$$

is an isomorphism, then if Φ has a right adjoint Θ, it follows that $\overline{\Phi}$ has a right adjoint $\overline{\Theta}$ and $U\overline{\Theta} \cong \Theta U'$.

<u>Proposition</u>: Let $U: \underline{B} \longrightarrow \underline{A}$ be a full and faithful indexed functor (each U^I is full and faithful) with a left adjoint F. If $\underline{A}$ is complete (at 1) then so is $\underline{B}$. If $\underline{A}$ is cocomplete (at 1) then so is $\underline{B}$.

<u>Proof</u>: The finite limits or colimits give no problem. The diagonal functors Δ_I commute with the U's and F's and so if $\underline{A}$ has Π_I or $\amalg_I$ then so will $\underline{B}$. That Δ_J preserves these in $\underline{B}$ follows from the fact that it does in $\underline{A}$. This proves the results at 1. The rest is obtained by localization. □

<u>Corollary</u>: Let A be any object of $\underline{A}$. If $\underline{A}$ has stable image factorizations and is complete (resp. cocomplete) then the indexed category $\underline{\mathrm{Sub}}(A)$ is complete (resp. cocomplete).

<u>Proof</u>: Immediate from the preceding proposition and (2.4.4). □

IV. The Adjoint Functor Theorems:

1. The General Adjoint Functor Theorem:

Before proving the general adjoint functor theorem we prove the following theorem, known as the initial object theorem, from which the adjoint functor theorem follows formally.

(1.1) Theorem: Let $\underline{A}$ be an $\underline{S}$-indexed category which is complete at 1 and has small homs at 1. $\underline{A}$ has an (indexed) initial object if and only if there is an object I of $\underline{S}$ and an object A of $\underline{A}^I$ such that for every A' in $\underline{A}^1$ there exist $i: 1 \to I$ and a morphism $A_i = i^*A \to A'$.

Proof: Let $W = \Pi_I A$. Then for every $i: 1 \to I$ we get a morphism $\Pi_I A = i^*\Delta_I \Pi_I A \xrightarrow{i^*(p)} i^*A$ where $p: \Delta_I \Pi_I A \to A$ is the projection morphism. Thus W is a weak initial object in $\underline{A}^1$.

We have two morphisms

$$W \overset{r}{\underset{s}{\rightrightarrows}} W^{\mathrm{Hom}(W,W)}$$

which correspond by the cotensor adjointness to

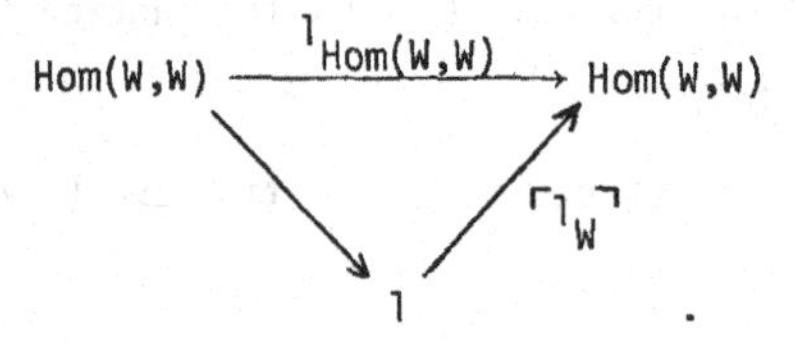

Any endomorphism $f: W \to W$ gives $\ulcorner f \urcorner: 1 \to \mathrm{Hom}(W,W)$ and a simple calculation shows the composites

$$W \overset{r}{\underset{s}{\rightrightarrows}} W^{\mathrm{Hom}(W,W)} \xrightarrow{W^{\ulcorner f \urcorner}} W^1 \cong W$$

to be f and 1_W . It follows that any morphism which equalizes r and s also equalizes f and 1_W .

Let $z: Z \to W$ be the equalizer of r and s . Z is also a weak initial object. Let $a,a': Z \rightrightarrows A'$ be any two morphisms out of Z and let $e: E \to Z$ be their equalizer. There exists $w: W \to E$ and so we get $W \xrightarrow{w} E \xrightarrow{e} Z \xrightarrow{z} W$. But $zewz = z$ so $ewz = 1_Z$, thus e which is a mono and a split epi is iso. It follows that $a = a'$ and so Z is initial in $\underline{A}^1$.

Δ_I has a right adjoint Π_I and so preserves initial objects. It follows that each $\underline{A}^I$ has an initial object and each α^* preserves it. □

The above proof is the same as the one in MacLane's book [ML].

We now give several versions of the general adjoint functor theorem, first of all at 1 and then localized. There are two cases, one where we consider solution sets of morphisms, the other where we consider solution sets of objects and where stronger hypotheses are needed.

(1.2) <u>Definition</u>: An indexed functor $U: \underline{B} \to \underline{A}$ satisfies the *solution set of morphisms* condition *at 1* (SSM1) if for every A in $\underline{A}^1$ there exist I in $\underline{S}$, an object B in $\underline{B}^I$, and a morphism $a: \Delta_I A \to U^I B$ in $\underline{A}^I$ such that for every B' in $\underline{B}^1$ and $a': A \to U^1 B'$ there exist $i: 1 \to I$ and $b: B_i = i^*B \to B'$ such that $a' = U^1 b \cdot i^* a$.

This is simply a translation of the usual solution set as found in MacLane [ML] .

(1.3) <u>Theorem</u>: Let $U: \underline{B} \to \underline{A}$ be an indexed functor where $\underline{B}$ is complete at 1 and $\underline{A}$ and $\underline{B}$ have small homs at 1 . Assume that U is continuous at at 1 . U has a left adjoint at 1 if and only if it satisfies SSM1 .

Proof: U has a left adjoint at 1 if and only if for each A in $\underline{A}^1$ the comma category (A,U) has a stable initial object (III, (2.2.2) and (2.2.3)). Since $\underline{B}$ is complete at 1 and U is continuous at 1 it follows that (A,U) is complete at 1 (corollary (III, 4.3.3)). Since $\underline{A}$ and $\underline{B}$ have small homs at 1 so does (A,U) by the results of (II, 3.5). The proof is complete when we notice that SSM1 is exactly the remaining condition of theorem (1.1) applied to the category (A,U) . □

The condition SSM1 is not stable under localization. The following condition is its localization.

(1.4) Definition: U: $\underline{B} \to \underline{A}$ satisfies the *solution set of morphisms* condition (SSM) if for every I in $\underline{S}$ and every A in $\underline{A}^I$ there exist $\alpha: J \to I$, B in $\underline{B}^J$ and $a: \alpha^*A \to U^J B$ such that for every B' in $\underline{B}^I$ and every $a': A \to U^I B'$ there exist $i: I \to J$ such that $\alpha i = 1_I$ and $b: i^*B \to B'$ such that $a' = U^I b \cdot i^*a$.

(1.5) Theorem: Let U: $\underline{B} \to \underline{A}$ be an indexed functor where $\underline{B}$ is complete and $\underline{A}$ and $\underline{B}$ have small homs. U has a left adjoint if and only if it is continuous and satisfies SSM.

Proof: Follows from the previous theorem by localization. □

Freyd's original adjoint functor theorem [Fr 1] had a solution set of objects rather than of morphisms.

(1.6) Definition: U: $\underline{B} \to \underline{A}$ satisfies the *solution set of objects* condition *at 1* (SSO1) if for every A in $\underline{A}^1$ there exist I in $\underline{S}$ and B in $\underline{B}^I$ such that for every B' in $\underline{B}^1$ and every morphism $a': A \to U^1B'$ there exist $i: 1 \to I$, $a: A \to U^1 i^*B$ and $b: i^*B \to B'$ such that $U^1b \cdot a = a'$.

(1.7) <u>Theorem</u>: Let $U: \underline{B} \to \underline{A}$ be an indexed functor where $\underline{B}$ is complete at 1, $\underline{A}$ has small homs, and $\underline{B}$ has small homs at 1. Assume that U is continuous at 1. Then U has a left adjoint at 1 if and only if it satisfies SSO1.

<u>Proof</u>: We shall show that when $\underline{A}$ has small homs, SSO1 implies SSM1. Let B in $\underline{B}^I$ be as in definition (1.6).

By small homs we have an object $h: \mathrm{Hom}^I(\Delta_I A, U^I B) \to I$ in $\underline{S}/I$ with the property that there is a natural bijection

$$\frac{\alpha \to h \quad \text{in } \underline{S}/I}{\Delta_J A \to U^J \alpha^* B \quad \text{in } \underline{A}^J}$$

where $\alpha: J \to I$ is any object of $\underline{S}/I$. Let $\bar{I} = \mathrm{Hom}^I(\Delta_I A, U^I B)$, $\bar{B} = h^*B$, and $\bar{a}: \Delta_{\bar{I}} A \to U^{\bar{I}} \bar{B}$ be the morphism corresponding to 1_h in the above bijection.

We claim that $(\bar{I}, \bar{B}, \bar{a})$ gives us a solution set of morphisms for U at A. Let B' be any object of $\underline{B}^1$ and $a': A \to U^1 B'$ any morphism of $\underline{A}^1$. Since (I,B) is a solution set of objects, there exist $i: 1 \to I$, $a: A \to U^1 i^*B$ and $b: i^*B \to B'$ such that $U^1 b \cdot a = a'$. The morphism $a: A \to U^1 i^*B$ corresponds by the above bijection to

$$\begin{array}{ccc} 1 & \xrightarrow{\bar{i}} & \mathrm{Hom}^I(\Delta_I A, U^I B) \\ & {}_{i}\searrow \quad \swarrow_{h} & \\ & I & \end{array}$$

and by naturality of that bijection in α we know that $\bar{i}^*(\bar{a}) = a$. Thus we have $\bar{i}: 1 \to \bar{I}$ and $\bar{b}: \bar{i}^* \bar{B} = i^*B \xrightarrow{b} B'$ such that $U^1 \bar{b} \cdot \bar{i}^* \bar{a} = U^1 b \cdot a = a'$. $\square$

We now give the localized version of the previous definition and theorem

(1.8) Definition: $U: \underline{B} \to \underline{A}$ satisfies the *solution set of objects* condition (SSO) if for every I in $\underline{S}$ and every A in $\underline{A}^I$ there exist J and B in $\underline{B}^J$ such that for every B' in $\underline{B}^I$ and every $a': A \to U^I B'$ there exist $i: I \to J$, $a: A \to U^I i^*B$, and $b: i^*B \to B'$ such that $U^I b \cdot a = a'$.

(1.9) Theorem: Let $U: \underline{B} \to \underline{A}$ be an indexed functor where $\underline{B}$ is complete and $\underline{A}$ and $\underline{B}$ have small homs. U has a left adjoint if and only if it is continuous and satisfies SSO. □

2. Generators:

(2.1) As before, we define generating families at 1, which are more intuitive and easier to work with, and then localize to get the notion of generating family which is the notion in which we are interested.

(2.1.1) Definition: A *generating family at 1* is an object G in $\underline{A}^I$ for some I, such that if $f,g: A \to A'$ in $\underline{A}^1$ with $f \neq g$, then there exist $\alpha: J \to I$ and $h: \alpha^*G \to \Delta_J A$ with $\Delta_J f \cdot h \neq \Delta_J g \cdot h$.

(2.1.2) Definition: A *generating family* is an object G in $\underline{A}^I$ for some I, such that for every J and every $f,g: A \to A'$ in $\underline{A}^J$ with $f \neq g$ there exist $(\alpha_1,\alpha_2): K \to I \times J$ and $h: \alpha_1^* G \to \alpha_2^* A$ with $\alpha_2^* f \cdot h \neq \alpha_2^* g \cdot h$.

It is easily seen that G is a generating family if and only if $p_2^* G$

generates $\underline{A}/J$ at 1 for every J (p_2: $I \times J \to I$ is the projection). Thus generating is the localization of generating at 1.

(2.1.3) Proposition: If $\underline{A}$ has small homs, then G in $\underline{A}^I$ is a generating family (at 1) if and only if the indexed functor $\text{Hom}^I(G,\Delta_I(\ )): \underline{A} \to \underline{S}^I$ is faithful (at 1).

Proof: Assume that G is a generating family at 1. For any $f,g: A \to A'$ in $\underline{A}^1$ with $f \neq g$ there exist $J \xrightarrow{\alpha} I$ and $h: \alpha^*G \to \Delta_J A$ with $\Delta_J f \cdot h \neq \Delta_J g \cdot h$. Now h corresponds to $\ulcorner h \urcorner: \alpha \to \text{Hom}^I(G,\Delta_I A)$ and we have $\text{Hom}^I(G,\Delta_I f) \cdot \ulcorner h \urcorner \neq \text{Hom}^I(G,\Delta_I g) \cdot \ulcorner h \urcorner$ and consequently $\text{Hom}^I(G,\Delta_I f) \neq \text{Hom}^I(G,\Delta_I g)$. It follows that $\text{Hom}^I(G,\Delta_I(\ ))$ is faithful at 1.

Conversely, if $\text{Hom}^I(G,\Delta_I(\ ))$ is faithful at 1 and $f \neq g$ (as above), then $\text{Hom}^I(G,\Delta_I f) \neq \text{Hom}^I(G,\Delta_I g)$. Now there exist an α and $\phi: \alpha \to \text{Hom}^I(G,\Delta_I A)$ such that $\text{Hom}^I(G,\Delta_I f) \cdot \phi \neq \text{Hom}^I(G,\Delta_I g) \cdot \phi$ (take ϕ to be the identity, for example). This ϕ corresponds to a morphism h as required for generators at 1.

The remainder of the proposition now follows by localization. □

Remark that a functor $\underline{A} \to \underline{S}^I$ is to be interpreted as an I-indexed family of functors $\underline{A} \to \underline{S}$. In particular, $\text{Hom}^I(G,\Delta_I(\ ))$ is the I-indexed family of functors obtained by homing out of each of the members of the I-indexed family G. The above proposition says that G is a generating family if and only if these hom functors are jointly faithful.

Let A be any object of $\underline{A}^1$ and let $\text{Hom}^I(G,\Delta_I A)$ be $x: X \to I$. By the universal property of $\text{Hom}^I(G,\Delta_I A)$ we have a bijection

$$\cong \frac{\alpha \longrightarrow \mathrm{Hom}^I(G,\Delta_I A) \text{ in } \underline{S}^I}{\alpha^* G \longrightarrow \Delta_J A \text{ in } \underline{A}^J}$$

where $\alpha: J \longrightarrow I$ is any object of $\underline{S}^I$. Corresponding to the identity morphism on top, we get the generic family of morphisms $h: x^*G \longrightarrow \Delta_X A$ (see II, 2.3).

For $f,g: A \longrightarrow A'$ in $\underline{A}^1$, $\mathrm{Hom}^I(G,\Delta_I f) = \mathrm{Hom}^I(G,\Delta_I g)$ if and only if the composites

$$x^*G \xrightarrow{h} \Delta_X A \overset{\Delta_X f}{\underset{\Delta_X g}{\rightrightarrows}} \Delta_X A'$$

are equal. Thus $\mathrm{Hom}^I(G,\Delta_I(\ ))$ is faithful at 1 if and only if the generic family of morphisms h is *jointly epic* (obvious definition) for all A in $\underline{A}^1$.

For A in $\underline{A}^J$, let $\mathrm{Hom}^{I\times J}(p_1^*G,p_2^*A)$ be $(x_1,x_2): X \longrightarrow I\times J$. Then there is a generic family of morphisms

$$h: x_1^*G \longrightarrow x_2^*A .$$

It is easily seen that $\mathrm{Hom}^I(G,\Delta_I(\ ))$ is faithful if and only if h is jointly epic (i.e. $x_2^*f\cdot h = x_2^*g\cdot h \Rightarrow f = g$) for all J and all A in $\underline{A}^J$.

(2.1.4) Proposition: Assume that $\underline{A}$ has small coproducts at 1 and let G be in $\underline{A}^I$. If for every A in $\underline{A}^1$ there exist $\alpha: J \longrightarrow I$ and an epimorphism $\amalg_J \alpha^*G \twoheadrightarrow A$ (not necessarily stable) then G is a generating family at 1. If $\underline{A}$ has small homs the converse is true.

Proof: Assume that the hypotheses of the first part hold and let $f,g: A \longrightarrow A'$ be in $\underline{A}^1$ with $f \neq g$. Then the composits

$$\amalg_J \alpha^*G \twoheadrightarrow A \overset{f}{\underset{g}{\rightrightarrows}} A'$$

are different and transposing this diagram through the adjointness $\amalg_J \dashv \Delta_J$

we get

$$\alpha^*G \xrightarrow{h} \Delta_J A \overset{\Delta_J f}{\underset{\Delta_J g}{\rightrightarrows}} \Delta_J A'$$

with $\Delta_J f\cdot h \neq \Delta_J g\cdot h$. Thus G is a generating family.

Assume now that $\underline{A}$ has small homs and that G is a generating family at 1 . From the above discussion, the generic family of morphisms

$$h: x^*G \longrightarrow \Delta_X A$$

is jointly epic. This is equivalent to saying that

$$\bar{h}: \amalg_X x^*G \longrightarrow A$$

is an epi. □

Localizing this proposition shows that if $\underline{A}$ has small homs and all $\amalg_\alpha$, then G in $\underline{A}^I$ is a generating family if and only if for every J and every A in $\underline{A}^J$ there exist $(\alpha_1,\alpha_2): K \longrightarrow I\times J$ and an epimorphism $\amalg_{\alpha_2}\alpha_1^*G \twoheadrightarrow A$.

Cogenerating families are defined dually.

(2.2) It is easily seen that when $\underline{S}$ is $\underline{\text{Set}}$ the above definitions are equivalent to the usual definition.

(2.2.1) It is always the case that 1 is a generator for $\underline{S}$ (a generator is a generating family indexed by 1). Indeed, $\text{Hom}^1(1,-) \cong 1_{\underline{S}}$.

(2.2.2) If $\underline{S}$ is a topos, then Ω is a cogenerator. It is well known that $\text{Hom}(-,\Omega): \underline{S}^{op} \longrightarrow \underline{S}$ is faithful (see for example [Pa]).

(2.2.3) If $p: \underline{E} \to \underline{S}$ makes $\underline{E}$ an $\underline{S}$-topos and if $\underline{E}$ as an $\underline{S}$-indexed category has a generating family G in $\underline{E}^I$, then $(\Delta_I\Omega)^G$ is a cogenerating family for $\underline{E}$ (the Ω is that of $\underline{E}$). $\mathrm{Hom}^I(\Delta_I A, (\Delta_I\Omega)^G) \cong \mathrm{Hom}^I(G, (\Delta_I\Omega)^{\Delta_I A})$ $\cong \mathrm{Hom}^I(G, \Delta_I(\Omega^A))$ which is a faithful functor of A since Ω^A is and $\mathrm{Hom}^I(G, \Delta_I(-))$ is faithful since G generates.

(2.2.4) Any small category has a generating family, namely the generic family of objects X. This follows immediately from the definitions.

(2.2.5) For any I, $\underline{S}^I$ has a generating family $\delta: I \to I \times I$ in $(\underline{S}^I)^I$. $\mathrm{Hom}^I(\delta, \Delta_I(\ )) \cong \mathrm{Hom}^I(\Sigma_I\delta, \ ) \cong \mathrm{Hom}^I(1_I, \ ) \cong 1_{\underline{S}^I}$ which is faithful.

(2.2.6) If $U: \underline{B} \to \underline{A}$ is a faithful indexed functor with a left adjoint F and $\underline{A}$ has a generating family G in $\underline{A}^I$, then $F^I(G)$ is a generating family for $\underline{B}$. Indeed, $\mathrm{Hom}^I(F^I(G), \Delta_I(\ )) \cong \mathrm{Hom}^I(G, U^I\Delta_I(\ )) \cong \mathrm{Hom}^I(G, \Delta_I U(\ ))$ which is clearly faithful.

In particular a reflective subcategory of a category with a generating family has one also.

(2.2.7) If $\mathbb{C}$ is a category object in $\underline{S}$, then $\underline{S}^{\mathbb{C}}$ has a generating family. It is well known that $\amalg_{\partial_1} \partial_0^*$ is a triple on $\underline{S}^{C_0}$ whose category of algebras is $\underline{S}^{\mathbb{C}}$. The result now follows from (2.2.5) and (2.2.6).

(2.2.8) If $\underline{S}$ is a topos with natural numbers object (see [J+W]) then the abelian groups $\mathbb{Z}$, $\mathbb{Q}$, $\mathbb{Q}/\mathbb{Z}$ can be constructed. It is known that $\mathbb{Z}$ is the free abelian group generated by 1 and consequently it is a generator for $\mathrm{Ab}(\underline{S})$. However, D. Van Osdol has shown that $\mathbb{Q}/\mathbb{Z}$ is not always a cogenerator for $\mathrm{Ab}(\underline{S})$ (take $\underline{S} = \underline{\mathrm{Set}}^{\bullet \leftarrow \bullet \rightarrow \bullet}$).

(2.3) Let $p: \underline{E} \to \underline{S}$ be a geometric morphism of topoi. In [Di2] R. Diaconescu defines a "set of" generators for $\underline{E}$ to be an object D of $\underline{E}$ such that for every X in $\underline{E}$

$$p^*p_*(\tilde{X}^D) \times D \xrightarrow{\varepsilon \times D} \tilde{X}^D \times D \xrightarrow{ev} \tilde{X}$$

is an epimorphism ($\tilde{X}$ is the partial morphism classifier [K+W] and ε is the counit for $p^* \dashv p_*$).

Proposition: $\underline{E}$ has generators as defined by Diaconescu if and only if $\underline{E}$ has a generating family at 1 as in definition (2.1.1).

Proof: Let D be a "set of" generators as defined by Diaconescu. Let $Sub(D)$ be the object of subobjects of D (II,4.6.1) and $G \rightarrowtail \Delta_{Sub(D)}D$ the generic family of subobjects of D (II,4.1). We shall show that G in $\underline{E}^{Sub(D)}$ is a generating family at 1 .

For any X we have

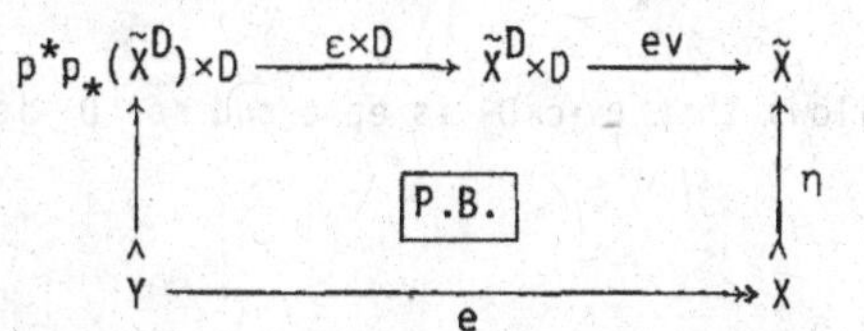

where e is an epimorphism (being the pullback of an epimorphism). Now Y determines a subobject $\bar{Y} \rightarrowtail \Delta_{p_*(\tilde{X}^D)}D$ in $\underline{E}^{p_*(\tilde{X}^D)}$ with $\amalg\bar{Y} = Y$. There exists $\phi: p_*(\tilde{X}^D) \to Sub(D)$ such that $\phi^*(G) \cong \bar{Y}$ and so we get $\amalg\phi^*(G) \cong Y \xrightarrow{e} X$. It follows by proposition (2.1.4) that G is a generating family at 1 .

Conversely, let G in $\underline{E}^I$ be a generating family at 1 and define $D = \amalg_I G$. We have $G \rightarrowtail \Delta_I D$. Since G generates at 1 , for any X there exist $\alpha: J \to I$ and an epimorphism $\amalg_J \alpha^*G \twoheadrightarrow \tilde{X}$. Since $\amalg_J \alpha^*G \rightarrowtail p^*J \times D$ and $\tilde{X}$ is injective (see [Fr2]) there exists ϕ making

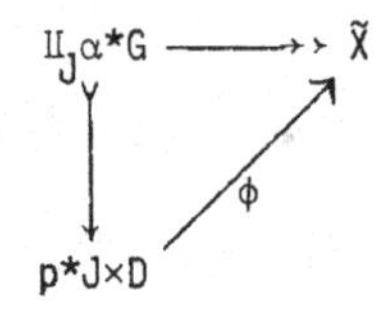

commute. ϕ is necessarily epic.

We have the bijections

$$\cong \frac{p^*J \times D \xrightarrow{\phi} \tilde{X}}{\cong \frac{p^*J \longrightarrow \tilde{X}^D}{J \xrightarrow[\psi]{} p_*(\tilde{X}^D)}}$$

and the composite

$$p^*J \times D \xrightarrow{p^*\psi \times D} p^*p_*(\tilde{X}^D) \times D \xrightarrow{\varepsilon \times D} \tilde{X}^D \times D \xrightarrow{ev} D$$

is equal to ϕ. It follows that $ev \cdot \varepsilon \times D$ is epic and so D is a "set of" Diaconescu generators. □

<u>Remark</u>: From the above discussion we see that D is a Diaconescu set of generators if and only if for every X there exist a J and an epimorphism $\amalg_J \Delta_J D \twoheadrightarrow \tilde{X}$. This property is stable under localization. Indeed, let X be in $\underline{E}^K$. Then $X \rightarrowtail \Delta_K \amalg_K X$ (because $\underline{E}$ is an $\underline{S}$-topos). There exist J and an epimorphism

$$\amalg_J \Delta_J D \twoheadrightarrow \widetilde{\amalg_K X}$$

in $\underline{E}$. Applying Δ_K we get an epimorphism

$$\Delta_K \amalg_J \Delta_J D \longrightarrow\!\!\!\!\rightarrow \Delta_K \widetilde{\amalg_K X} .$$

The following diagram commutes up to isomorphism

$$\begin{array}{ccccc} \underline{E} & \xrightarrow{\Delta_J} & \underline{E}^J & \xrightarrow{\amalg_J} & \underline{E} \\ \downarrow \Delta_K & & \downarrow p_1^* & & \downarrow \Delta_K \\ \underline{E}^K & \xrightarrow[p_2^*]{} & \underline{E}^{J\times K} & \xrightarrow[\amalg p_2]{} & \underline{E}^K \end{array}$$

and so $\Delta_K \amalg_J \Delta_J D \cong \amalg_{p_2} p_2^*(\Delta_K D)$. Since Δ_K is logical, $\Delta_K \widetilde{\amalg_K X} \cong \widetilde{\Delta_K \amalg_K X}$ and $\tilde{X} \rightarrowtail \widetilde{\Delta_K \amalg_K X}$ but since $\tilde{X}$ is injective this mono has a retract $\widetilde{\Delta_K \amalg_K X} \longrightarrow\!\!\!\!\rightarrow \tilde{X}$. Putting all this together gives an epimorphism

$$\amalg_{p_2} p_2^*(\Delta_K D) \longrightarrow\!\!\!\!\rightarrow \tilde{X} .$$

Therefore having Diaconescu generators is also equivalent to having a generating family (2.1.2).

3. The Special Adjoint Functor Theorem:

The following theorem is the indexed version of Freyd's Special Adjoint Functor Theorem at 1 . The proof is the same as Freyd's, suitably internalized.

(3.1) Theorem: Let $\underline{B}$ be complete at 1 and well-powered at 1 and assume that it has a cogenerating family at 1 and small homs. Let $\underline{A}$ be a category with small homs and $U: \underline{B} \rightarrow \underline{A}$ an indexed functor which is continuous at 1 , then U has a left adjoint at 1 .

Proof: Let $C \in \underline{B}^I$ be the cogenerating family for $\underline{B}$ and A any object of $\underline{A}^1$. We denote $\mathrm{Hom}^I(\Delta_I A, U^I C)$ by $x: X \rightarrow I$ and the corresponding generic family of morphisms by $g\colon \Delta_X A \to x^*U^I C$. We shall show that $\mathrm{Sub}(\Pi_X x^*C)$ with the generic family of subobjects is a solution set of objects at 1 for U at A .

Let B be any object of $\underline{B}^1$ and denote $\mathrm{Hom}^I(\Delta_I B, C)$ by $y\colon Y \to I$ and the generic family of morphisms by $h\colon \Delta_Y B \to y^*C$. For any $f\colon A \to UB$,

$$\Delta_Y A \xrightarrow{\Delta_Y f} \Delta_Y UB \cong U^Y \Delta_Y B \xrightarrow{U^Y h} U^Y y^*C \cong y^*U^I C$$

induces, by the universal property of $\mathrm{Hom}^I(\Delta_I A, U^I C)$, a unique

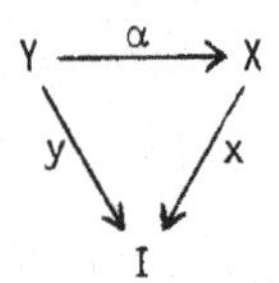

such that $\alpha^*(g) = U^Y h \cdot \Delta_Y f$.

If $p\colon \Delta_Y \Pi_Y y^*C \to y^*C$ and $q\colon \Delta_X \Pi_X x^*C \to x^*C$ denote the respective projection morphisms (i.e. counits for the adjunction $\Delta \dashv \Pi$) , then α induces a unique $s\colon \Pi_X x^*C \to \Pi_Y y^*C$ such that

$$\begin{array}{ccc}
\Delta_Y \Pi_X x^*C & \xrightarrow{\alpha^*(q)} & \alpha^*x^*C \\
{\scriptstyle \Delta_Y s}\downarrow & & \wr\| \\
\Delta_Y \Pi_Y y^*C & \xrightarrow[p]{} & y^*C
\end{array}$$

commutes.

From $h\colon \Delta_Y B \to y^*C$ we get $\bar{h}\colon B \to \Pi_Y y^*C$ which is monic since C is a cogenerating family. Define $B_0 \rightarrowtail \Pi_X x^*C$ by the pullback

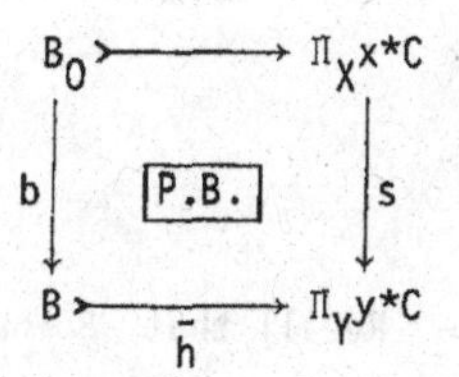

Now $g\colon \Delta_X A \longrightarrow x^*U^I C$ induces $\bar{g}\colon A \longrightarrow \Pi_X x^*U^I C \cong U\Pi_X x^*C$ such that $U^X q\cdot\Delta_X \bar{g} = g$ (since U preserves Π_X (including q)) . We claim that $Us\cdot\bar{g} = U\bar{h}\cdot f$. Since U preserves Π_Y , it is sufficient to show that $U^Y p\cdot\Delta_Y(Us\cdot\bar{g}) = U^Y p\cdot\Delta_Y(U\bar{h}\cdot f)$.

$$U^Y p\cdot\Delta_Y(Us\cdot\bar{g}) = U^Y p\cdot U^Y\Delta_Y s\cdot\Delta_Y\bar{g} = U^Y(p\cdot\Delta_Y s)\cdot\Delta_Y\bar{g} = U^Y\alpha^* q\cdot\Delta_Y\bar{g} = \alpha^*(U^X q\cdot\Delta_X\bar{g})$$

$$= \alpha^* g = U^Y h\cdot\Delta_Y f = U^Y(p\cdot\Delta_Y\bar{h})\cdot\Delta_Y f = U^Y p\cdot\Delta_Y(U\bar{h}\cdot f) .$$

Since U preserves pullbacks

$$\begin{array}{ccc} UB_0 & \rightarrowtail & U\Pi_X x^*C \\ {\scriptstyle Ub}\downarrow & & \downarrow{\scriptstyle Us} \\ UB & \xrightarrow[U\bar{h}]{} & U\Pi_Y y^*C \end{array}$$

is a pullback, and so $f\colon A \longrightarrow UB$ factors through Ub with $B_0 \rightarrowtail \Pi_X x^*C$. This proves the theorem. □

Localizing this theorem, we get the special adjoint functor theorem.

(3.2) <u>Corollary</u>: Let $\underline{B}$ be a complete, well-powered category with a cogenerating family and small homs. Let $\underline{A}$ be a category with small homs and $U\colon \underline{B} \longrightarrow \underline{A}$ an indexed functor. Then U has a left adjoint if and only if it is continuous. □

V. Applications:

1. Small Categories:

Let $\underline{B}$ be a small category. Recall that $\underline{B}$ has a generic family of objects (II,2.1), i.e. an object I in $\underline{S}$ and an object B in $\underline{B}^I$ such that for every J in $\underline{S}$ and B' in $\underline{B}^J$ there exists a unique $\alpha: J \to I$ such that $\alpha^*B \cong B'$.

Let $U: \underline{B} \to \underline{A}$ be any indexed functor and A an object of $\underline{A}^1$. For any B' in $\underline{B}^1$ and any morphism $a': A \to UB'$ there exists $i: 1 \to I$ such that $i^*B = B'$ and thus we get a factorization

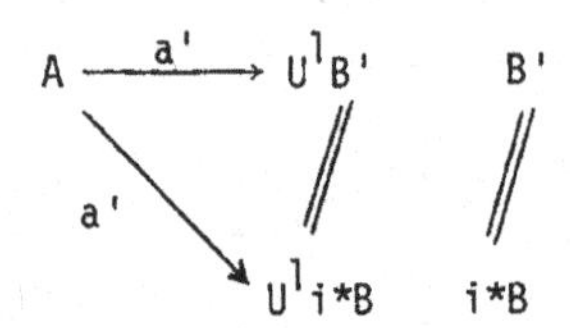

This shows that (I,B) is a solution set of objects for U at A. Thus U satisfies SSO1. But the hypotheses are stable under localization so U also satisfies SSO.

It follows that if $\underline{B}$ is complete at 1 and $\underline{A}$ has small homs, then any functor U which is continuous at 1 has a left adjoint at 1. If $\underline{B}$ is complete and $\underline{A}$ has small homs, then any continuous functor has a left adjoint.

As an example of this, let $\underline{S}$ be a topos, $Gr(\underline{S})$ the category of group objects in $\underline{S}$, and $U: Gr(\underline{S}) \to \underline{S}$ the forgetful functor. $\underline{S}$ is well-powered and it is easily seen that $Gr(\underline{S})$ is also. U induces a functor $u: \underline{Sub}(G) \to \underline{Sub}(UG)$ which is indexed. Now $\underline{Sub}(G)$ and $\underline{Sub}(UG)$ are both reflective subcategories of complete categories and so are themselves complete. Furthermore since U is continuous, so is u. Since $\underline{Sub}(G)$ is small, we get a left adjoint v for u. v applied to a subobject of UG gives the subgroup of G generated by that subobject.

Remark: It is not known whether a small category which is complete must be a preordered object, even if we assume the base category $\underline{S}$ to be a topos.

(1.1) Proposition: Assume that $\underline{S}$ has small homs and that $\underline{B}$ is small. Then $\underline{B}$ is complete if and only if it is cocomplete.

Proof: Assume that $\underline{B}$ is complete. Since $\underline{S}$ has small homs it is cartesian closed and so far any category object $\mathbb{C}$ in $\underline{S}$, $\underline{B}^{\mathbb{C}}$ is also small and consequently has small homs. $\Delta_{\mathbb{C}}: \underline{B} \to \underline{B}^{\mathbb{C}}$ is continuous and therefore has a left adjoint. Finite colimits are similar. Therefore $\underline{B}$ is cocomplete at 1. Since the hypotheses are stable under localization it follows that $\underline{B}$ is cocomplete.

The other half is dual. □

There is another kind of completeness for partially ordered objects in a topos (see [Mk2]). Let $X_1 \rightarrowtail X_0 \times X_0$ be a poset object in the topos $\underline{S}$. We denote this poset by $\mathbb{X}$ and we shall define an indexed functor $\text{seg}: [\mathbb{X}] \to \underline{\text{Sub}}(X_0)$. For any $\phi: I \to X_0$ we send the object $[\phi]$ of $[\mathbb{X}]^I$ to $\text{seg}[\phi] \rightarrowtail \Delta_I X_0$ defined by the pullback

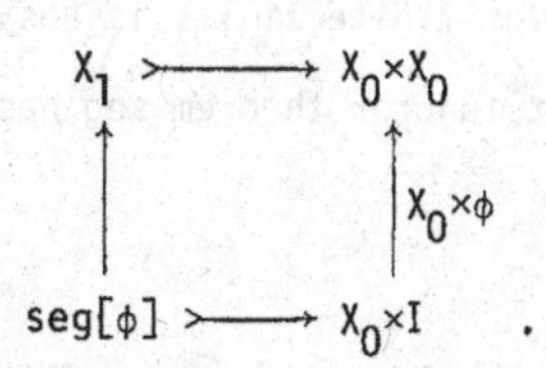

From the universal property of the pullback and the way $[\mathbb{X}]$ is an indexed category we see that $\text{seg}[\phi]$ is characterized by the following property: for any $(\alpha,\psi): J \rightarrowtail I \times X_0$, $(\alpha,\psi) \subseteq \text{seg}[\phi]$ (as subobjects of $\Delta_I X_0$) if and only if

$[\psi] \leq [\phi\alpha]$ in $[\mathbb{X}]^J$. In particular we see that $(1_I,\phi) \subseteq \text{seg}[\phi]$.

Suppose $\phi,\psi: I \to X_0$ are such that $[\psi] \leq [\phi]$. Then for any α we have $[\psi\alpha] = \alpha^*[\psi] \leq \alpha^*[\phi] = [\phi\alpha]$. Now for any subobject (α,θ) of $\Delta_I X_0$ we have $(\alpha,\theta) \subseteq \text{seg}[\psi] \Rightarrow [\theta] \leq [\psi\alpha] \Rightarrow [\theta] \leq [\phi\alpha] \Rightarrow (\alpha,\theta) \subseteq \text{seg}[\phi]$, so seg is functorial. That seg is indexed follows from its definition as a pullback.

$\mathbb{X}$ is said to be *seg-complete* if seg has a left adjoint
sup: $\underline{\text{Sub}}(X_0) \to [\mathbb{X}]$.

(1.2) <u>Proposition</u>: A poset $\mathbb{X}$ is seg-complete if and only if it is complete.

<u>Proof</u>: Since we are dealing with posets seg is clearly faithful. It is also full, for if $\text{seg}[\psi] \subseteq \text{seg}[\phi]$ then $(1_I,\psi) \subseteq \text{seg}[\psi] \subseteq \text{seg}[\phi]$ so $[\psi] \leq 1_I^*[\phi] = [\phi]$. Thus if $\mathbb{X}$ is seg-complete, $\mathbb{X}$ is a full reflective subcategory of $\text{Sub}(X_0)$ which is complete, so $\mathbb{X}$ is complete.

Now assume that $\mathbb{X}$ is complete. We will show that seg preserves Π_I. Let $\phi: I \to X_0$, then $\Pi_I[\phi]: 1 \to X_0$. For any $J \xrightarrow{(\tau,\psi)} 1\times X_0$ we have that $(\tau,\psi) \subseteq \text{seg}\ \Pi_I[\phi] \Leftrightarrow [\psi] \leq \Delta_J \Pi_I[\phi] \Leftrightarrow [\psi] \leq \Pi_I^J\ \Delta_J^I[\phi]$
$\Leftrightarrow \Delta_I^J[\psi] \leq \Delta_J^I\ [\phi] \Leftrightarrow [\psi p_2] \leq [\phi p_1] \Leftrightarrow (p_1,\psi p_2) \subseteq \text{seg}[\phi] \Leftrightarrow \Delta_I(\tau,\psi) \subseteq \text{seg}[\phi]$
$\Leftrightarrow (\tau,\psi) \subseteq \Pi_I \text{seg}[\phi]$, where $p_1: I\times J \to I$ and $p_2: I\times J \to J$ are the projections. It follows that $\text{seg}\ \Pi_I[\phi] \cong \Pi_I \text{seg}[\phi]$. That seg preserves the Π_α follows by localization. That seg preserves finite limits is easy, since they are external.

Therefore, by the adjoint functor theorem seg has a left adjoint, i.e. $\mathbb{X}$ is seg-complete. □

The seg defined above is the lower segment. The upper segment is contravariant and its adjoint on the right (if it has one) is called inf. By duality, the existence of inf is equivalent to $\mathbb{X}$ being cocomplete. But since $\mathbb{X}$ is cocomplete if and only if it is complete we see that all four completeness concepts coincide.

Bénabou also had this result [Bn2] but his methods were different. We have given it here mainly to illustrate how the adjoint functor theorem can be used on small categories.

Remark: In view of (IV, 2.2.4), any small category has a generating (and cogenerating) family, so we could have used the special adjoint functor theorem to prove propositions (1.1) and (1.2).

2. Free Φ-algebras:

(2.1) Let $\Phi\colon \underline{A} \to \underline{A}$ be an indexed functor. A Φ-*algebra* is a pair (A,a) where A is an object of $\underline{A}$ and $a\colon \Phi A \to A$. A homomorphism $f\colon(A,a) \to (B,b)$ is a morphism $f\colon A \to B$ in $\underline{A}$ such that $f\cdot a = b\cdot\Phi f$. We denote the category of Φ-algebras by $(\Phi;\underline{A})$. This is clearly an indexed category with $(\Phi;\underline{A})^I = (\Phi^I;\underline{A}^I)$. There is an obvious forgetful functor $U\colon (\Phi;\underline{A}) \to \underline{A}$ which is also indexed. The constructions giving $(\Phi;\underline{A})$ and U are clearly stable under localization.

The comma category $(\Phi,1_{\underline{A}})$ has two canonical projections $(\Phi,1_{\underline{A}}) \rightrightarrows \underline{A}$ which give a functor $(\Phi,1_{\underline{A}}) \to \underline{A}\times\underline{A}$. Then

$$\begin{array}{ccc} (\Phi;\underline{A}) & \longrightarrow & (\Phi,1_{\underline{A}}) \\ \downarrow & & \downarrow \\ \underline{A} & \xrightarrow{\ \Delta\ } & \underline{A}\times\underline{A} \end{array}$$

is a pullback diagram.

It follows that if $\underline{A}$ has small homs (at 1) then so has $(\Phi;\underline{A})$. We can also conclude that if $\underline{A}$ is small then so is $(\Phi;\underline{A})$ since if $\underline{A}$ is small so is $\underline{A}^{2}$ and pullbacks of small categories are small.

(2.1.1) Proposition: If $\underline{A}$ is complete (at 1) then $(\Phi;\underline{A})$ is complete (at 1) and $U: (\Phi;\underline{A}) \longrightarrow \underline{A}$ is continuous (at 1).

Proof: Finite limits are treated as in the usual (Set-indexed) case.

Let $a: \Phi^{I}A \longrightarrow A$ be an object of $(\Phi;\underline{A})^{I}$. We claim that the Φ-algebra

$$\Phi\Pi_{I}A \xrightarrow{\lambda A} \Pi_{I}\Phi^{I}A \xrightarrow{\Pi_{I}a} \Pi_{I}A$$

is the product of the family (A,a), where $\lambda: \Phi\Pi_{I} \longrightarrow \Pi_{I}\Phi^{I}$ is the canonical morphism.

Let $b: \Phi B \dashrightarrow B$ be any Φ-algebra and $f: (B,b) \longrightarrow (\Pi_{I}A, \Pi_{I}a\cdot\lambda A)$ a homomorphism, i.e. the following diagram commutes

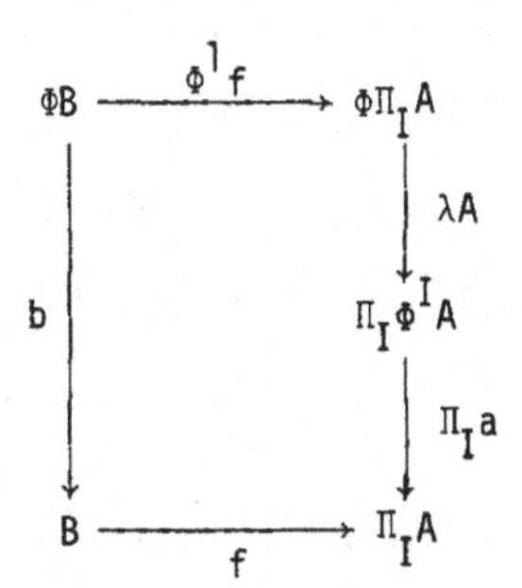

Now morphisms $f: B \longrightarrow \Pi_{I}A$ are in bijection with morphisms $\bar{f}: \Delta_{I}B \longrightarrow A$ where $\bar{f}$ is given by $\bar{f} = p\cdot\Delta_{I}f$ ($p: \Delta_{I}\Pi_{I}A \longrightarrow A$ is the projection morphism). The commutivity of the above diagram, when transformed by the adjunction $\Delta_{I} \dashv \Pi_{I}$ becomes equivalent to the commutivity of ★ below

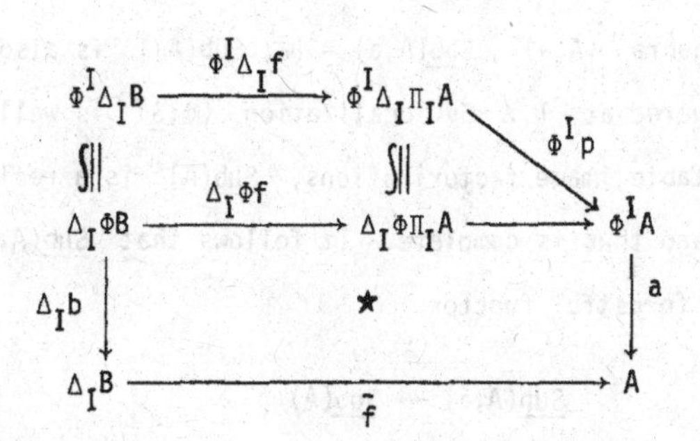

i.e. $\bar{f}: \Delta_I(B,b) \to (A,a)$ is a homomorphism.

That Δ_J preserves Π_I is clear since Δ_J preserves everything in the construction of Π_I. It is also clear that U preserves Π_I.

The rest of the proposition follows by localization. □

If $\underline{A}$ has (stable) finite limits then $(\Phi;\underline{A})$ also has and U preserves them. It follows that U preserves monos. Thus $f: (A,a) \to (B,b)$ is monic if and only if $f: A \to B$ is monic in $\underline{A}$.

Assume that $\underline{A}$ has stable image factorizations and let (A,a) be a Φ-algebra. If $\underline{\mathrm{Sub}}(A)$ is the indexed poset of subobjects of A we get an indexed functor

$$\phi: \underline{\mathrm{Sub}}(A) \to \underline{\mathrm{Sub}}(A)$$

defined at I by sending $m: A_0 \rightarrowtail \Delta_I A$ to the image of

$$\Phi^I A \xrightarrow{\Phi^I m} \Phi^I \Delta_I A = \Delta_I \Phi A \xrightarrow{\Delta_I a} \Delta_I A .$$

It is an easy calculation to see that $\underline{\mathrm{Sub}}(A,a) = (\phi;\underline{\mathrm{Sub}}(A))$.

Now let us assume that $\underline{S}$ is an elementary topos and take $\underline{A} = \underline{S}$.

Since $\underline{S}$ is well-powered and has small homs, $\underline{\mathrm{Sub}}(A)$ is small for any A in $\underline{S}$. For any Φ-algebra (A,a) , $\underline{Sub}(A,a) = (\phi; \underline{Sub}(A))$ is also small and so $(\Phi;\underline{S})$ is well-powered at 1 . By localization $(\Phi;\underline{S})$ is well powered.

Since we have stable image factorizations, $\underline{Sub}(A)$ is a reflective subcategory of $\underline{S}/A$ and thus is complete. It follows that $\underline{Sub}(A,a)$ is also complete and that the forgetful functor

$$\underline{Sub}(A,a) \longrightarrow \underline{Sub}(A)$$

is continuous (prop. (2.1)).

By the results of section 1, we conclude that the forgetful functor $\underline{Sub}(A,a) \rightarrow \underline{Sub}(A)$ has a left adjoint which we denote by $\langle - \rangle$. Given a subobject $A_0 \rightarrowtail A$, $\langle A_0 \rangle \rightarrowtail (A,a)$ is the subalgebra of (A,a) generated by A_0 . It has the property that for any subalgebra $(B,b) \rightarrowtail (A,a)$, $A_0 \subseteq B$ if and only if $\langle A_0 \rangle \subseteq (B,b)$. In particular if (B,b) is a subalgebra of $\langle A_0 \rangle$ containing A_0 then $(B,b) = \langle A_0 \rangle$.

(2.2) We must recall Freyd's unique existentiation proposition ([Fr2], proposition 2.2.1). Given any morphism $f: C \longrightarrow A$ in a topos there exists a monomorphism $Q \rightarrowtail A$ such that

$$\begin{array}{ccc} Q & = & Q \\ \downarrow & & \downarrow \\ C & \xrightarrow[f]{} & A \end{array}$$

is a pullback and for any other pullback

$$\begin{array}{ccc} X & = & X \\ {\scriptstyle g}\downarrow & & \downarrow \\ C & \xrightarrow[f]{} & A \end{array}$$

g factors through Q .

Note that f is a monomorphism if and only if $Q \rightarrowtail C$ is an isomorphism. Thus if $F: \underline{S} \to \underline{S}$ preserves those pullbacks obtained by unique existentiation, it also preserves monomorphisms. Indeed, if f is a monomorphism, then $Q \rightarrowtail C$ is an isomorphism and so is $FQ \to FC$. Since

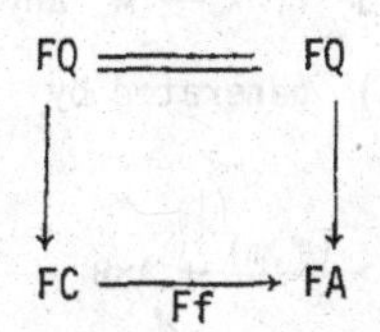

is a pullback by hypothesis, $FQ \to FC$ factors through $Q' \rightarrowtail FC$, the unique existentiation subobject corresponding to Ff. Therefore $Q' \rightarrowtail FC$ is an isomorphism and so Ff is monic.

(2.2.1) <u>Proposition</u>: Let $F: \underline{S} \to \underline{S}$ be any (not necessarily indexed) functor and G a subfunctor of F. If F preserves the pullbacks obtained by unique existentiation, then so does G.

<u>Proof</u>: This follows by a simple diagram chase. □

(2.2.2) <u>Theorem</u>: Let $\Phi: \underline{S} \to \underline{S}$ be an indexed functor such that Φ^1 preserves the pullbacks of unique existentiation. Assume that for every X in $\underline{S}$ there exists an object B and monomorphisms

(i) $X \overset{m}{\rightarrowtail} B$

(ii) $\Phi B \overset{b}{\rightarrowtail} B$

(iii) $B+B \rightarrowtail B$

then $U: (\Phi;\underline{S}) \to \underline{S}$ has a left adjoint at 1.

<u>Proof</u>: We shall show that U satisfies SSO1 and then the result will follow from theorem (IV, 1.7).

Let X be any object of $\underline{S}$. There exists B as above. Because of (iii) we can assume that m and b are disjoint. We will show that the solution set for X is $\mathrm{Sub}(B,b)$ with the generic family of subobjects $(B_0,b_0) \rightarrowtail \Delta(B,b)$ in $(\Phi;\underline{S})^{\mathrm{Sub}\,(B,b)}$.

Let (A,a) be any algebra and $f\colon X \to A$ any morphism in $\underline{S}$. Let $\langle X\rangle$ be the subalgebra of $(A,a) \times (B,b)$ generated by $X \xrightarrow{(f,m)} A\times B$

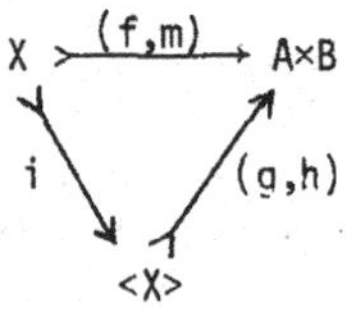

$$\begin{array}{ccc} X & \xrightarrow{(f,m)} & A\times B \\ & {\scriptstyle i}\searrow \quad \nearrow{\scriptstyle (g,h)} & \\ & \langle X\rangle & \end{array}$$

$g\colon \langle X\rangle \to A$ and $h\colon \langle X\rangle \to B$ are homomorphisms. We want to show that h is monic.

Let the structure morphism of $\langle X\rangle$ be $x\colon \Phi\langle X\rangle \to \langle X\rangle$. Then $\binom{i}{x}\colon X + \Phi\langle X\rangle \to \langle X\rangle$ is epi. Indeed, if

$$X + \Phi\langle X\rangle \xrightarrow{\binom{u}{v}} Y \xrightarrow{w} \langle X\rangle$$

is the image factorization of $\binom{i}{x}$, then $(Y, v\cdot\Phi w)$ is an algebra and w a homomorphism so that $(Y, v\cdot\Phi w)$ is a subalgebra of $\langle X\rangle$ containing X and therefore w must be an isomorphism.

Let

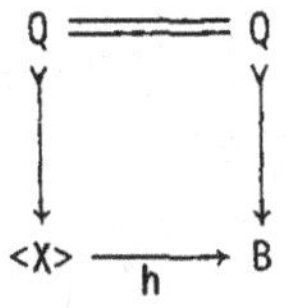

$$\begin{array}{ccc} Q & = & Q \\ \downarrow & & \downarrow \\ \langle X\rangle & \xrightarrow{h} & B \end{array}$$

be the unique existentiation pullback. We want to show that Q contains X and is a subalgebra of $\langle X\rangle$.

The following diagrams are pullbacks

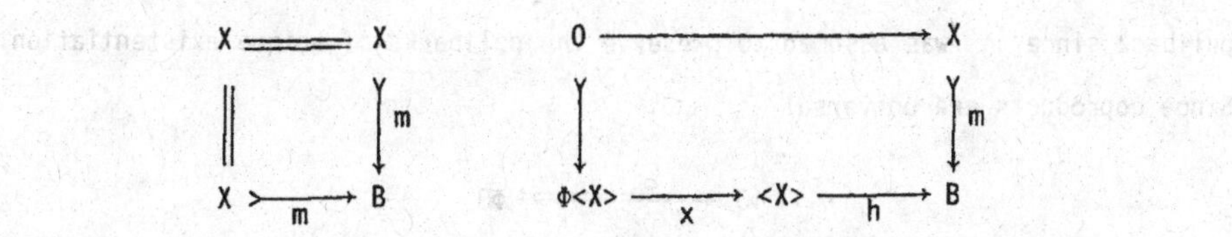

the second one since $hx = \beta \cdot \Phi h$ and m and β are disjoint. Since coproducts are universal in a topos

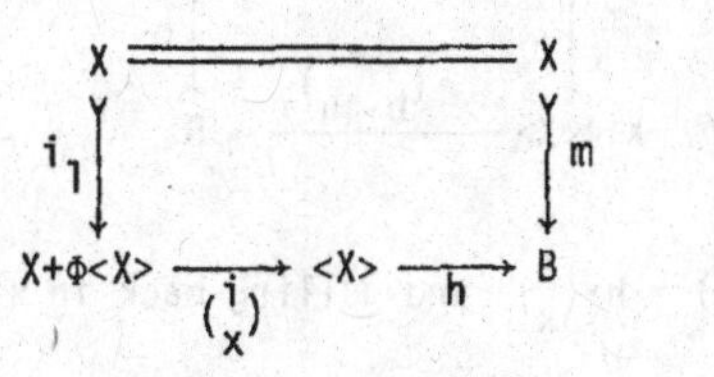

is a pullback. Pulling back in stages we get

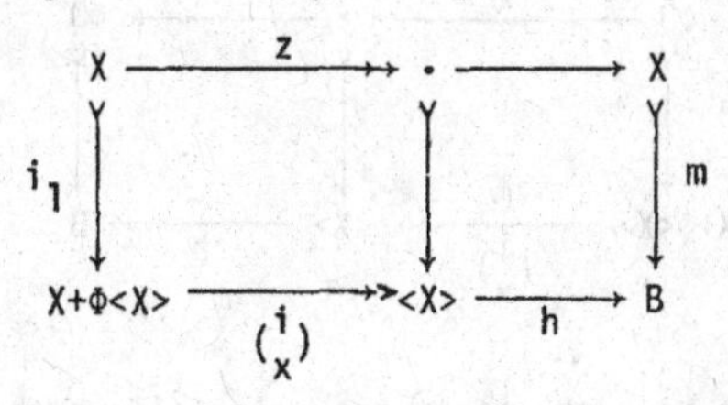

where z is epi (the pullback of an epi) and split mono, therefore iso. It follows that

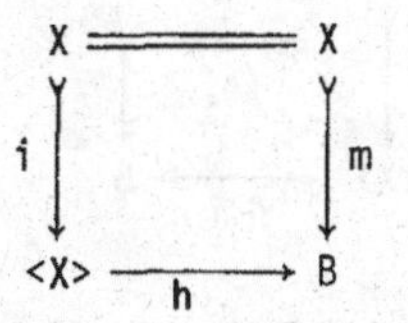

is a pullback and so we get $X \subseteq Q$.

The following (outside) squares are pullbacks

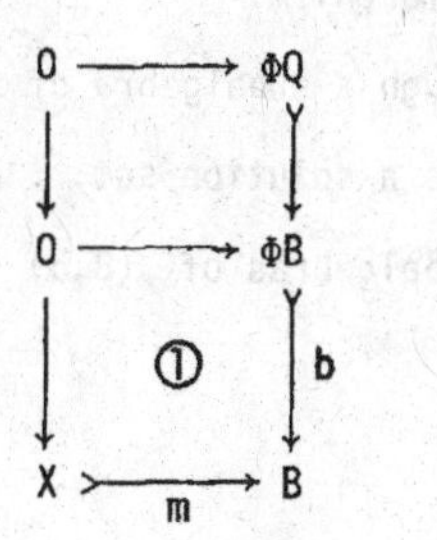

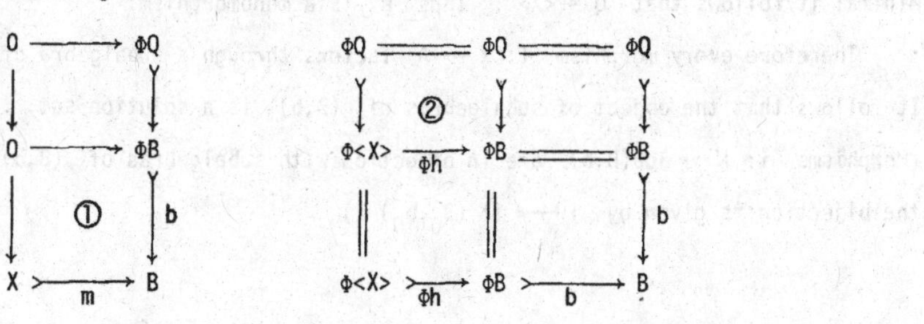

where ① is a pullback since m and b were chosen disjoint and ② is a pullback since Φ was assumed to preserve the pullbacks of unique existentiation. Since coproducts are universal

$$\begin{array}{ccc} \Phi Q & = & \Phi Q \\ \downarrow & & \downarrow \\ \Phi\langle X\rangle & & \Phi B \\ \downarrow & & \downarrow b \\ X+\Phi\langle X\rangle & \xrightarrow{\binom{m}{b\cdot\Phi h}} & B \end{array}$$

is a pullback. But $\binom{m}{b\cdot\Phi h} = h\cdot\binom{i}{x}$ and pulling back in stages we get

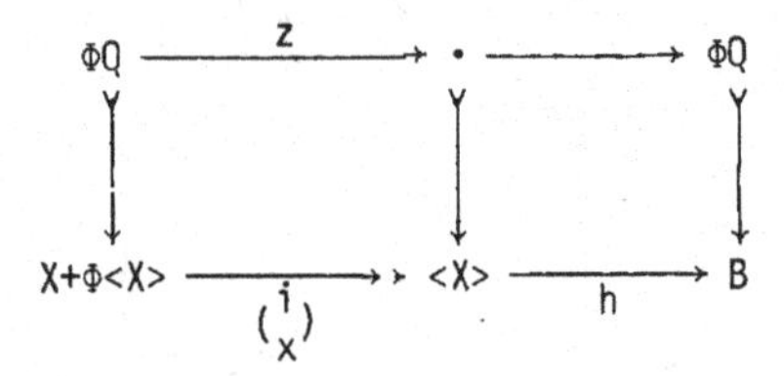

where z is epi and split mono therefore iso. It follows that

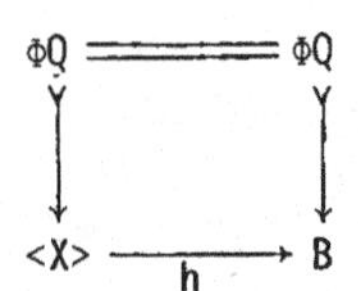

is a pullback and so $\Phi Q \subseteq Q$ as subobjects of $\langle X\rangle$.

Therefore Q is a subalgebra of $\langle X\rangle$ containing X and since $\langle X\rangle$ is minimal it follows that $Q = \langle X\rangle$. Thus h is a monomorphism.

Therefore every morphism $f: X \to A$ factors through a subalgebra of (B,b). It follows that the object of subalgebras of (B,b) is a solution set. (Morphisms $i: 1 \to \mathrm{Sub}(B,b)$ are in bijection with subalgebras of (B,b) and the bijection is given by $i \longmapsto i^*(B_0,b_0)$.) □

The above proof was adapted from the one in [Sc] .

(2.2.3) Remark: In view of results of Barr [Br1], the above theorem says that Φ generates a free triple.

(2.3) As a first application of this theorem we have the following.

(2.3.1) Proposition: A topos $\underline{S}$ has a natural numbers object if and only if there exists an object A such that there are monos $A+A \rightarrowtail A$ and $1 \rightarrowtail A$.

Proof: Take $\Phi = 1_{\underline{S}}$ in the above theorem. Then $(\Phi;\underline{S})$ is the category of objects of $\underline{S}$ equipped with an endomorphism. For any X conditions (i), (ii), and (iii) are satisfied by taking $B = X \times A$. The value of the left adjoint at the object 1 is the natural numbers object.

Conversely if there is a natural numbers object $\mathbb{N}$ then it is well known that $A = \mathbb{N}$ satisfies the conditions in the proposition. □

(2.3.2) Remark: Condition (iii) in the theorem is there only so that we can choose m and b disjoint. Thus in the above proposition we can replace the conditions by the existence of disjoint monos $A \rightarrowtail A$ and $1 \rightarrowtail A$. We then get Freyd's theorem 5.44 [Fr2].

(2.3.3) Lemma: Let $\underline{S}$ be a topos with natural numbers object $\mathbb{N}$. For any K in $\underline{S}$, the object $B = \Omega^{K^{\mathbb{N}}}$ has the following properties:

(i) there exists a monomorphism $K \rightarrowtail B$

(ii) for any monomorphism $L \rightarrowtail K$ there exists a monomorphism $B^L \rightarrowtail B$.

Proof: The unique $\mathbb{N} \to 1$ is epic and so induces a mono $K \rightarrowtail K^{\mathbb{N}}$. The singleton morphism $K^{\mathbb{N}} \rightarrowtail \Omega^{K^{\mathbb{N}}}$ is always monic and thus we get $K \rightarrowtail \Omega^{K^{\mathbb{N}}}$.

The direct image morphism induced by a mono is monic and so for any mono $L \rightarrowtail K$ we have

$$(\Omega^{K^{\mathbb{N}}})^L \cong \Omega^{K^{\mathbb{N}} \times L} \rightarrowtail \Omega^{K^{\mathbb{N}} \times K} \cong \Omega^{K^{\mathbb{N}+1}} \cong \Omega^{K^{\mathbb{N}}} \quad . \qquad \square$$

(2.3.4) Proposition: Let $\underline{S}$ be a topos with natural numbers object. For any I in $\underline{S}$ the functor $\Phi = (\ )^I$ satisfies the hypotheses of theorem (2.2.2) and consequently free Φ-algebras exist.

Proof: $(\ )^I$ has a left adjoint and so preserves all pullbacks.

For any X in $\underline{S}$ let $B = \Omega^{(I+X+2)^{\mathbb{N}}}$. Then (using lemma 2.3.3)

(i) $X \rightarrowtail I+X+2 \rightarrowtail B$

(ii) $I \rightarrowtail I+X+2$ therefore $B^I \rightarrowtail B$

(iii) $2 \rightarrowtail I+X+2$ therefore $2 \rightarrowtail B$ and $B^2 \rightarrowtail B$ so we have

$$B+B \cong 2\times B \rightarrowtail B\times B \cong B^2 \rightarrowtail B \ . \qquad \square$$

This proposition says that for any object I we have free algebras with one I-ary operation and no equations. We wish to generalize this to the case where we have a family of arities. Let $t\colon I \to J$ be a J-indexed family of arities. An algebra of type t should then be an object A in $\underline{S}$ together with a morphism $(\Delta_J A)^t \to \Delta_J A$ in $\underline{S}^J$. Such a morphism is equivalent to a morphism $a\colon \Pi_J[(\Delta_J A)^t] \to A$ in $\underline{S}$. Thus if we take $\Phi A = \Pi_J[(\Delta_J A)^t]$ then an algebra of type t is a Φ-algebra.

(2.3.5) Proposition: Let $\underline{S}$ be a topos with natural numbers object. For any $t\colon I \to J$ the functor $\Phi = \Pi_J[(\Delta_J -)^t]$ satisfies the hypotheses of theorem (2.2.2).

Proof: Δ_J , $(\ )^t$, and Π_J all preserve pullbacks so Φ does also.

Now for any X in $\underline{S}$ we let $B = \Omega^{(I+J+X+2)^{\mathbb{N}}}$. As before $X \rightarrowtail B$ and $B+B \rightarrowtail B$. Now Δ_J is a logical functor so

$$\Delta_J[\Omega^{(I+J+X+2)^{\mathbb{N}}}] \cong \Omega^{(\Delta_J I+\Delta_J J+\Delta_J X+2)^{\mathbb{N}}}$$

where the Ω, 2, $\mathbb{N}$ on the right are in $\underline{S}^J$. Now $t \rightarrowtail \Delta_J I$ in $\underline{S}^J$ and so by the lemma applied to $\underline{S}^J$ we have that $(\Delta_J B)^t \rightarrowtail \Delta_J B$. Now $\amalg_J$ preserves monos so $\amalg_J[(\Delta_J B)^t] \rightarrowtail J\times B$. $J \rightarrowtail B$ so $J\times B \rightarrowtail B\times B \cong B^2 \rightarrowtail B$ and by composing we get $\amalg_J[(\Delta_J B)^t] \rightarrowtail B$. □

(2.3.6) Proposition: Let Ψ be a subfunctor of Φ. If Φ satisfies the hypotheses of theorem (2.2.2), then so does Ψ.

Proof: That Ψ preserves the pullbacks of unique existentiation follows from (2.2.1). The other hypotheses are obvious. □

Another functor which satisfies the hypotheses of theorem (2.2.2) is $\Phi X = (J+X)^I$.

The question of imposing equations on Φ-algebras is closely related to the existence of coequalizers. If Φ does not preserve epimorphisms (and $(\ )^I$ does not in general) these questions are difficult. They have been studied extensively by R. Rosebrugh [Rs2] (see also [J+W]).

3. Existence of Colimits:

(3.1) Proposition: Let $\underline{A}$ be complete at 1, well-powered at 1, have a cogenerating family at 1, and have small homs. Then $\underline{A}$ has finite colimits at 1.

Proof: If $\underline{X}$ is a finite category, then $\underline{A}^{\underline{X}}$ has small homs (II, 3.9). It is easily seen that $\Delta: \underline{A} \to \underline{A}^{\underline{X}}$ is continuous at 1 and so by theorem (IV, 3.1) it has a left adjoint at 1 . □

(3.2) Corollary: If $\underline{S}$ is cartesian closed and has a subobject classifier, then it has finite colimits at 1 .

Proof: By (III, 4.2.1), $\underline{S}$ is complete at 1 . $\underline{S}$ is well-powered at 1 by (II, 4.2.1), and (II, 3.9) tells us that $\underline{S}$ has small homs. If Ω is the subobject classifier, (IV, 2.2.2) says that Ω is a cogenerator. □

(3.3) Proposition: Let $\underline{A}$ be complete, well-powered, have a cogenerating family and small homs. If $\underline{S}$ has small homs, then $\underline{A}$ is cocomplete.

Proof: By proposition (3.1), $\underline{A}$ has finite colimits.

For any small category $\mathbb{C}$, $\underline{A}^{\mathbb{C}}$ has small homs (corollary III, 1.4). It is easily seen that $\Delta: \underline{A} \to \underline{A}^{\mathbb{C}}$ is continuous and thus by corollary (IV, 3.2) has a left adjoint. Thus $\underline{A}$ has $\mathbb{C}$-colimits. Since the hypotheses are stable under localization, the result follows. □

4. Logical Functors:

As a final application of the special adjoint functor theorem, we prove the following theorem due to Mikkelsen [Mk2].

Theorem: If a logical functor between topoi has a right adjoint, then it also has a left adjoint.

Proof: Let the logical functor be $L: \underline{S} \to \underline{E}$ (where $\underline{S}$ and $\underline{E}$ are topoi) and assume that L has a right adjoint thus giving a geometric morphism

$p: \underline{E} \to \underline{S}$ with $p^* = L$. By (I, 2.4), $\underline{E}$ becomes an $\underline{S}$-indexed category and (I, 2.4.4) says that L is an $\underline{S}$-indexed functor.

As in corollary (3.2) above, $\underline{S}$ is complete at 1 , well-powered at 1 , has small homs and Ω is a cogenerator. By (II, 3.9), $\underline{E}$ has small homs. Since L is logical it preserves finite limits at 1 . From the proof of proposition (III, 3.5.1) we see that Π_I is constructed from finite limits and exponentiation and so is preserved by L , in the sense that $L\Pi_I \cong \Pi_{LI}L$. From (III, 4.2.5) we see that this means that L preserves products at 1 and so by (III, 4.3.1), L is continuous at 1 . By theorem (IV, 3.1), L has a left adjoint at 1 . □

Remark: In the above proof, it was not necessary that L be logical but only that it preserve products at 1 and be left exact.

References:

[Br 1] Barr, M., Coequalizers and free triples, Math. Zeit. 116 (1970), 307-322.

[Br 2] Barr, M., The Point of the Empty Set, Cahiers de Topologie et Géométrie Différentielle, Vol. XIII - 4, 357-368.

[Bn 1] Bénabou, J., Introduction to bicategories, Reports of the Midwest Category Seminar, Lecture Notes in Mathematics 47, Springer, Berlin-Heidelberg-New York 1967.

[Bn 2] Bénabou, J., Logique catégorique, Séminaire de Mathématiques Supérieures, Université de Montréal, 1974. To appear.

[Bn 3] Bénabou, J., Théories relatives à un corpus, C.R. Acad. Sc. Paris, 281 (1975), 831-834.

[Bn 4] Bénabou, J., Fibrations petites et localement petites, C.R. Acad. Sc. Paris, 281 (1975), 897-900.

[Cl] Celeyrette, J., Thesis.

[Da] Day, B.J., An adjoint-functor theorem over topoi, Preprint, Univ. of Sydney, 1976.

[Di 1] Diaconescu, R., Change of base for some toposes, Ph.D. thesis, Dalhousie 1973.

[Di 2] Diaconescu, R., Change of base for toposes with generators, J. Pure and Applied Alg. 6 (1975), 191-218.

[Db] Dubuc, E., Kan extensions in enriched category theory, Lecture Notes in Mathematics, 145, Berlin-Heidelberg-New York, Springer 1970.

[Fr 1] Freyd, P., Abelian categories: An introduction to the theory of functors, New York, Harper and Row 1964.

[Fr 2] Freyd, P., Aspects of topoi, Bull. Austral. Math. Soc. 7 (1972), 1-76.

[Gi 1] Giraud, J., Méthode de la descente, Bull. Soc. Math. France, Mémoire 2, 1964.

[Gi 2] Giraud, J., Cohomologie non abélienne, Grundlehren 179, Berlin-Heidelberg-New York, Springer, 1971.

[Gy 1] Gray, J., Fibred and Cofibred Categories, Proc. Conf. on Categorical Algebra, La Jolla 1975, 21-83.

[Gy 2] Gray, J., Formal Category Theory: Adjointness for 2-Categories, Lecture Notes in Mathematics 391, Springer, Berlin-Heidelberg-New York, 1974.

[Gk] Grothendieck, A., Catégories fibrées et descente, Séminaire de géométrie algébrique de l'Institut des Hautes Etudes Scientifiques. Paris 1961.

[Jn] Johnstone, P., Topos theory, L.M.S. Mathematical Monographs no. 10, Academic Press 1977.

[J+W] Johnstone, P. and Wraith, G., Algebraic theories and recursion in elementary topos theory, this volume.

[Kl] Kelly, G.M., Adjunction for enriched categories, Reports of the Midwest Category Seminar III, Lecture Notes in Math. 106, Springer, Berlin-Heidelberg-New York, 1969.

[K+W] Kock, A. and Wraith, G., Elementary toposes, Lecture Notes No. 30, Aarhus Univ. (1971).

[La] Lawvere, F.W., Theory of Categories over a Base Topos, Lecture Notes, Perugia, 1972-73.

[LS] LeSaffre, B., Structures algébriques dans les topos élémentaires, Thèse, Paris VII, 1974.

[ML] MacLane, S., Categories for the Working Mathematician, Graduate Texts in Mathematics 5, New York Heidelberg Berlin, Springer (1971).

[MK 1] Mikkelsen, C., On the internal completeness of elementary topoi, Oberwolfach Tagungsbericht 30, 1973.

[MK 2] Mikkelsen, C., Lattice theoretic and logical aspects of elementary topoi. Thesis. Aarhus Universitet Various Publications Series No. 25 (1976).

[Pa] Paré, R., Colimits in Topoi, Bull. Amer. Math. Soc. 80 (1974), 556-561.

[Pn 1] Penon, J., Catégories localement internes, C.R. Acad. Sc. Paris, 278, 1577-1580 (1974).

[Pn 2] Penon, J., Locally internal categories, Lecture at the Mulvey-Tierney topos seminar, Columbia, Feb. 1976.

[Rn] Reynolds, G.D., Tensor and Cotensor Products of Bifunctors, Thesis, Wesleyan University, 1973.

[Rs 1] Rosebrugh, R., Abstract Families of Algebras, Thesis, Dalhousie Univ., 1977.

[Rs 2] Rosebrugh, R., Coequalizers in Algebras for an Internal Type, this volume.

[Sc] Schumacher, D., Absolutely free algebras in a topos containing an infinite object, Canad. Math. Bull., Vol. 19(3) (1976), 323-328.

[St] Stout, L.N., Quels sont les espaces topologiques dans les topos? Preprint, McGill Univ., 1975.

[Wd 1] Wood, R.J., Indicial Methods for Relative Categories, Thesis, Dalhousie Univ., 1976.

[Wd 2] Wood, R.J., V-Indexed Categories, this volume.

$\underline{V}$-INDEXED CATEGORIES

R.J. Wood

§0 INTRODUCTION

For $\underline{S}$ a category with finite limits and $\underline{A}$ an $\underline{S}$-indexed category with small homs, [P&S] , the categories $\underline{A}^I$, for $I\epsilon\underline{S}$, become $\underline{S}/I$-categories (in the sense of [E&K]) in a compatible way. It seems natural to ask whether the notion of indexing can be profitably applied to the study of $\underline{V}$-categories for arbitrary monoidal $\underline{V}$.

If $\underset{\sim}{A}$ is a $\underline{V}$-category and I an object of $\underline{V}$, it is not immediately clear how to define a category $\underset{\sim}{A}^I$ of I-indexed families of objects of $\underset{\sim}{A}$. However, for A and B objects of $\underset{\sim}{A}$ it is reasonable to consider a morphism $I \longrightarrow [AB]$ in $\underline{V}$ as an I-indexed family of morphisms from A to B . Furthermore, a cotensor product [IB] in $\underset{\sim}{A}$ has long been regarded as a sort of "product" of an I-indexed family of B's .

Thus a $\underline{V}$-category may be regarded as a category equipped with I-indexed families of morphisms for all $I\epsilon\underline{V}$. If $\underline{V}$ is not closed, $\underline{V}$ itself is not a $\underline{V}$-category, but we may view a morphism $V\otimes I \longrightarrow W$ in $\underline{V}$ as an I-indexed family of morphisms from V to W . A similar remark applies to any $\underline{V}$-tensored or $\underline{V}$-cotensored category in the sense of [Ln] .

Taking the notion of $\underline{V}$-indexed families as basic also sheds light on some of the familiar difficulties with $\underline{V}$-categories. While general $\underline{V}$-categories and the like do not come equipped with a non-trivial notion of I-indexed family of objects, certain categories constructed from them do. For example, if $\underset{\sim}{A}$ is a $\underline{V}$-category and $A_o\epsilon\underset{\sim}{A}$, we might define an I-indexed family of objects of the comma category $A_o/\underset{\sim}{A}$ to be a pair $\langle A,\alpha \rangle$ where $A\epsilon\underset{\sim}{A}$ and $I \xrightarrow{\alpha} [A_oA]$ is in $\underline{V}$. If $\underline{V}$ has pullbacks we can form the comma object $A_o/\underset{\sim}{A}$ in $\underline{\underline{V\text{-}Cat}}$, but for a general $\underline{V}$ the construction takes no account of the above "families" of objects. In short, information is lost in the passage from $\underset{\sim}{A} \xrightarrow{[A_o-]} \underline{V}$ to $A_o/\underset{\sim}{A} \longrightarrow \underset{\sim}{A}$ and similarly for related examples. The $\underline{V}$-indexed categories that we introduce allow various types of families of objects and morphisms so that family-theoretic information is retained in such constructions.

§1 LARGE $\underline{V}$-CATEGORIES

Throughout, $\underline{V} = \langle \underline{V},U,\otimes \rangle$ will denote a fixed strict monoidal category with unit object U . Strictness has been assumed here only for ease of exposition. The reader is referred to [Wd] for details in the general case, although a brief account of some of the necessary modifications is given here. We assume universes $\mathcal{U}_1$ and $\mathcal{U}_2$ with $\mathcal{U}_1\epsilon\mathcal{U}_2$ and denote the category of $\mathcal{U}_1$, respectively $\mathcal{U}_2$, sets by $\underline{set}$, respectively $\underline{SET}$. We write $\underline{\underline{cat}}$, respectively $\underline{\underline{CAT}}$, for the 2-category of category objects in $\underline{set}$, respectively $\underline{SET}$. $\underline{V}$ is an object of $\underline{\underline{CAT}}$ which has $\mathcal{U}_1$ homs, i.e. $\underline{V}^{op}\times\underline{V} \xrightarrow{(-,-)} \underline{set}$. $\underline{\underline{V\text{-}Cat}}$ denotes the 2-category of $\underline{V}$-categories whose set of objects is a $\mathcal{U}_2$ set. The reader is assumed to be familiar with the biclosed monoidal structure of $\underline{SET}^{\underline{V}^{op}}$ as found in [Dy] . $\underline{SET}^{\underline{V}^{op}}$ -$\underline{\underline{cat}}$ denotes the 2-category of $\underline{SET}^{\underline{V}^{op}}$ -categories whose set of objects is a $\mathcal{U}_2$ set. It follows from [Dy], [D&K], [Wf] etc.

that $\underline{SET}^{\underline{V}^{op}}$ -$\underline{\underline{cat}}$ is U_2 -bicomplete as a 2-category and, if $\underline{V}$ is symmetric, is symmetric monoidal closed.

A *large* $\underline{V}$*-category* $\underset{\sim}{A}$ consists of a U_2 set O, of objects, together with, for each pair $A,B\in O$, a functor $\underline{V}^{op} \xrightarrow{[\![AB]\!]} \underline{SET}$, and functions $1 \xrightarrow{\iota} (U)[\![AA]\!]$ and $(I)[\![AB]\!]\times(J)[\![BC]\!] \xrightarrow{\gamma} (I\otimes J)[\![AC]\!]$, the latter natural in I and J, subject to unitary and associative axioms that we will presently display.

An element of $(I)[\![AB]\!]$ will be referred to as an I-*indexed family of morphisms from* A *to* B or simply as an I-*morphism from* A *to* B. We write $A \xrightarrow{(f;I)} B$ for such an element. If $J \xrightarrow{u} I$ is in $\underline{V}$, we write $A \xrightarrow{(fu^*;J)} B$ for $f((u)[\![AB]\!])$ and think of u^* as *substitution along* u. For each A, ι gives a distinguished U-morphism $A \xrightarrow{(A;U)} A$, the identity on A. If $A \xrightarrow{(f;I)} B$ and $B \xrightarrow{(g;J)} C$, γ gives their composite, which is an $I\otimes J$-morphism $A \xrightarrow{(fg;I\otimes J)} C$. Commutative diagrams involving I-morphisms have the obvious meaning, and the axioms for a large $\underline{V}$-category are simply that,

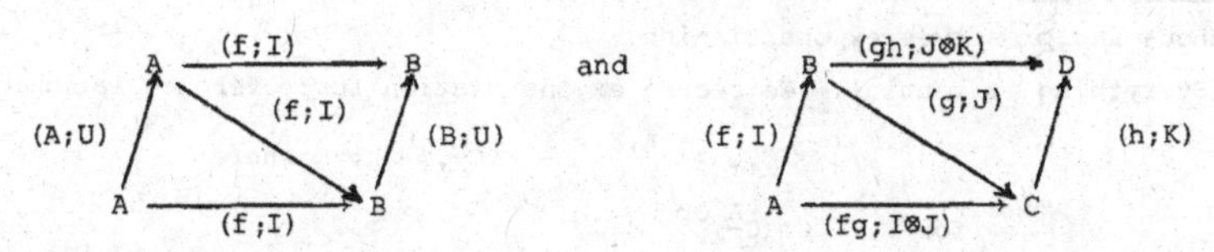

commute for all f,g and h.

Given large $\underline{V}$-categories $\underset{\sim}{A}$ and $\underset{\sim}{A}'$, a *large* $\underline{V}$*-functor* $\underset{\sim}{A} \xrightarrow{F} \underset{\sim}{A}'$ is an assignment that preserves the above structure. A *large* $\underline{V}$*-natural transformation* $\underset{\sim}{A} \underset{G}{\overset{F}{\underset{\Downarrow t}{\rightrightarrows}}} \underset{\sim}{A}'$ is a collection of U-morphisms $AF \xrightarrow{(At;U)} AG$ in $\underset{\sim}{A}'$, parametrized by the objects of $\underset{\sim}{A}$, such that for every I, for every I-morphism $A \xrightarrow{(f;I)} B$ in $\underset{\sim}{A}$,

$$\begin{array}{ccc} AF & \xrightarrow{(fF;I)} & BF \\ {\scriptstyle (At;U)}\downarrow & & \downarrow{\scriptstyle (Bt;U)} \\ AG & \xrightarrow[(fG;I)]{} & BG \end{array}$$

commutes.

If $\underset{\sim}{A}$ is such that $[\![AB]\!]$ is representable for all $A,B\in O$, we write $[AB]$ for the representing object and say that $\underset{\sim}{A}$ has *small homs*.

Proposition 1. $\underline{V}$-$\underline{\underline{CAT}}$ is 2-equivalent to $\underline{SET}^{\underline{V}^{op}}$ -$\underline{\underline{cat}}$, and the 2-full sub 2-category of $\underline{V}$-$\underline{\underline{CAT}}$ determined by the objects with small homs is 2-equivalent to $\underline{V}$-$\underline{\underline{Cat}}$.

Proof: (Sketch) The following bijections indicate that to give internal composition for a $\underline{SET}^{\underline{V}^{op}}$ -category is to give a γ as in the definition above.

$$([\![AB]\!]\otimes[\![BC]\!],[\![AC]\!])$$
$$\cong (\int^{I,J}(I)[\![AB]\!]\times(J)[\![BC]\!]\times(-,I\otimes J),[\![AC]\!])$$
$$\cong \int_{I,J}((I)[\![AB]\!]\times(J)[\![BC]\!]\times(-,I\otimes J),[\![AC]\!])$$
$$\cong \int_{I,J}((I)[\![AB]\!]\times(J)[\![BC]\!],(I\otimes J)[\![AC]\!])$$

<u>Corollary 2</u>. $\underline{V}$-$\underline{\underline{CAT}}$ is $\mathcal{U}_2$-bicomplete as a 2-category and, if $\underline{V}$ is symmetric, is monoidal closed. ∎

Large $\underline{V}$-categories are easier to work with than $\underline{SET}^{\underline{V}^{op}}$-categories in the same way that graded categories (large $\mathbb{N}$-categories) are easier to work with than $\underline{SET}^{\mathbb{N}}$-categories. We will demonstrate this elsewhere. For our present purposes large $\underline{V}$-categories are the most important example of $\underline{V}$-indexed categories. We introduce them separately for clarity and motivation.

§2 $\underline{V}$-INDEXED CATEGORIES

Let $\underline{\underline{SET}}^{()op} \xrightarrow{\mathbb{P}} \underline{\underline{CAT}}$ denote the fibration for which the fibre over $\underline{A}$ is $\underline{SET}^{\underline{A}^{op}}$. Thus $\underline{\underline{SET}}^{()op}$ has as objects, pairs $<\underline{A},\Gamma>$, where $\underline{A}^{op} \xrightarrow{\Gamma} \underline{SET}$. A morphism $<\underline{A},\Gamma> \xrightarrow{<F,\alpha>} <\underline{B},\Phi>$ consists of a functor $\underline{A} \xrightarrow{F} \underline{B}$ together with a natural transformation $\Gamma \xrightarrow{\alpha} F^{op}\Phi$. A 2-cell $<F,\alpha> \xrightarrow{t} <G,B>$ is a natural transformation $F \xrightarrow{t} G$ such that $\alpha = \beta . t^{op}\Phi$.

<u>Proposition 3</u>. $\underline{\underline{SET}}^{()op}$ is bicomplete as a 2-category and cartesian closed. $\mathbb{P}$ is 2-bicontinuous and preserves exponentiation.

<u>Proof:</u> Everything is routine. We record exponentiation for reference later.

$$<\underline{C},\psi>^{<\underline{A},\Gamma>} \simeq <\underline{C}^{\underline{A}},\psi^{\Gamma}>, \quad \text{where}$$

$$\text{for} \quad G \in (\underline{C}^{\underline{A}})^{op} .$$

$$(G)\psi^{\Gamma} \simeq (\Gamma, G^{op}\psi)\underline{SET}^{\underline{A}^{op}}$$

$$\simeq \int_A (A\Gamma, AG^{op}\psi)$$

∎

In particular we can carry out category theory in $\underline{\underline{SET}}^{()op}$. $\mathbb{P}$ applied to a category object in $\underline{\underline{SET}}^{()op}$ yields a category object in $\underline{\underline{CAT}}$. Thus given a category object $\underline{\mathbb{C}}$ in $\underline{\underline{CAT}}$, we can speak of "categories" in $\underline{\underline{SET}}^{()op}$ over $\underline{\mathbb{C}}$. We will write

$\underline{\mathbb{V}}$ for the category object $\underline{V}\times\underline{V} \overset{P_0}{\underset{P_1}{\overset{\otimes}{\rightrightarrows}}} \underline{V} \overset{\longrightarrow}{\underset{\longrightarrow}{\xleftarrow{U}}} \mathbb{1}$ in $\underline{\underline{CAT}}$.

<u>Proposition 4</u>. Large $\underline{V}$-categories are precisely category objects in $\underline{\underline{SET}}^{()op}$ over $\underline{\mathbb{V}}$.

<u>Proof:</u> Given a large $\underline{V}$-category $\underset{\sim}{A}$, let $M = \sum_{A,B\in O} [\![AB]\!] : \underline{V}^{op} \longrightarrow \underline{SET}$. Regarding the set of objects of $\underset{\sim}{A}$, O, as $\mathbb{1}^{op} \xrightarrow{O} \underline{SET}$, define $<\underline{V},M> \overset{<!,\partial_0>}{\underset{<!,\partial_1>}{\rightrightarrows}} <\mathbb{1},O>$ in $\underline{\underline{SET}}^{()op}$ by $IM \xrightarrow{I\partial_i} O$, $i = 0,1, I\in\underline{V}$, where the A,B'th component of $I\partial_0$ is A and the A,B'th component of $I\partial_1$ is B. Now form the pullback of $<!,\partial_0>$ and $<!,\partial_1>$ in $\underline{\underline{SET}}^{()op}$. This is given by $<\underline{V}\times\underline{V}, M\underset{O}{\times}M>$, where $<I,J> M\underset{O}{\times}M$ is given by the following pullback in $\underline{SET}$:

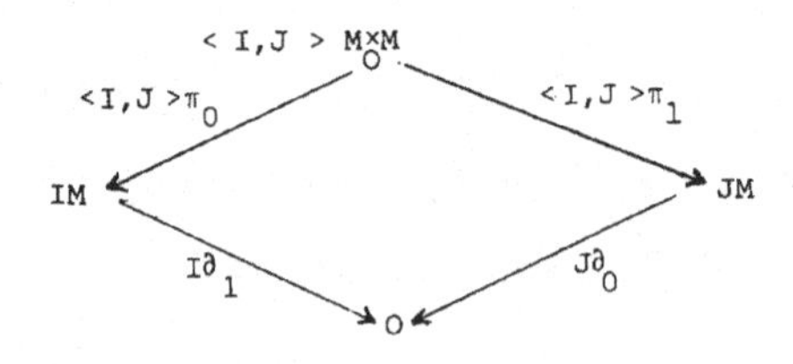

Note that the π's above also give $<\underline{V}\times\underline{V}, M\underset{O}{\times}M> \underset{<P_1,\pi_1>}{\overset{<P_0,\pi_0>}{\rightrightarrows}} <\underline{V},M>$ in $\underline{\underline{SET}}^{()op}$, where the P_i, $i = 0,1$, are the projections from $\underline{V}\times\underline{V}$ to $\underline{V}$. $<I,J> M\underset{O}{\times}M$ is also given by $\sum_{A,B,C\in O} (I)[\![AB]\!]\times(J)[\![BC]\!]$, from which it follows that we can define $<\underline{V}\times\underline{V}, M\underset{O}{\times}M> \xrightarrow{<\otimes,\gamma'>} <\underline{V},M>$ by $<(I)[\![AB]\!]\times(J)[\![BC]\!] \xrightarrow{\gamma} (I\otimes J)[\![AC]\!] \xrightarrow{i} (I\otimes J)M>_{A,B,C}$ where the γ's are those given in the structure of $\underset{\sim}{A}$ and i is the injection into the coproduct. The remaining structure and the axioms for a category object follow easily.

Conversely, given a category object $<\underline{V},M> \underset{<!,\partial_1>}{\overset{<!,\partial_0>}{\rightrightarrows}} <\mathbb{1},O>$ etc. over $\underline{\mathbb{V}}$, we can define a functor $[\![AB]\!] : \underline{V}^{op} \longrightarrow \underline{SET}$, for each pair $A,B\in O$, by the following pullback in $\underline{SET}^{\underline{V}^{op}}$:

$$\begin{array}{ccc} [\![AB]\!] & \longrightarrow & M \\ \downarrow & & \downarrow{\scriptstyle <\partial_0,\partial_1>} \\ 1 & \xrightarrow[<A,B>]{} & !O\times!O \end{array} .$$

It follows then that the front and back faces of the following cube are pullbacks:

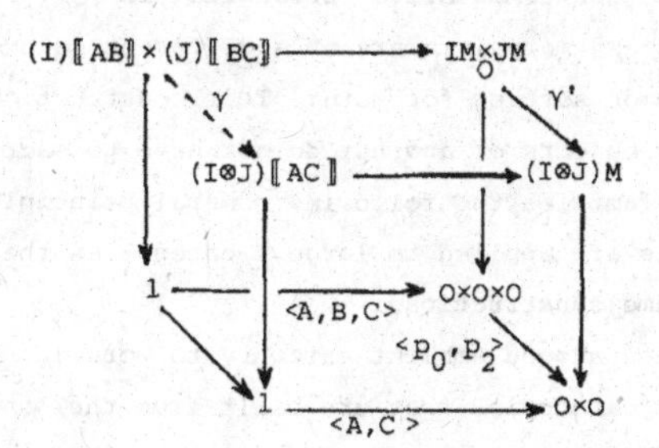

and we define the γ's for a large $\underline{V}$-category structure as the morphisms induced by the given internal composition morphism γ' above.

It is clear that the above constructions are essentially inverse to each other. ∎

For $\underset{\sim}{A}$ a large $\underline{V}$-category and $A_o\in\underset{\sim}{A}$ we exhibit now the comma object $A_o/\underset{\sim}{A}$ in $\underline{\underline{cat}}(\underline{\underline{SET}}^{()op})$. Here we identify A_o with the functorial morphism from $<\mathbb{1},1> \rightrightarrows <\mathbb{1},1>$ (the terminal object in $\underline{\underline{cat}}(\underline{\underline{SET}}^{()op})$) to $\underset{\sim}{A}$ determined by A_o.

$A_o/\underset{\sim}{A}$ is necessarily a category object over $0/\underline{\mathbb{V}}$, where 0 denotes the unique "object" of $\underline{\mathbb{V}}$. $0/\underline{\mathbb{V}}$ is the diagram:

$$\underline{V}\times\underline{V}\times\underline{V} \quad \begin{array}{c} \xrightarrow{<P_0,P_1>} \\ \xrightarrow{<P_0,P_1\otimes P_2>} \\ \xrightarrow{<P_0\otimes P_1,P_2>} \end{array} \quad \underline{V}\times\underline{V} \quad \begin{array}{c} \xrightarrow{P_0} \\ \xleftarrow{<\underline{V},U>} \\ \xrightarrow{\otimes} \end{array} \quad \underline{V} ,$$

in $\underline{\underline{CAT}}$. The P_i denote projections. Direct calculation shows that the object of objects of $A_o/\underset{\sim}{A}$ is $<\underline{V}, \sum_{A\in\underset{\sim}{A}} [\![A_oA]\!]>$ and the object of morphisms is

$< \underline{V}\times\underline{V} , \sum_{A,B\in\mathbb{A}} [\![A_oA]\!]\times[\![AB]\!] >$.

We prefer to work with a family-theoretic interpretation, however, so we describe $A_o/\mathbb{A}$ as suggested in the introduction. For each $I\in\underline{V}$ it has I-*indexed families of objects* or simply I-*objects* and these are given by I-morphisms in $\mathbb{A}$ with domain A_o . In the event that $\mathbb{A}$ has small homs (i.e. $\mathbb{A}$ is an ordinary $\underline{V}$-category) such an I-object is just a pair $< A,\alpha >$ with $A\in\mathbb{A}$ and $I \xrightarrow{\alpha} [A_oA]$ in $\underline{V}$. For each pair $I,J\in\underline{V}$, $A_o/\mathbb{A}$ has I,J-*indexed families of morphisms,* a typical one being

$$\begin{array}{ccc} & & A_o \\ & \swarrow^{(a;I)} & \\ A & \xrightarrow[(f;J)]{} & B \end{array} ,$$

where the components are $\underline{V}$-morphisms of $\mathbb{A}$. The domain of such an I,J-morphism is (a;I) , while the codomain is the I⊗J-object, (af;I⊗J) . There are of course obvious substitution operations for both objects and morphisms, inherited from those of $\mathbb{A}$.

In short, $A_o/\mathbb{A}$ in $\underline{\underline{cat}}(\underline{\underline{SET}}^{()op})$ looks like an ordinary comma category in $\underline{\underline{CAT}}$, in that its "morphisms" are commutative "triangles" in $\mathbb{A}$. For large (ordinary) $\underline{V}$-categories the objects and morphisms are of a different nature but by working in $\underline{\underline{SET}}^{()op}$ we provide a common setting for both. Thus constructions which tend to make morphisms of one category objects of another do not have to sacrifice information. From the point of view of families the following general principle is suggested: When categorical constructions are applied to large $\underline{V}$-categories the families involved should be subjected to the same constructions.

$\underline{\underline{SET}}^{()op}$ provides a good ambient category to work in, however, we wish to deal only with objects over categories that are built from the monoidal data of $\underline{V}$. We now give an approach that allows for the development of a theory of $\underline{V}$-indexed categories closed under finite limits. Other possibilities will be briefly discussed in §5 .

Let $\underline{T}$ denote the category of finitely presentable monoids.

Lemma 5 $\underline{T}$ is finitely cocomplete and any left exact functor $\underline{T}^{op} \longrightarrow \underline{S}$, where $\underline{S}$ has finite limits, is essentially determined by a monoid in $\underline{S}$. ∎

Corollary 6 Any left exact functor $\underline{T}^{op} \longrightarrow \underline{\underline{CAT}}$ is essentially of the form $[-\underline{V}]$ for $\underline{V}$ a strict monoidal category, where, for $T\in\underline{T}$, $[T\underline{V}]$ denotes the category whose objects are strict monoidal functors from T , regarded as a discrete monoidal category, to $\underline{V}$. ∎

If f is a morphism in $\underline{T}$ we write $f^{\#}$ for $[f\underline{V}]$. Now define $\underline{V}$-$\underline{\text{ind-SET}}$, the category of $\underline{V}$-*indexed sets,* by the following pullback:

$$\begin{array}{ccc} \underline{V}\text{-}\underline{\text{ind-SET}} & \longrightarrow & \underline{\underline{SET}}^{()op} \\ {\scriptstyle P_{\underline{V}}}\downarrow & & \downarrow{\scriptstyle \mathbb{P}} \\ \underline{T}^{op} & \xrightarrow[{[-\underline{V}]}]{} & \underline{\underline{CAT}} \end{array}$$

Lemma 7 $P_{\underline{V}}$ is a fibration and $\underline{V}$-$\underline{\text{ind-SET}}$ has finite limits which are preserved by $P_{\underline{V}}$.

Proof: $P_{\underline{V}}$ is actually a split normal fibration since it is a pullback of $\mathbb{P}$. The second assertion follows from a general lemma: If $\underline{E} \xrightarrow{P} \underline{B}$ is a fibration (not necessarily split normal) and for some class of limits C, $\underline{B}$ has C-limits, $(I)P^{-1}$ has C-limits for all I in $\underline{B}$, and uP^{-1} preserves them, for all $J \xrightarrow{u} I$ in $\underline{B}$, then $\underline{E}$ has C-limits and P preserves them. In the present situation $(T)P_{\underline{V}}^{-1} = \underline{SET}^{[T\underline{V}]^{op}}$. ∎

Remark 8 If $\underline{V}$ is not strict, the construction of $\underline{V\text{-ind-SET}}$ is somewhat more complicated. In that case one constructs a pseudo functor from the category of finitely presented monoids to $\underline{CAT}$ using the monoidal data of $\underline{V}$ and a variant of the subequalizers of [Lk]. The resulting $P_{\underline{V}}$ is not split in the general case. Note that in any event the 2-structure of $\underline{\underline{SET}}^{()op}$ is forgotten.

Note also that given "change of base" data $\underline{V} \xrightarrow{F} \underline{W}$, where $\underline{W}$ is also monoidal, we obtain a corresponding natural transformation $[-\underline{V}] \xrightarrow{[-F]} [-\underline{W}]$ which induces a functor over $\underline{T}^{op}$ as indicated:

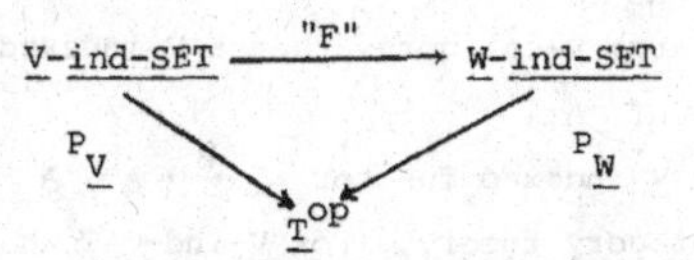

.

Definition 9 A $\underline{V}$-*indexed category* is a category object in $\underline{V\text{-ind-SET}}$. Other terminology is similarly determined by:

$$\underline{V\text{-ind-CAT}} \xrightarrow{\mathbb{P}_{\underline{V}}} \underline{\underline{cat}}(\underline{T}^{op})$$

$$\underset{\text{Def.}}{=\!=} \underline{\underline{cat}}(\underline{V\text{-ind-SET}}) \xrightarrow{\underline{\underline{cat}}(P_{\underline{V}})} \underline{\underline{cat}}(\underline{T}^{op}) \quad .$$

As an immediate consequence of the definition we have:

Proposition 10 $\underline{V\text{-ind-CAT}}$ is finitely complete as a 2-category and $\mathbb{P}_{\underline{V}}$ is left exact as a 2-functor. ∎

If $\langle \bar{T}, \underset{\sim}{A} \rangle$ is an object of $\underline{V\text{-ind-CAT}}$, where $\bar{T}$ is a category object in $\underline{T}^{op}$, we say that $\underset{\sim}{A}$ is a $\underline{V}$-indexed category of *type* $\bar{T}$. It will sometimes be convenient to write just $\underset{\sim}{A}$ for $\langle \bar{T}, \underset{\sim}{A} \rangle$. If when this is done we later have to refer to the type of $\underset{\sim}{A}$ we will denote it by $\bar{\underset{\sim}{A}}$. An analogous convention will apply to $\underline{V}$-indexed functors etc.

Before considering examples let us note in concrete terms what a $\underline{V}$-indexed category $\underset{\sim}{A}$ of type $\bar{T} : T_2 \underset{p_1}{\overset{p_0}{\underset{c}{\leftleftarrows}}} T_1 \underset{d_1}{\overset{d_0}{\underset{i}{\rightleftarrows}}} T_0$ "looks like". It has for each $\vec{J} \in [T, \underline{V}]$, $\vec{J}$-*indexed families of morphisms* or $\vec{J}$-*morphisms* and these we denote by expressions such as $(A; \vec{I}_0) \xrightarrow{(f; \vec{J})} (B; \vec{I}_1)$, where $\vec{I}_0 = \vec{J}d_0^{\#}$, $\vec{I}_1 = \vec{J}d_1^{\#}$. If T_1 can be presented with n generators and m relations, then $\vec{J}$ may be thought of as an n-tuple of objects of $\underline{V}$, $\langle J_0, J_1, \ldots, J_{n-1} \rangle$, for which m equations involving the J_i, ⊗, and U are valid. For example we might have $J_0 \otimes J_1 = J_2 \otimes J_3$. Note that for not necessarily strict $\underline{V}$ such equations must be replaced by specified "compatible" isomorphisms. Compatibility means roughly that all well formed diagrams involving such isomorphisms and the monoidal data of $\underline{V}$ commute.

§3 EXAMPLES AND APPLICATIONS

Large $\underline{V}$-categories are precisely $\underline{V}$-indexed categories of type $\bar{\mathbb{N}} : \mathbb{N}+\mathbb{N} \overset{x}{\underset{y}{\overset{xy}{\leftrightarrows}}} \mathbb{N} \leftrightarrows 0$. (We write x,y etc. for the generators of $\mathbb{N}+\mathbb{N}$ etc.) This follows immediately from proposition 4 and the observation that $[\bar{\mathbb{N}}\underline{V}] = \underline{V}$. If $\underset{\sim}{A}$ and $\underset{\sim}{B}$ are large $\underline{V}$-categories, a $\underline{V}$-indexed functor $\underset{\sim}{A} \xrightarrow{F} \underset{\sim}{B}$ has a type, $\bar{\mathbb{N}} \xleftarrow{t} \bar{\mathbb{N}}$. It follows that t is either the identity or the zero morphism. ($\underline{\underline{\text{cat}}}(\underline{T}^{op})$ has a zero object since $\underline{T}$ does.) If F is of identity type, it is a large $\underline{V}$-functor, otherwise it is an assignment that takes objects to objects and I-morphisms to U-morphisms, for all $I \in \underline{V}$, "functorially". If $\underset{\sim}{A} \overset{F}{\underset{G}{\rightrightarrows}} \underset{\sim}{B}$ are large $\underline{V}$-functors, a $\underline{V}$-indexed natural transformation $F \xrightarrow{t} G$ is just a large $\underline{V}$-natural transformation.

Any $\underline{V}$-indexed set $< T,\Gamma >$ $([T\underline{V}]^{op} \xrightarrow{\Gamma} \underline{\text{SET}})$ gives rise to a discrete $\underline{V}$-indexed category $\Gamma : \Gamma \rightrightarrows \Gamma \leftrightarrows \Gamma$ of type $T : T \leftrightarrows T \rightrightarrows T$. In particular, for $I \in \underline{V}$ we get $\underset{\sim}{I}$, the discrete $\underline{V}$-indexed category on $< \mathbb{N},(-,I) >$.

Proposition 11 If $\underset{\sim}{A}$ is a large $\underline{V}$-category, then a $\underline{V}$-indexed functor $\underset{\sim}{I} \longrightarrow \underset{\sim}{A}$ is determined by an object, B say, of $\underset{\sim}{A}$. ∎

To take the limit of a $\underline{V}$-indexed functor $\underset{\sim}{I} \xrightarrow{B} \underset{\sim}{A}$, $\underset{\sim}{A}$ a large $\underline{V}$-category, we can proceed as in ordinary category theory. For $\underline{V}$-$\underline{\underline{\text{ind-CAT}}}$ has a terminal object, $\mathbb{1}$, which is the discrete $\underline{V}$-indexed category on $< 0,1 >$, and a $\underline{V}$-indexed functor $\mathbb{1} \longrightarrow \underset{\sim}{A}$ is also determined by an object, A say, of $\underset{\sim}{A}$. Thus we can define a cone from A to B to be a $\underline{V}$-indexed natural transformation,

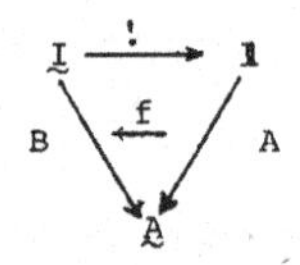

Such an f itself has a type and a universal cone must necessarily have a universal type. Direct calculation shows that there are types corresponding to each natural number n and that there is no universal type. Thus $\underset{\sim}{I} \xrightarrow{B} \underset{\sim}{A}$ does not have a limit in the usual sense. However, for a fixed n we can ask for a universal cone of "type n", which we denote by $\underset{\longleftarrow n}{\lim} B \xrightarrow{p_n} B$. A cone f of type n from A to B is easily seen to be an $I^{(n)}$-morphism $A \xrightarrow{(f;I^{(n)})} B$ in $\underset{\sim}{A}$, where $I^{(0)} = U$ and $I^{(n)} = I^{(n-1)} \otimes I$ for $n \geq 1$.

Proposition 12 $\underset{\longleftarrow n}{\lim} B$ exists if and only if the weak cotensor $[I^{(n)}B]$ exists, in which case they are isomorphic. If $\underset{\sim}{A}$ is an ordinary $\underline{V}$-category, $\underset{\longleftarrow}{\lim} B \xrightarrow{(p_n;I^{(n)})} B$ "is" the morphism $I^{(n)} \longrightarrow [[I^{(n)}B],B]$ in $\underline{V}$ corresponding to the identity $[I^{(n)}B] \longrightarrow [I^{(n)}B]$ in $\underset{\sim}{A}$. ∎

We can state a similar result about (strong) cotensors in a large $\underline{V}$-category but first we make a general definition of a right Kan extension notion in $\underline{V}$-$\underline{\underline{\text{ind-CAT}}}$.

Definition 13 For ε as shown:

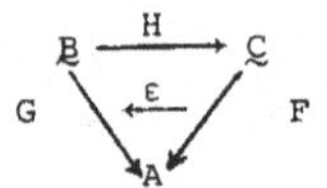

of type t in V-ind-CAT , we say that ε exhibits F as a *right* t,s - *extension of* G *along* H if for any $\underset{\sim}{C} \xrightarrow{F'} \underset{\sim}{A}$ and any $HF' \xrightarrow{\tau} G$ of type $\bar{H}s.t$, there exists a unique $F' \xrightarrow{\tau'} F$ of type s such that

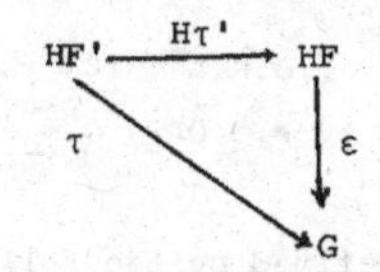

commutes.

True Kan extensions in V-ind-CAT would be Kan extensions at the level of types and these are too rare to be of interest.

Proposition 13 A weak cotensor $[IB] \xrightarrow{(p;I)} B$, in a large V-category $\underset{\sim}{A}$, is a cotensor if and only if, for all $J \in \underline{V}$, $P_1 p$ exhibits $![IB]$ as a right t,s-extension of $P_1 B$ along P_0 ; where the P_i are the projections from the product as indicated below:

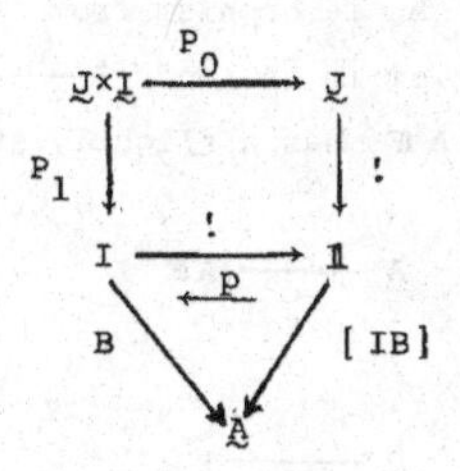

;

t (the type of $P_1 p$) is, in abbreviated form, $\mathbb{N} + \mathbb{N} \xleftarrow{y} \mathbb{N}$; and s abbreviated is $\mathbb{N} \xleftarrow{x} \mathbb{N}$. ∎

The above proposition is an application of the pointwise Kan-extension concept introduced by Street, [S]. Note that here this notion of "pointwise" agrees with that of Dubuc, [Dc] . From a family theoretic point of view $P_1 p$ in the above diagram is really a J-indexed family of cones from [IB] to B .

The "comma category" $A_0/\underset{\sim}{A}$ that we discussed in §2 is a V-indexed category of type $\bar{\mathbb{N}}^* : \mathbb{N} + \mathbb{N} + \mathbb{N} \; \overset{\binom{x}{y}}{\longleftarrow}\;\overset{\binom{x}{yz}}{\longleftarrow}\;\overset{\binom{xy}{z}}{\longleftarrow} \; \mathbb{N} + \mathbb{N} \; \overset{x}{\longleftarrow}\;\overset{\binom{x}{1}}{\longrightarrow}\;\overset{xy}{\longleftarrow} \; \mathbb{N}$, for $[\bar{\mathbb{N}}^* \underline{V}] = 0/\underline{V}$. Viewed as such it is a true comma object in V-ind-CAT . We will return to V-indexed categories of this type later. A similar example is provided by again considering a large V-category $\underset{\sim}{A}$ and forming $\underset{\sim}{A}^2$ in V-ind-CAT . Its object of objects is of type $\mathbb{N}$ and thus $\underset{\sim}{A}^2$ has, for each $I \in \underline{V}$, I-indexed families of objects, these being given by I-morphisms in $\underset{\sim}{A}$. The type of its object of morphisms is given by the following pushout in $\underline{T}$:

$$\begin{array}{ccc} \mathbb{N} & \xrightarrow{xy} & \mathbb{N}+\mathbb{N} \\ {\scriptstyle xy}\downarrow & & \downarrow \\ \mathbb{N}+\mathbb{N} & \longrightarrow & P \end{array} \quad .$$

It follows that a "morphism" in $\underset{\sim}{A}^2$ is just a "commutative square":

$$\begin{array}{ccc} A & \xrightarrow{(f;K)} & B \\ {\scriptstyle (a;I)}\downarrow & & \downarrow{\scriptstyle (b;L)} \\ A' & \xrightarrow[(g;J)]{} & B' \end{array}$$

in $\underset{\sim}{A}$. The above "morphism" is an $< I,J,K,L \mid I\otimes J = K\otimes L >$ - morphism.

A $\underline{V}$-indexed category of trivial type, $0 : 0 \leftleftarrows 0 \rightleftarrows 0$, is just an ordinary category. In fact:

Proposition 14 The 2-functor $\mathbb{I}$ defined by the following pullback is 2-full and faithful.

$$\begin{array}{ccc} \underline{\underline{CAT}} & \xrightarrow{\mathbb{I}} & \underline{V}\text{-}\underline{\underline{ind\text{-}CAT}} \\ \downarrow & & \downarrow{\scriptstyle \mathbb{P}_{\underline{V}}} \\ \mathbf{1} & \xrightarrow[0]{} & \underline{\underline{cat}}(\underline{T}^{op}) \end{array}$$

■

If $\underline{A}$ is an ordinary category let $\underline{A}\,\mathbb{F}$ denote the free large $\underline{V}$-category on $\underline{A}$. (The objects of $\underline{A}\,\mathbb{F}$ are those of $\underline{A}$. An I-morphism from A to B in $\underline{A}\,\mathbb{F}$ is a pair <f,u> , where $A \xrightarrow{f} B$ is a morphism in $\underline{A}$ and $I \xrightarrow{u} U$ is a morphism in $\underline{V}$). Regarded as an object of $\underline{V}$-$\underline{\underline{ind\text{-}CAT}}$ $A\,\mathbb{F}$ has a slightly stronger universal property.

Proposition 15

$$\begin{array}{ccc} \underline{A} & \longrightarrow & A\,\mathbb{F} \\ \vdots & & \vdots \\ 0 & \xleftarrow[!]{} & \bar{\mathbb{N}} \end{array}$$

■

is cocartesian relative to $\mathbb{P}_{\underline{V}}$.

Cocartesian morphisms relative to $\mathbb{P}_{\underline{V}}$ are in general quite complicated. Suppose however that $\underline{V}$ is symmetric and that $\underline{T}$ in our construction of $\underline{V}$-$\underline{\underline{ind\text{-}CAT}}$ is replaced by $\underline{T}_c$, the category of finitely presentable commutative monoids. If $\underset{\sim}{A}$ and $\underset{\sim}{B}$ are large $\underline{V}$-categories their product in $\underline{V}$-$\underline{\underline{ind\text{-}CAT}}$ is of type $\bar{\mathbb{N}}\oplus\bar{\mathbb{N}}$ $(\bar{\mathbb{N}} + \bar{\mathbb{N}})$ and we now have a morphism of types $\bar{\mathbb{N}}\oplus\bar{\mathbb{N}} \xleftarrow{\Delta} \bar{\mathbb{N}}$ constructed from the diagonal $\mathbb{N} \xrightarrow{\Delta} \mathbb{N}\oplus\mathbb{N}$ in $\underline{T}_c$.

Proposition 16 For $\underline{V}$ symmetric, $\underline{T}_c$ as above, and $\underset{\sim}{A}$, $\underset{\sim}{B}$ large $\underline{V}$-categories,

$$\begin{array}{ccc} \underset{\sim}{A}\times\underset{\sim}{B} & \longrightarrow & \underset{\sim}{A}\otimes\underset{\sim}{B} \\ \vdots & & \vdots \\ \bar{\mathbb{N}}\oplus\bar{\mathbb{N}} & \xleftarrow[\Delta]{} & \bar{\mathbb{N}} \end{array}$$

is cocartesian relative to $\mathbb{P}_{\underline{V}}$, where $\underset{\sim}{A}\otimes\underset{\sim}{B}$ denotes the usual tensor product of large $\underline{V}$-categories.

Proof: (Sketch) The proposition follows from the fact that for $F,G \in \underline{SET}^{\underline{V}^{op}}$, $F\otimes G$ is the left Kan extension of $(\underline{V}\times\underline{V})^{op} \xrightarrow{F\times G} \underline{SET}$ along $(\underline{V}\times\underline{V})^{op} \xrightarrow{\otimes^{op}} \underline{V}^{op}$. Here $F\times G$ is the product of F and G in $\underline{\underline{SET}}^{()^{op}}$. I.e. $< I,J > F\times G \cong IF\times JG$.

■

If $\underline{V}$ is not symmetric $\underline{V}$-$\underline{\underline{CAT}}$ does not admit a tensor product, yet the product of large $\underline{V}$-categories, as $\underline{V}$-indexed categories, contains the information necessary for a

$\otimes$ were $\underline{V}$ symmetric.

If $\underset{\sim}{A}$ is a $\underline{V}$-indexed category of type $\bar{T}$, $\underset{\sim}{A}^{op}$ is a $\underline{V}$-indexed category of type $\bar{T}_{op}$. The latter is obtained by interchanging the roles of $T_1 \underset{d_1}{\overset{d_0}{\leftleftarrows}} T_0$ in $\bar{T}$. In particular if $\underset{\sim}{A}$ is a large $\underline{V}$-category, $\underset{\sim}{A}^{op}$ is a $\underline{V}$-indexed category of type:

$\bar{\mathbb{N}}_{op} : \mathbb{N}+\mathbb{N} \underset{x}{\overset{y}{\underset{xy}{\leftleftarrows}}} \mathbb{N} \rightleftarrows 0$. Equivalently, $\underset{\sim}{A}^{op}$ is a large $\underline{V}^{rev}$-category, where $\underline{V}^{rev}$ is the monoidal category whose underlying category is $\underline{V}$ and for which $I\otimes^{rev}J = J\otimes I$ (c.f.[R]). Now for $\underset{\sim}{A}$ and $\underset{\sim}{B}$ large $\underline{V}$-categories, $\underset{\sim}{A}^{op}\times\underset{\sim}{B}$ is of type $\bar{\mathbb{N}}_{op}+\bar{\mathbb{N}}$, but furthermore $\underline{V}$ itself is also. Let $\underset{\sim}{V}$ denote the $\underline{V}$-indexed category of type $\bar{\mathbb{N}}_{op}+\bar{\mathbb{N}}$ whose objects are those of $\underline{V}$ and for which an I,J-morphism from V to W is a morphism $I\otimes V\otimes J \longrightarrow W$ in $\underline{V}$.

<u>Proposition 17</u> For $\underset{\sim}{A}$ and $\underset{\sim}{B}$ (large) $\underline{V}$-categories a $\underline{V}$-indexed functor from $\underset{\sim}{A}^{op}\times\underset{\sim}{B}$ to $\underset{\sim}{V}$ of identity type is the same thing as a "$\underline{V}$-valued $\underline{V}$-bifunctor of mixed variance" as defined in [Ln]. ∎

§4 V-DISCRETE V-FIBRATIONS

Let $\underset{\sim}{V}_*$ denote the comma object $U/\underset{\sim}{V}$ in V-ind-CAT where $\underset{\sim}{V}$ is regarded as an object of type $\bar{\mathbb{N}}$. Thus $\underset{\sim}{V}_*$ is of type $\bar{\mathbb{N}}^*$ as in §3. Explicitly, $\underset{\sim}{V}_*$ has, for all $I\in\underline{V}$, I-objects and these are I-morphisms with domain U, $U \xrightarrow{(x;I)} V$, in $\underset{\sim}{V}$ or simply morphisms $U\otimes I = I \xrightarrow{x} V$ in $\underline{V}$. We regard $(x;I)$ as an object of $\underset{\sim}{V}$, V, together with an I-indexed family of its "elements". As a comma category $\underset{\sim}{V}_*$ comes equipped with a projection, $\underset{\sim}{V}_* \xrightarrow{Q} \underset{\sim}{V}$, which simply sends $U \xrightarrow{(x;I)} V$ to V. We refer to Q as the *universal* $\underline{V}$*-discrete* $\underline{V}$*-cofibration*. Our point of view is that $\underline{V}$-discrete V-cofibrations "are" pullbacks of Q in V-ind-CAT and we show that these admit a reasonable classification. It is convenient to deal with the types first.

<u>Lemma 18</u> If $\underset{\sim}{B} \xrightarrow{F} \underset{\sim}{V}$ is a $\underline{V}$-indexed functor and $\underset{\sim}{B}$ is of type

$\bar{M} : M_2 \overset{p_0}{\underset{p_1}{\overset{c}{\leftleftarrows}}} M_1 \overset{d_0}{\underset{d_1}{\overset{i}{\rightleftarrows}}} M_0$, the type of F is determined by an element $m\in M_1$ satisfying $mi = 1$ (in M_0) and $mc = (mp_0).(mp_1)$ (in M_2). ∎

We will refer to the type of such an F simply by m.

The pullback of Q along F is necessarily a comma object together with one of its projections, $U/F \xrightarrow{Q_F} \underset{\sim}{B}$. We denote the type of $U/F \xrightarrow{Q_F} \underset{\sim}{B}$ by $!/m \xleftarrow{\hat{m}} \bar{M}$.

<u>Lemma 19</u> $!/m \xleftarrow{\hat{m}} \bar{M}$ is given explicitly by the following diagram in $\underline{T}$:

$$\begin{array}{ccccc}
!/m : \mathbb{N}+M_2 & \overset{\mathbb{N}+p_0,\ \mathbb{N}+c,\ \binom{x(mp_0)}{p_1 j_1}}{\longleftarrow} & \mathbb{N}+M_1 & \overset{\mathbb{N}+d_0,\ \mathbb{N}+i,\ \binom{xm}{d_1 j_1}}{\rightleftarrows} & \mathbb{N}+M_0 \\
\hat{m}\uparrow \quad j_1\uparrow & & \uparrow j_1 & & \uparrow j_1 \\
\bar{M} \quad M_2 & \overset{p_0,\ c,\ p_1}{\longleftarrow} & M_1 & \overset{d_0,\ i,\ d_1}{\rightleftarrows} & M_0
\end{array},$$

where the j_1's are coproduct injections. ∎

We call a type $\bar{P} \longleftarrow \bar{M}$ a *type cofibration* if it is of the form $\hat{m}$ for some $m\in M_1$ as above.

Proposition 20 $\bar{P} \longleftarrow \bar{M}$ is a type cofibration if and only if its "domain" data is of the form:

$$\begin{array}{ccc} \mathbb{N}+M_1 & \xleftarrow{N + d_0} & \mathbb{N}+M_0 \\ {\scriptstyle j_1}\uparrow & & \uparrow{\scriptstyle j_1} \\ M_1 & \xleftarrow[d_0]{} & M_0 \end{array} ,$$

and the "codomain" operation of $\bar{P}$ is of the form $\begin{pmatrix} xm \\ d_1 j_1 \end{pmatrix}$ for some $m \in M_1$ with $mi = 1$ $mc = (mp_0).(mp_1)$. ∎

If $\bar{M} \underset{m_1}{\overset{m_0}{\longleftarrow\!\!\!\longleftarrow}} \bar{\mathbb{N}}$ (with $\downarrow t$) is the type of a $\underline{V}$-indexed natural transformation between $\underline{V}$-valued $\underline{V}$-indexed functors it gives rise to a type $\hat{t}$ as shown below:

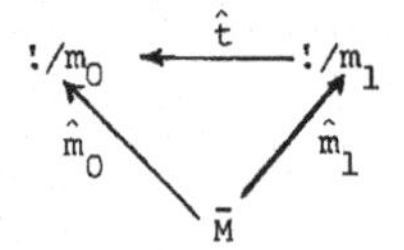

We call such a $\hat{t}$ a *morphism of type cofibrations*. If $\hat{m}_1 s = \hat{m}_0$ it does not follow that s is of the form $\hat{t}$ for some t as above, but a characterization of morphisms of type cofibrations is straightforward and left to the reader. We do have however:

Proposition 21 If $\underset{\sim}{B} \underset{F_1}{\overset{F_0}{\rightrightarrows}} \underset{\sim}{V}$ are of types m_0 and m_1 respectively and G as below:

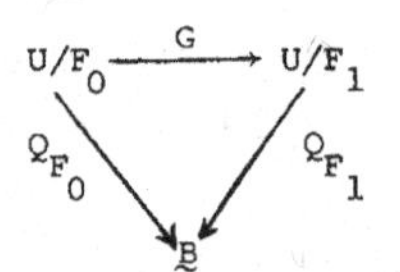

is of type $\hat{t}$ for some $m_0 \xrightarrow{t} m_1$, then there exists a unique $F_0 \xrightarrow{\tau} F_1$ of type t such that $G = U/\tau$. ∎

If $\underset{\sim}{E} \xrightarrow{G} \underset{\sim}{B}$ is a $\underline{V}$-indexed functor whose type is a type cofibration, say $!/m \xleftarrow{\hat{m}} \bar{M}$, we can ask whether it satisfies the following:

Definition 22 *Lifting condition*. For any $< I, I_0 > \in \underline{V} \times [M_0 \underline{V}]$, any I, I_0-object of $\underset{\sim}{E}$, $(X; I, I_0)$, any $I_1 \in [M_1 \underline{V}]$ with $I_1 d_0^{\#} = I_0$, and any I_1-morphism of $\underset{\sim}{B}$, $(f; I_1)$ whose domain is $(XG; I_0)$; there exists a unique I, I_1-morphism of $\underset{\sim}{E}$ whose domain is $(X; I, I_0)$ and G of which is $(f; I_1)$.

Diagramatically:

$$\begin{array}{ccc} (X; I, I_0) & \dashrightarrow & (Y, I \otimes I_1 m^{\#}, J_0) \\ \vdots & & \vdots \\ (XG; I_0) & \xrightarrow[(f; I_1)]{} & (B; J_0) \end{array}$$

(Note that the codomain of the lift is necessarily an $I \otimes I_1 m^{\#}, J_0$-object.)

It is routine to show that the lifting condition is stable under substitution and that if $\underset{\sim}{B} \xrightarrow{F} \underset{\sim}{V}$, then Q_F satisfies the lifting condition.

For $\underline{E} \xrightarrow{G} \underline{B}$ of type $!/m \xleftarrow{\hat{m}} \bar{M}$, we can define, for each $I_0 \epsilon [M_0\underline{V}]$ and each I_0-object A of $\underline{B}$, a functor:

$$\underline{V}^{op} \xrightarrow{(A;I_0)G^{-1}} \underline{SET} ,$$

whose value at $I \epsilon \underline{V}$ is the set of all I,I_0-objects of $\underline{E}$, $(X;I,I_0)$, such that $XG = A$. We can ask whether G satisfies the following:

Definition 23 *Smallness condition*. For each $I_0 \epsilon [M_0\underline{V}]$ and each I_0-object, A of $\underline{B}$; $(A;I_0)G^{-1}$ is representable and further for any $I'_0 \xrightarrow{u_0} I_0$ in $[M_0\underline{V}]$, $(Au_0^*;I'_0)G^{-1} \simeq (A;I_0)G^{-1}$.

For F as before it is easy to see that Q_F satisfies the smallness condition.

Definition 24 A $\underline{V}$-indexed functor $\underline{E} \xrightarrow{G} \underline{B}$, whose type is a type cofibration, is a *$\underline{V}$-discrete $\underline{V}$-cofibration* if it satisfies:

1) the lifting condition
2) the smallness condition.

$\underline{V}\text{-}d_0fib_{\underline{B}}$ will denote the category of $\underline{V}$-discrete $\underline{V}$-cofibrations over $\underline{B}$. $\underline{V}\text{-}d_0fib_{\underline{B}}$ and $(\underline{B},\underline{V})$, the category of $\underline{V}$-valued $\underline{V}$-indexed functors on $\underline{B}$, are both defined over $(\bar{M},\bar{N})$, the obvious type category.

Theorem 25 $\underline{V}\text{-}d_0fib_{\underline{B}}$ and $(\underline{B},\underline{V})$ are equivalent categories over $(\bar{M},\bar{N})$.

Proof: (Sketch) $(\underline{B},\underline{V}) \longrightarrow \underline{V}\text{-}d_0fib_{\underline{B}}$ is of course given by pulling back $\underline{V}_* \xrightarrow{Q} \underline{V}$. In the other direction, given $\underline{E} \xrightarrow{G} \underline{B}$ of type $\hat{m}$, we proceed as follows. For each $I_0 \epsilon [M_0\underline{V}]$ and each I_0-object A in $\underline{B}$, choose a representing object $AF_G \epsilon \underline{V}$ for $(A;I_0)G^{-1}$, in such a way that for $I'_0 \xrightarrow{u_0} I_0$ in $[M_0\underline{V}]$, $Au_0^*F_G = AF_G$. This can be done since G satisfies the smallness condition. Let $(-,AF_G) \xrightarrow[\sim]{\varphi} (A;I_0)G^{-1}$ denote such a representation. Then $g = (1)(AF_G\varphi)$ is a "generic" AF_G,I_0-object in $\underline{E}$ over $(A;I_0)$. Now for an I_1-morphism $(A;I_0) \xrightarrow{(f;I_1)} (B;J_0)$ in $\underline{B}$ consider its "lift" to a morphism in $\underline{E}$ with domain $(g;AF_G,I_0)$:

$$\begin{array}{ccc} (g;AF_G,I_0) & \dashrightarrow & (Y;AF_G \otimes I_1m^{\#},J_0) \\ \vdots & & \vdots \\ (A;I_0) & \xrightarrow[(f;I_1)]{} & (B;J_0) \end{array}$$

The codomain of the lift is an element of $(AF_G \otimes I_1m^{\#})(B;J_0)G^{-1}$ and if we denote the representation of $(B;J_0)G^{-1}$ by $(-,BF_G) \xrightarrow[\sim]{\varphi} (B;J_0)G^{-1}$, then $fF_G \overset{Def}{=\!=} (Y)(AF_G \otimes I_1m^{\#})\varphi^{-1}$ is a morphism, $AF_G \otimes I_1m^{\#} \xrightarrow{fF_G} BF_G$ in $\underline{V}$, that is an $I_1m^{\#}$-morphism from AF_G to BF_G in $\underline{V}$. This defines a $\underline{V}$-indexed functor $\underline{B} \xrightarrow{F_G} \underline{V}$ of type m . The rest of the proof is straightforward and left to the reader.

∎

An analogous theorem can of course be stated for $\underline{V}$-discrete $\underline{V}$-fibrations which are defined as expected.

If $\underline{B}$ is a $\underline{V}$-category we may consider the functors $\underline{B} \xrightarrow{[B_0-]} \underline{V}$. For these, $U/[B_0-] \longrightarrow \underline{B}$ is just $B_0/\underline{B} \longrightarrow \underline{B}$. Now if $\underline{B}$ is only a large $\underline{V}$-category, the latter still makes sense and satisfies the lifting condition but $[B_0-]$ may fail to exist.

From the preceeding theorem we see that $B_0/\underset{\sim}{B} \longrightarrow \underset{\sim}{B}$ contains the information that $[B_0-]$ would, were it available. Thus $\underline{V}$-indexed categories of type $\bar{\mathbb{N}}^*$ etc. are very natural objects to consider.

We give two simple examples.

If $\underline{V} = \underline{ab}$ and $\underset{\sim}{B} = R$ is a ring regarded as an $\underline{ab}$-category with one object then an $\underline{ab}$-functor $R \rightarrow \underline{ab}$ is just a right R-module M . The associated total category $\mathcal{I}/M$ has group homomorphisms $I \xrightarrow{x} M$ as I-objects and pairs of homomorphisms $< I \xrightarrow{x} M$, $J \xrightarrow{f} M >$ as I,J-morphisms. The codomain of such a morphism is $I \otimes J \xrightarrow{x \otimes f} M \otimes R \xrightarrow{a} M$, where a is the action of R on M .

Let $\underline{V} = \mathbb{R}$ as in [Le] . Let $B \xrightarrow{f} \mathbb{R}$ be a distance decreasing map between generalized metric spaces (i.e. an $\mathbb{R}$-functor). For $i \in \mathbb{R}$ the i-objects of $0/f$ are given by $\{b \in B \mid i \geq bf\}$. For $i,j \in \mathbb{R}$ an i,j-morphism is a pair $<b,c> \in B \times B$ with $i \geq bf$ and $j \geq [bc]$ ($[bc]$ = distance from b to c). It's codomain is the i+j object c . If bf is the distance from b to some closed subset B' of B , then $0/f$ can be viewed as B equipped with an $\mathbb{R}$-indexed family (in the usual sense) of concentric subsets, the innermost being B'.

Let $\underset{\sim}{A}$ be a large $\underline{V}$-category. If $\underset{\sim}{E} \xrightarrow{H} \underset{\sim}{A}$ has the type of $\underset{\sim}{V}_* \xrightarrow{Q} \underset{\sim}{V}$, we may consider various "$\underleftarrow{\lim}$ H" as in §3. For the remainder of this paper, however, we make the following definitions. For $A \in \underset{\sim}{A}$, a *cone* c *from* A *to* H consists of: for each $I \in \underline{V}$ and each I-object X of $\underset{\sim}{E}$ an I-morphism $A \xrightarrow{(Xc;I)} XH$ in $\underset{\sim}{A}$ subject to the requirement that for every I,J-morphism $X \xrightarrow{(x;I,J)} X'$ in $\underset{\sim}{E}$,

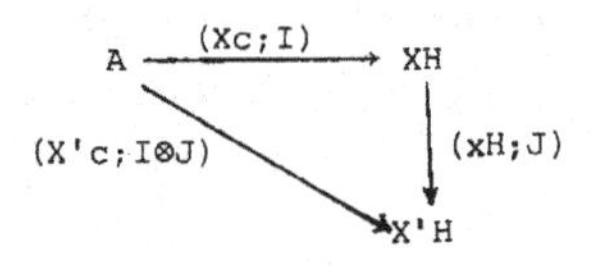

commutes and $(Xc)u^* = (Xu^*)c$ for all $I' \xrightarrow{u} I$ in $\underline{V}$. We write $A \xrightarrow{(c;)} H$. For $K \in \underline{V}$ we say that c is a *K-indexed family of cones from* A *to* H or simply that c is a K-*cone from* A *to* H, if c consists of a $K \otimes I$-morphism $A \xrightarrow{(Xc;K \otimes I)} XH$ in $\underset{\sim}{A}$, for each I-object X in $\underset{\sim}{E}$, with

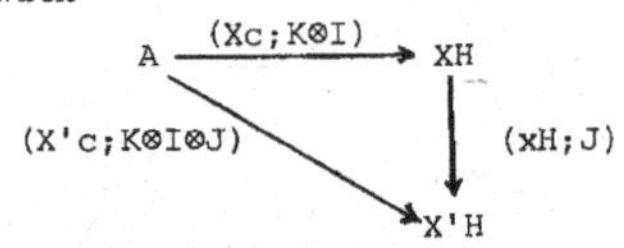

for every $(x;I,J)$ in $\underset{\sim}{E}$ and substitution respected as above. We denote such c by $A \xrightarrow{(c;K)} H$.

Definition 26 For $\underset{\sim}{E} \xrightarrow{H} \underset{\sim}{A}$ as above, the limit of H(if it exists) is an object $\underleftarrow{\lim}$ H of $\underset{\sim}{A}$ together with a cone, $\underleftarrow{\lim}\, H \xrightarrow{(p;)} H$, such that for any $K \in \underline{V}$, for any K-cone $A \xrightarrow{(c;K)} H$, there exists a unique K-morphism $A \xrightarrow{(\bar{c};K)} \underleftarrow{\lim}\, H$ in $\underset{\sim}{A}$ such that

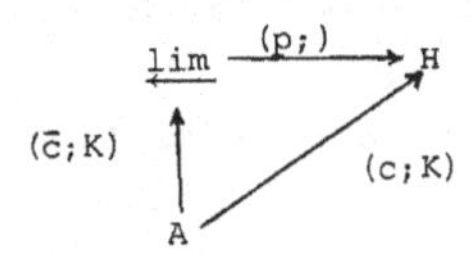

commutes.

Recalling the mean cotensor product $\{F,G\}$ of $\underset{\sim}{V}$-functors:

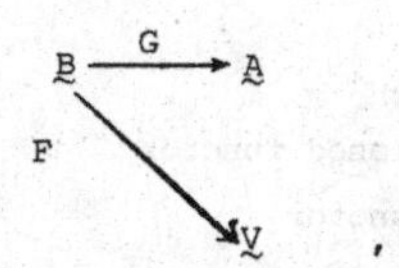

as defined in [B&K] ($\{F,G\} \simeq \int_{B}[BF,BG]$ when the indicated cotensors exist), and writing Q_F for the $\underline{V}$-discrete $\underline{V}$-cofibration corresponding to F, we can state:

Theorem 27 For F,G as above, $\{F,G\}$ exists if and only if $\underleftarrow{\lim}\, Q_F G$ exists in which case they are isomorphic. ∎

§5 APPENDIX

The category $\underline{T}$ that we have used to define types may be altered in at least two ways.

On one hand if the $\otimes$ of $\underline{V}$ satisfies certain "equations" it may be desirable to replace $\underline{T}$ by a "quotient theory" as we did in §3 for a symmetric $\underline{V}$.

On the other hand we do not have to restrict ourselves to a $\underline{V}$-ind-CAT which is only finitely complete. Our ambient category, $\underline{SET}^{()op}$, is considerably richer (proposition 3) so we give a variant of $\underline{V}$-ind-SET (and hence $\underline{V}$-ind-CAT) which takes advantage of this.

We define $\underline{set}^{\underline{T}} \xrightarrow{\underline{V}\otimes-} \underline{CAT}$ by left Kan extension:

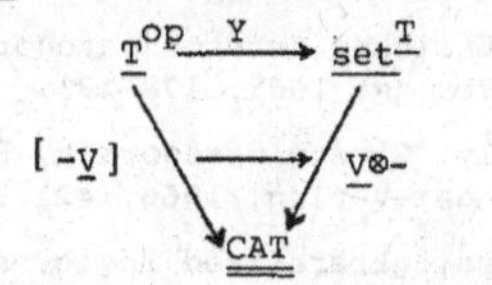

Thus for $F \in \underline{set}^{\underline{T}}$, $\underline{V}\otimes F \simeq \int^{T} TF.[T\underline{V}]$, the coend taken over $T \in \underline{T}$.

Proposition 28 $\underline{V}\otimes$-preserves colimits and finite products. ∎

Now redefine $\underline{V}$-ind-SET $\xrightarrow{P_{\underline{V}}} \underline{set}^{\underline{T}}$ as the pullback of $\underline{SET}^{()op} \xrightarrow{P} \underline{CAT}$ along $\underline{V}\otimes-$.

Proposition 29 $\underline{V}$-ind-SET is $\mathcal{U}_1$-bicomplete and cartesian closed. $P_{\underline{V}}$ is bicontinuous and preserves exponentiation.

Proof: The statement about exponentials follows from:

Lemma 30 Let

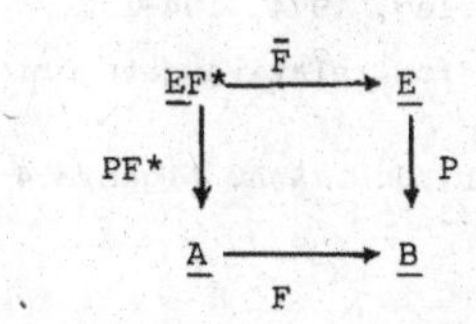

be a pullback in $\underline{CAT}$.

If 1) P is a fibration

2) $\underline{A}$, $\underline{E}$, and $\underline{B}$ are cartesian closed

3) P is a cartesian closed functor

4) F is a cartesian functor,

then 1) PF* is a fibration

2) $\underline{E}F^*$ is cartesian closed

3) PF* is a cartesian closed functor

4) $\bar{F}$ is a cartesian functor.

∎

Of course V-ind-SET above contains our earlier category of V-indexed sets as a full subcategory.

It is fairly difficult to explicitly calculate functor categories in our new V-ind-CAT, however, the situation is somewhat simpler for symmetric $\underline{V}$ and $\underline{T}_c$ as before. In that situation one can also show that $\underline{V}^{\underline{I}}$ externalized is $\underline{V}[U \xrightarrow{x} I]$; i.e. $\underline{V}$ with an indeterminant U section of I adjoined, universally for symmetric monoidal categories. It is interesting to compare this with the corresponding situation for S-indexed categories.

REFERENCES

[B&K] F. Borceux and G.M. Kelly, A notion of limit for enriched categories, Bull. Austral. Math. Soc. 12 (1975), 49-72.

[Dc] E.J. Dubuc, Kan extensions in enriched category theory, Lecture Notes in Math. 145, Springer-Verlag, 1970.

[Dy] B.J. Day, On closed categories of functors, Lecture Notes in Math. 137, Springer-Verlag, 1970, 1-38.

[D&K] B.J. Day and G.M. Kelly, Enriched functor categories, Lecture Notes in Math. 106, Springer-Verlag, 1969, 178-191.

[E&K] S. Eilenberg and G.M. Kelly, Closed categories, Proc. Conf. Categorical Algebra, La Jolla, Springer-Verlag, 1966, 421-562.

[Le] F.W. Lawvere, Metric spaces, generalized logic, and closed categories, Preprint from Institute di Mathematica, Universita di Perugia, 1973.

[Lk] J. Lambek, Subequalizers, Canad. Math. Bull. 13 (1970), 337-349.

[Ln] F.E.J. Linton, The multilinear Yoneda lemmas, Lecture Notes in Math. 195, Springer-Verlag, 1971, 209-229.

[P&S] R. Paré and D. Schumacher, S-indexed categories, Various unpublished lectures in Halifax.

[R] G.D. Reynolds, Tensor and cotensor products of bifunctors, Thesis, Wesleyan University, 1973.

[S] R. Street, Fibrations and Yoneda's lemma in a 2-category, Lecture notes in Math. 420, Springer-Verlag, 1974, 104-133.

[Wd] R.J. Wood, Indicial methods for relative categories, Thesis, Dalhousie University, 1976.

[Wf] H. Wolff, V-cat and V-graph, J.P. & A. Algebra 4 (1974), 123-135.

ALGEBRAIC THEORIES IN TOPOSES

by

Peter T. Johnstone and Gavin C. Wraith

INTRODUCTION

If $\mathcal{E}$ is a topos, we have a notion of $\mathcal{E}$-based algebraic theory. When $\mathcal{E}$ is $\mathcal{S}$, the category of sets and functions, this notion reduces to the usual one [29, 38].

Let us give a simple example when $\mathcal{E}$ is Shv(X), the topos of sheaves on a topological space X. If R is a sheaf of rings on X, the theory of left R-modules is a Shv(X)-based algebraic theory. Instead of a set of unary operations, as we would have in a $\mathcal{S}$-based algebraic theory, we have a sheaf (on X) of unary operations.

We must distinguish between finitary algebraic theories and infinitary ones. We use the term "infinitary" to mean that we permit, but do not insist upon, infinitary operations. Finitary algebraic theories are geometric, in the sense that their models are preserved by inverse image functors, and so are part of the general study of geometric theories and classifying toposes [28, 36].

Our approach to finitary algebraic theories is rather unusual; it is based on the concept of the "object classifier". We shall show that if $\mathcal{E}$ is a topos satisfying the axiom of infinity, there is an $\mathcal{E}$-topos $\mathcal{E}[U]$, together with a distinguished object U of $\mathcal{E}[U]$, which has the following universal property:

Given any object X in an $\mathcal{E}$-topos $\mathcal{F}$, there exists (uniquely up to canonical isomorphism) a map of $\mathcal{E}$-toposes

$$\tilde{X}: \mathfrak{J} \longrightarrow \mathcal{E}[U]$$

such that $\tilde{X}^*(U) \cong X$. Moreover, any morphism $X \xrightarrow{a} Y$ in $\mathfrak{J}$ induces a unique natural transformation $\tilde{a}: \tilde{X}^* \longrightarrow \tilde{Y}^*$ such that $\tilde{a}_U \cong a$. We refer to U as the generic object and $\tilde{X}$ as the classifying map of X. We have deliberately used a notation reminiscent of that for polynomial rings: one should picture a typical object of $\mathcal{E}[U]$ as a "polynomial" in the "indeterminate object" U. The objects and maps of $\mathcal{E}$ are "constants", "addition" is given by ($\mathcal{E}$-indexed) colimits and "multiplication" by finite limits. We shall see that "composition" of "polynomials" induces a (non-symmetric) monoidal structure $(\mathcal{E}[U], \otimes, U)$ on $\mathcal{E}[U]$, with U as unit object.

We shall identify finitary $\mathcal{E}$-based algebraic theories with monoids in $(\mathcal{E}[U], \otimes, U)$. It will turn out that this is equivalent to two other definitions:

(a) mimic Lawvere's definition of a finitary algebraic theory as a category with objects indexed by the natural numbers, satisfying appropriate conditions [22];

(b) give a monad on every $\mathcal{E}$-topos, so as to commute up to isomorphism with inverse image parts of maps of $\mathcal{E}$-toposes.

We shall define an infinitary $\mathcal{E}$-based algebraic theory to be a locally internal (= indexed) monad on $\mathcal{E}$ [33, 34]. In the $\mathcal{S}$-based case, it follows from the work of Linton [26] that this is equivalent to a definition in terms of operations and equations, but in general it gives rise to a number of interesting new phenomena which cannot occur over $\mathcal{S}$. For example, we shall see that the free model functor for an $\mathcal{E}$-based algebraic theory preserves monomorphisms if $\mathcal{E}$ is Boolean or if the theory is finitary, but not in the general case.

We have not investigated the relationship between our $\mathcal{E}$-based algebraic theories and the "algebraic theories enriched in a closed category" studied by F. Borceux and B. Day [7]. The basic difference is that Borceux and Day work with an external category of finite sets, and hence require external cocompleteness assumptions on their base category $\mathcal{V}$, whereas we are able to avoid any such assumptions by the systematic use of internal and indexed categories. Doubtless it would be possible to produce a synthesis of the two approaches, to handle algebraic theories enriched in a closed category which is itself locally internal over a topos, using the techniques of R. Wood [40].

This article has had a long gestation, and really arose out of conjectures about recursive definitions. In 1972 the second author delivered a lecture on the object classifier for at the University of Montreal, and at the same time learnt from A. Joyal and J. Bénabou how to construct internal full subcategories. The same summer, at the Oberwolfach category theory conference, he presented the section criterion (2.1 below) and the uniqueness theorem for recursion (2.2), and raised the following general question:

> Given an object X in a topos $\mathcal{E}$ satisfying the axiom of infinity, and a functor $T: \mathcal{E} \longrightarrow \mathcal{E}$, does there exist an object F in $\mathcal{E}/N$ such that $o^*(F) \cong X$ and $s^*(F) \cong T^N(F)$, where T^N denotes the functor T "applied fibrewise"?

We think of X and T as recursion data; T is a "machine" for manufacturing new objects out of old, and X is the initial data. In the case $\mathcal{E} = \mathcal{S}$ the answer is yes, because we simply take F to be

$$\coprod_{n \geq 0} T^n(X) \quad .$$

The uniqueness theorem asserts that if T is an indexed functor, then the solution F is unique up to canonical isomorphism if it exists.

We shall see that for an arbitrary topos, the answer to the

general question must be no. However, the second author conjectured that the answer was yes if T was induced by an object of $\mathcal{E}[U]$, because of the suggestive nature of the conjecture when phrased in terms of the following diagram:

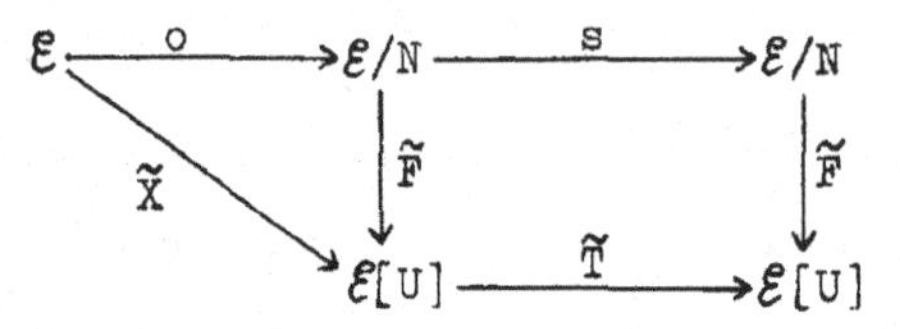

This conjecture was resolved positively by both authors independently early in 1974. The first author's method [15] amounted in effect to constructing free finitary algebraic theories, while the second author's involved iterating profunctor composition to show that $\mathcal{E}/N$ is a natural number object in the category of bounded $\mathcal{E}$-toposes.

At the Bangor conference in September 1973 [14], the first author showed how to construct the internal category of finitely-presented models of a finitely-presented, finitary $\mathcal{S}$-based algebraic theory. (B. Lesaffre [25] had shown that free functors for such theories exist in any topos satisfying the axiom of infinity.) The point is that one can then prove that the topos of $\mathcal{E}$-valued functors on this internal category is a classifying topos for the theory - thereby generalizing M. Hakim's original result [13] for the $\mathcal{S}$-based theory of commutative rings. The methods of [14] apply equally well to $\mathcal{E}$-based finitary algebraic theories. More recently, the first author has extended them to construct classifying toposes for such things as the finitely-generated models of a theory.

While these constructions involve rather more work than the general techniques of Joyal, Tierney [36] and Bénabou [5] for proving the existence of classifying toposes, they are still valuable in that they give us a much more explicit description of the classifying topos and the generic model than the general method can hope to do.

This is essential if we are to make any worthwhile calculations involving the "theory of the generic model". (See, for example, Kock [20].)

The first suggestion that our work on the subject of algebraic theories and recursion should be written up as a joint paper was made by the first author at the time of his Ph.D. examination in July 1974. However, no action on this suggestion was taken for some time. In the autumn of 1974, the second author produced a manuscript of some 30 pages entitled "Finitary Algebraic Theories in Topoi"; a much more comprehensive draft of about 70 pages, entitled "Algebraic Theories and Recursion in Elementary Topos Theory", was written by the first author in the spring and summer of 1975. The latter contained a good deal of detailed exposition of background material on toposes, which was later incorporated into the first author's book [17].

When we revived the idea of writing a joint paper in the autumn of 1976, it therefore seemed desirable to produce a new draft, omitting the background material but including some of our more recent work (notably the chapter on infinitary theories). The present text is the result; it was written between November 1976 and March 1977 by the two authors.

Here is a brief description of the contents of the paper. Chapters I and II are concerned with topos-theoretic preliminaries and with the setting-up of notation; we refer the reader to [17] for proofs of most of the results stated here. In chapter III, we develop the technique of iterating profunctor composition, and use it to prove that $\mathcal{E}/N$ is a natural number object in the category of bounded $\mathcal{E}$-toposes. Chapter IV introduces the object classifier, studies its basic properties and proves the existence theorem for recursively defined objects. In chapter V we reach the definition of a finitary algebraic theory in a topos; we study the consequences of this definition in some detail, and we extend the methods of chapter IV to construct a classifying topos for such a theory. The last two chapters investigate possible extensions of the basic definition of chapter V: to many-sorted theories such as the theory of categories (chapter VI) and to theories whose operations may have arities which are not finite cardinals (chapter VII). We shall see that in each case some of our results for single-sorted finitary theories fail to generalize; so both these chapters are rather more tentative in nature than what has gone before.

Although it is described in detail in chapter I, it is perhaps appropriate to say something here about our custom of using "variables" to handle indexed families of objects. Since the idea of associating a formal language with an elementary topos was first popularized by J. Bénabou, there has been a tendency for writers on category theory, and particularly on topos theory, to move away from the traditional "diagrammatic" style exemplified by [11] towards a much more formal "quasi-set-theoretical" notation as used in [31]. We believe strongly that there is a need for an intermediate style of presentation, in which set-theoretic notations are used informally in such a way as (we hope) to enlighten the reader without overwhelming him. In fact our notation is very similar to that which has been evolved (for similar reasons) by A. Kock [21, 20].

We should also perhaps mention that the symbol $\Box$ is normally used in the text to mark the end of a proof. When it appears at the end of the statement of a theorem, it indicates either that no proof is given, or that the proof is contained in the informal discussion which precedes the statement.

We are conscious of our debt to many other workers in category theory whose ideas we have used in our own research. Foremost, of course, we are indebted to Bill Lawvere as the originator of both the categorical approach to algebraic theories and the elementary theory of toposes. Others whose work has contributed significantly to our understanding of the subject-matter of these notes include Jean Bénabou, Robert Coates, André Joyal, Anders Kock, Brigitte Lesaffre, Christian Juul Mikkelsen, Robert Paré and Jacques Penon: our thanks to all of them. In addition, the second author is indebted to the first for typing the final manuscript, and both authors are indebted to Robert Paré for agreeing to the suggestion that our work be published together with his.

CHAPTER I : INTERNAL CATEGORIES AND PROFUNCTORS

We begin with a few words about prerequisites. We shall assume that the reader is acquainted with the basic ideas of elementary topos theory; our references for this material are generally to the recent book of the first author [17], but most of the results we require may also be found in the second author's lecture notes [39]. Some acquaintance with the "classical" theory of $\mathcal{S}$-based algebraic theories is desirable but not essential; the standard references for this are Manes [29] and Wraith [38].

We shall make fairly systematic use of the theory of indexed categories developed by R. Paré and D. Schumacher [33] in the first half of this volume: our notation and terminology generally follows theirs, but we shall follow J. Penon [34] in using the term <u>locally internal category</u> for what Paré and Schumacher would call an indexed category with small homs. Thus a locally internal category $\mathcal{C}$ over a topos $\mathcal{E}$ is specified by giving an $(\mathcal{E}/I)$-category $\mathcal{C}^I$ for each object I of $\mathcal{E}$, together with functors $f^*: \mathcal{C}^J \longrightarrow \mathcal{C}^I$ for each $I \xrightarrow{f} J$ which are full embeddings of $(\mathcal{E}/I)$-categories and satisfy the appropriate coherence condition. The enriched structure on the categories $\mathcal{C}^I$ is generally essential from our point of view, since it is required for the uniqueness theorem (2.2) from which a great many of our arguments stem.

We shall often require the following lemma, which is mentioned in [33], II.3.2.

<u>1.1 Lemma</u> Let $\mathcal{C}$ and $\mathcal{D}$ be locally internal categories over $\mathcal{E}$. Then any indexed functor $T: \mathcal{C} \longrightarrow \mathcal{D}$ is in fact locally internal.

<u>Proof</u> We have to show that $T^I: \mathcal{C}^I \longrightarrow \mathcal{D}^I$ is enriched over $\mathcal{E}/I$; i.e. given two objects A, B of $\mathcal{C}^I$, we have to produce a morphism

$$\mathcal{C}^I(A,B) \longrightarrow \mathcal{D}^I(T^IA,T^IB)$$

in $\mathcal{E}/I$ which "internalizes" the effect of T^I on morphisms of $\mathcal{C}^I$. Write $(X \xrightarrow{\varphi} I)$ for the object $\mathcal{C}^I(A,B)$. Then the diagonal map $X \longrightarrow X \times_I X$ can be regarded as a global element of the object

$$\varphi^*(\mathcal{C}^I(A,B)) \cong \mathcal{C}^X(\varphi^*A,\varphi^*B)$$

of $\mathcal{E}/X$, i.e. a morphism $\varphi^*A \longrightarrow \varphi^*B$ in $\mathcal{C}^X$. Applying T^X to this, we obtain a morphism

$$\varphi^*T^IA \cong T^X\varphi^*A \longrightarrow T^X\varphi^*B \cong \varphi^*T^IB \quad ,$$

and hence an element of $\varphi^*(\mathcal{D}^I(T^IA,T^IB))$ in $\mathcal{E}/X$. But this in turn corresponds to a morphism

$$\mathcal{C}^I(A,B) = \Sigma_\varphi(1_X) \longrightarrow \mathcal{D}^I(T^IA,T^IB)$$

in $\mathcal{E}/I$, as required. □

If $\mathcal{E}$ is a topos, an <u>$\mathcal{E}$-topos</u> means a topos $\mathcal{F}$ equipped with a geometric morphism $\mathcal{F} \xrightarrow{f} \mathcal{E}$. Normally we shall abuse notation in the standard way, by omitting to mention the morphism f and simply saying "$\mathcal{F}$ is an $\mathcal{E}$-topos". If I is an object of $\mathcal{E}$, we shall refer to $f^*(I)$ as "the (constant) object I in $\mathcal{F}$", and when it is possible to do so without ambiguity, we shall simply denote it by I. (This follows the tradition of denoting by the same letter an element of a ring and the constant polynomial which it determines.) We shall frequently consider $\mathcal{F}$ as a locally internal category over $\mathcal{E}$: the indexing is defined by $\mathcal{F}^I = \mathcal{F}/f^*I$, and the enriched structure using the functor f_* (cf. [17], Example A.3(ii)).

We shall employ a notation half way between the formal language of Bénabou, Osius [31] et al., and the language of category theory. If I is an object of a topos $\mathcal{E}$, we use notations such as $X(\underset{\sim}{i})$ or $X_{\underset{\sim}{i}}$ for objects of $\mathcal{E}/I$. This emphasises the interpretation of $\mathcal{E}/I$ as the category of I-indexed families of objects of $\mathcal{E}$, and we shall use the same notation for more general categories indexed over $\mathcal{E}$. One should think of the symbol $\underset{\sim}{i}$ as a variable of type I - its role is to point out which $\mathcal{E}$-topos $X(\underset{\sim}{i})$ belongs to. Similarly, if I and J are two objects of $\mathcal{E}$, an expression $X(\underset{\sim}{i},\underset{\sim}{j})$ refers to an object of $\mathcal{E}/I\times J$; and if $\underset{\sim}{i}_1$, $\underset{\sim}{i}_2$ are both variables of type I, then $Y(\underset{\sim}{i}_1,\underset{\sim}{i}_2)$ refers to an object of $\mathcal{E}/I\times I$. (The expression $Y(\underset{\sim}{i},\underset{\sim}{i})$ would then denote the pullback of $Y(\underset{\sim}{i}_1,\underset{\sim}{i}_2)$ along the diagonal map $I \longrightarrow I\times I$.)

More generally, for any morphism $I \xrightarrow{f} J$ of $\mathcal{E}$ and object $X(\underset{\sim}{j})$ of $\mathcal{E}/J$, the pullback of $X(\underset{\sim}{j})$ along f will be denoted by $X(f\underset{\sim}{i})$. As a particular case, we do not use any symbol to denote a variable of type 1, and so if $1 \xrightarrow{j} J$ is a global element of J, then $X(j)$ denotes the pullback of $X(\underset{\sim}{j})$ along j.

An expression of the form $X(\underset{\sim}{i}) \times Y(\underset{\sim}{j})$ must be interpreted as the object of $\mathcal{E}/I\times J$ given by $\pi_1{}^*(X(\underset{\sim}{i})) \times \pi_2{}^*(Y(\underset{\sim}{j}))$, where

$$I \xleftarrow{\pi_1} I\times J \xrightarrow{\pi_2} J$$

are the product projections. Similar remarks apply to expressions like

$$X(\underset{\sim}{i}) \amalg Y(\underset{\sim}{j}) \quad , \quad X(\underset{\sim}{i}_1,\underset{\sim}{i}_2)^{Y(\underset{\sim}{i}_2,\underset{\sim}{j})} \quad ,$$

and so on. In each case one determines the $\mathcal{E}$-topos in which the interpretation should live by counting up the free variables in the expression, and then re-interprets the sub-expressions, which generally live in other $\mathcal{E}$-toposes, by pulling back along the appropriate projection maps. This process may sound clumsy, but will be found to be quite easy in practice.

The functors

$$\Sigma_{\pi_1} \ , \ \Pi_{\pi_1} \ : \ \mathcal{E}/I\times J \longrightarrow \mathcal{E}/I$$

(i.e. the left and right adjoints of pullback along π_1) will be seen to behave like quantifiers on variables of type J; to emphasise this resemblance, we shall write them as $\Sigma_{\underset{\sim}{j}}$ and $\Pi_{\underset{\sim}{j}}$. Thus $\Sigma_{\underset{\sim}{i}} X(\underset{\sim}{i})$ denotes the object of $\mathcal{E}$ obtained from $X(\underset{\sim}{i})$ by forgetting the structure map into I.

We shall demonstrate this notation by unfolding some of the fundamental notions of category theory within a topos. Detailed proofs of the results which follow will be found in chapter 2 of [17].

An <u>internal category</u> $\underline{A}$ in $\mathcal{E}$ is given by an object A_o of $\mathcal{E}$, called the object of objects of $\underline{A}$, and an object $\underline{A}(\underset{\sim}{a}_1,\underset{\sim}{a}_2)$ of $\mathcal{E}/A_o \times A_o$, together with maps

$$\underline{A}(\underset{\sim}{a}_1,\underset{\sim}{a}_2)\times \underline{A}(\underset{\sim}{a}_2,\underset{\sim}{a}_3) \xrightarrow{c} \underline{A}(\underset{\sim}{a}_1,\underset{\sim}{a}_3) \qquad \text{in } \mathcal{E}/A_o\times A_o\times A_o$$

and

$$1 \xrightarrow{e} \underline{A}(\underset{\sim}{a},\underset{\sim}{a}) \qquad \text{in } \mathcal{E}/A_o \quad ,$$

satisfying commutative diagrams which say that c is associative and e is a two-sided unit for it. We shall occasionally write A_1 for the object $\Sigma_{\underset{\sim}{a}_1,\underset{\sim}{a}_2}\underline{A}(\underset{\sim}{a}_1,\underset{\sim}{a}_2)$ of $\mathcal{E}$, and call it the object of morphisms of $\underline{A}$.

An <u>internal functor</u> $f\colon \underline{A} \longrightarrow \underline{B}$ between internal categories consists of a morphism $f_o\colon A_o \longrightarrow B_o$, together with a morphism

$$\underline{A}(\underset{\sim}{a}_1,\underset{\sim}{a}_2) \longrightarrow \underline{B}(f_o\underset{\sim}{a}_1,f_o\underset{\sim}{a}_2)$$

in $\mathcal{E}/A_o\times A_o$ which is compatible in the obvious sense with composition and unit maps. We write $\underset{\approx}{\mathrm{cat}}(\mathcal{E})$ for the category of internal categories and functors in $\mathcal{E}$. We recall that $\underset{\approx}{\mathrm{cat}}(\mathcal{E})$ is an $\mathcal{E}$-category (in fact a locally internal category, with indexing given by

$$(\underset{\approx}{\mathrm{cat}}(\mathcal{E}))^I = \underset{\approx}{\mathrm{cat}}(\mathcal{E}/I) \quad),$$

and that it is cartesian closed. There is also a definition of <u>internal natural transformation</u> which makes $\underset{\approx}{\mathrm{cat}}(\mathcal{E})$ into a 2-category.

The notion of internal functor should not be confused with that of <u>internal diagram</u>, which we may think of as a functor from an internal category to the topos in which it lives. An internal diagram F on $\underline{A}$ consists of an object $F(\underset{\sim}{a})$ of $\mathcal{E}/A_0$, together with an <u>action map</u>

$$F(\underset{\sim}{a}_1) \times \underline{A}(\underset{\sim}{a}_1, \underset{\sim}{a}_2) \xrightarrow{\alpha} F(\underset{\sim}{a}_2)$$

in $\mathcal{E}/A_0 \times A_0$, which is associative and unitary relative to c and e. We write $\mathcal{E}^{\underline{A}}$ for the category of internal diagrams on $\underline{A}$ (we leave it to the reader to define morphisms of internal diagrams); we recall that $\mathcal{E}^{\underline{A}}$ is an $\mathcal{E}$-topos, and that the forgetful functor

$$\mathcal{E}^{\underline{A}} \longrightarrow \mathcal{E}/A_0 \ ; \quad F \longmapsto F(\underset{\sim}{a})$$

is both monadic and comonadic (so that it is the inverse image of an essential, surjective map of $\mathcal{E}$-toposes). More generally, any internal functor $\underline{A} \xrightarrow{f} \underline{B}$ induces an essential map of $\mathcal{E}$-toposes

$$\mathcal{E}^{\underline{A}} \longrightarrow \mathcal{E}^{\underline{B}} \quad ,$$

so that the assignment $\underline{A} \longmapsto \mathcal{E}^{\underline{A}}$ becomes a (pseudo)functor

$$\underset{\sim\sim}{\text{cat}}(\mathcal{E}) \longrightarrow \underline{\underline{\text{Top}}}/\mathcal{E} \quad .$$

An internal diagram on the opposite category $\underline{A}^{op}$ is called an <u>internal presheaf</u> on $\underline{A}$. If F is an internal presheaf on $\underline{A}$, we customarily write the action of $\underline{A}^{op}$ on $F(\underset{\sim}{a})$ as a morphism

$$\underline{A}(\underset{\sim}{a}_1, \underset{\sim}{a}_2) \times F(\underset{\sim}{a}_2) \longrightarrow F(\underset{\sim}{a}_1) \quad .$$

If $X(\underset{\sim}{i})$ is an object of $\mathcal{E}/I$ for some I, we can define an internal category $\text{Full}_{\mathcal{E}}(X)$, the <u>internal full subcategory</u> of $\mathcal{E}$ generated by $X(\underset{\sim}{i})$, as follows: $\text{Full}_{\mathcal{E}}(X)_0 = I$, and

$$\text{Full}_{\mathcal{E}}(X)(\underset{\sim}{i}_1, \underset{\sim}{i}_2) = X(\underset{\sim}{i}_2)^{X(\underset{\sim}{i}_1)} \quad .$$

The composition and unit maps are defined in the obvious way. Note that if $\underline{A}$ is a category and $F(\underset{\sim}{a})$ an object of $\mathcal{E}/A_0$, then actions of $\underline{A}$ on $F(\underset{\sim}{a})$ are in 1-1 correspondence with internal functors $\underline{A} \longrightarrow \text{Full}_{\mathcal{E}}(F)$ which are the identity on objects. (The correspondence is given by transposing a map

$$F(\underset{\sim}{a}_1) \times \underline{A}(\underset{\sim}{a}_1, \underset{\sim}{a}_2) \longrightarrow F(\underset{\sim}{a}_2)$$

to obtain

$$\underline{A}(a_1,a_2) \longrightarrow F(a_2)^{F(a_1)} = \mathrm{Full}_{\mathcal{E}}(F)(a_1,a_2) \quad .)$$

In particular, the identity functor on $\mathrm{Full}_{\mathcal{E}}(X)$ corresponds to an internal diagram structure on the object $X(i)$; if we think of internal diagrams as functors into $\mathcal{E}$, this diagram is the "inclusion functor" $\mathrm{Full}_{\mathcal{E}}(X) \longrightarrow \mathcal{E}$. We denote it simply by X.

More generally, J. Penon [34] has pointed out that the internal full subcategory construction works for any locally internal category over $\mathcal{E}$: if $\mathcal{C}$ is such a category and $A(i)$ is an object of $\mathcal{C}^I$, we can construct an internal category $\mathrm{Full}_{\mathcal{C}}(A)$ with object of objects I. The "inclusion functor" mentioned above becomes a full embedding of locally internal categories

$$\mathrm{Full}_{\mathcal{C}}(A) \xrightarrow{\ A\ } \mathcal{C} \quad ,$$

where we identify the internal category $\mathrm{Full}_{\mathcal{C}}(A)$ with the locally internal category which it generates as in [34].

Now let $\underline{A}$ and $\underline{B}$ be two internal categories. An <u>internal profunctor</u> T from $\underline{A}$ to $\underline{B}$ (written $\underline{A} \overset{T}{- - \to} \underline{B}$) is an internal diagram on $\underline{A}^{op} \times \underline{B}$: equivalently, it is given by an object $T(a,b)$ of $\mathcal{E}/A_0 \times B_0$ equipped with left and right action maps

$$\underline{A}(a_1,a_2) \times T(a_2,b) \xrightarrow{\ \alpha\ } T(a_1,b)$$

and

$$T(a,b_1) \times \underline{B}(b_1,b_2) \xrightarrow{\ \beta\ } T(a,b_2)$$

which are unitary and associative and commute with each other. Equivalently again, we may regard T as an internal presheaf on $\underline{A}$ in the $\mathcal{E}$-topos $\mathcal{E}^{\underline{B}}$, or an internal diagram on $\underline{B}$ in $\mathcal{E}^{\underline{A}^{op}}$ - in particular, internal diagrams and presheaves on $\underline{A}$ may be regarded as profunctors $\underline{1} - - - \to \underline{A}$ and $\underline{A} - - - \to \underline{1}$ respectively, where $\underline{1}$ is the terminal object of $\mathrm{cat}(\mathcal{E})$. We write $\mathrm{Prof}(\underline{A},\underline{B})$ for $\mathcal{E}^{\underline{A}^{op} \times \underline{B}}$, the category of profunctors from $\underline{A}$ to $\underline{B}$.

Now suppose we are given profunctors

$$\underline{A} \overset{S}{- - \to} \underline{B} \overset{T}{- - \to} \underline{C} \quad .$$

Then we can define a profunctor

$$S\otimes_{\underline{B}}T\colon \underline{A} - - - \to \underline{C}$$

(called the <u>composite</u> or <u>tensor product</u> of S and T) as follows: $S\otimes_{\underline{B}}T(\underset{\sim}{a},\underset{\sim}{c})$ is defined by the coequalizer diagram

$$\Sigma_{\underset{\sim}{b}_1,\underset{\sim}{b}_2} S(\underset{\sim}{a},\underset{\sim}{b}_1)\times\underline{B}(\underset{\sim}{b}_1,\underset{\sim}{b}_2)\times T(\underset{\sim}{b}_2,\underset{\sim}{c}) \underset{1\times\alpha_T}{\overset{\beta_S\times 1}{\rightrightarrows}} \Sigma_{\underset{\sim}{b}} S(\underset{\sim}{a},\underset{\sim}{b})\times T(\underset{\sim}{b},\underset{\sim}{c}) \longrightarrow S\otimes_{\underline{B}}T(\underset{\sim}{a},\underset{\sim}{c})$$

in $\mathcal{E}/A_o\times C_o$, and the actions of $\underline{A}$ and $\underline{C}$ are induced by α_S and β_T respectively. We note for future reference that the above diagram is actually a reflexive coequalizer diagram; that is to say, the map

$$\Sigma_{\underset{\sim}{b}} S(\underset{\sim}{a},\underset{\sim}{b})\times T(\underset{\sim}{b},\underset{\sim}{c}) \longrightarrow \Sigma_{\underset{\sim}{b}_1,\underset{\sim}{b}_2} S(\underset{\sim}{a},\underset{\sim}{b}_1)\times\underline{B}(\underset{\sim}{b}_1,\underset{\sim}{b}_2)\times T(\underset{\sim}{b}_2,\underset{\sim}{c})$$

induced by the unit map $1 \xrightarrow{e} \underline{B}(\underset{\sim}{b},\underset{\sim}{b})$ splits both the maps $\beta_S\times 1$ and $1\times\alpha_T$.

It is easy to show that $\otimes_{\underline{B}}$ is a bifunctor

$$\mathrm{Prof}(\underline{A},\underline{B}) \times \mathrm{Prof}(\underline{B},\underline{C}) \longrightarrow \mathrm{Prof}(\underline{A},\underline{C}) \quad ,$$

and that it is associative up to coherent natural isomorphism. Moreover, there is a profunctor

$$Y(\underline{B})\colon \underline{B} - - - \to \underline{B}$$

which is a two-sided unit for tensor product: specifically, $Y(\underline{B})(\underset{\sim}{b}_1,\underset{\sim}{b}_2) = \underline{B}(\underset{\sim}{b}_1,\underset{\sim}{b}_2)$, with both left and right actions of $\underline{B}$ given by the composition map. (We call $Y(\underline{B})$ the <u>Yoneda</u> or <u>unit profunctor</u> on $\underline{B}$.) Thus internal profunctors in $\mathcal{E}$ form the 1-arrows of a bicategory, whose objects are the internal categories. An important property of profunctors in a topos is the fact that this bicategory is <u>biclosed</u>, i.e. for a fixed $\underline{B} - \overset{T}{-} - \to \underline{C}$, the functors

$$(-)\otimes_{\underline{B}}T\colon \mathrm{Prof}(\underline{A},\underline{B}) \longrightarrow \mathrm{Prof}(\underline{A},\underline{C})$$

and

$$T\otimes_{\underline{C}}(-)\colon \mathrm{Prof}(\underline{C},\underline{D}) \longrightarrow \mathrm{Prof}(\underline{B},\underline{D})$$

have right adjoints. (See [17], Corollary 2.49.)

It follows that the functor $(-)\otimes_{\underline{B}}T$ preserves arbitrary colimits. We shall frequently be interested in the case when this functor preserves finite limits; if this happens we shall say that the left action of $\underline{B}$ on $T(\underset{\sim}{b},\underset{\sim}{c})$ is flat, or simply that T is left flat. In particular, we say that a presheaf on $\underline{B}$ is flat if it is left flat as a profunctor $\underline{B} \dashrightarrow \underline{1}$, and we write $\mathrm{Flat}(\underline{B}^{op},\mathcal{E})$ for the full subcategory of flat presheaves in $\mathcal{E}^{\underline{B}^{op}}$.

We can now state the important theorem of R. Diaconescu [11], which we shall use repeatedly in later chapters.

1.2 Theorem Let $\underline{A}$ be an internal category in a topos $\mathcal{E}$, and let $\mathcal{F}$ be an $\mathcal{E}$-topos. Then there is an equivalence (natural in $\mathcal{F}$) between the category $\underline{\underline{\mathrm{Top}}}/\mathcal{E}\ (\mathcal{F},\mathcal{E}^{\underline{A}})$ of geometric morphisms over $\mathcal{E}$ from $\mathcal{F}$ to $\mathcal{E}^{\underline{A}}$, and the category $\mathrm{Flat}(\underline{A}^{op},\mathcal{F})$ of flat presheaves on $\underline{A}$ in $\mathcal{F}$. □

The equivalence is obtained as follows: let G be a flat presheaf on $\underline{A}$ in $\mathcal{F}$. Then the functor

$$(-)\otimes_{\underline{A}}G\colon \mathcal{F}^{\underline{A}} \longrightarrow \mathcal{F}$$

is the inverse image of a geometric morphism $h(G)\colon \mathcal{F} \longrightarrow \mathcal{F}^{\underline{A}}$, which is in fact a map of $\mathcal{F}$-toposes. But we have a commutative diagram of geometric morphisms (in fact a pullback in $\underline{\underline{\mathrm{Top}}}$)

$$\begin{array}{ccc} \mathcal{F}^{\underline{A}} & \xrightarrow{f^{\underline{A}}} & \mathcal{E}^{\underline{A}} \\ \downarrow & & \downarrow \\ \mathcal{F} & \xrightarrow{f} & \mathcal{E} \end{array}$$

where the inverse image $(f^{\underline{A}})^*$ is the functor "apply f^* to all parts of an internal diagram on $\underline{A}$". Thus we have a functor

$$\mathrm{Flat}(\underline{A}^{op},\mathcal{F}) \longrightarrow \underline{\underline{\mathrm{Top}}}/\mathcal{E}\ (\mathcal{F},\mathcal{E}^{\underline{A}});\ G \longmapsto f^{\underline{A}}.h(G) \quad .$$

Conversely, if we are given $\mathcal{F} \xrightarrow{g} \mathcal{E}^{\underline{A}}$ over $\mathcal{E}$, we may define the presheaf G to be $(g^{\underline{A}^{op}})^*(Y(\underline{A}))$, where the unit profunctor $Y(\underline{A})$ is regarded as a presheaf on $\underline{A}$ in $\mathcal{E}^{\underline{A}}$. The fact that the two constructions are inverse up to natural isomorphism follows from the fact that $Y(\underline{A})$ is a unit for $\otimes_{\underline{A}}$.

There is an alternative definition of flatness for presheaves involving filtered categories, which we shall require in later chapters. Recall that an internal category $\underline{F}$ is said to be <u>filtered</u> ([17], Definition 2.51) if it satisfies the formal versions (in the internal language of $\mathcal{E}$) of the following three statements:

(a) F_0 is nonempty.

(b) Given $U, V \in F_0$, there exists $W \in F_0$ and a diagram

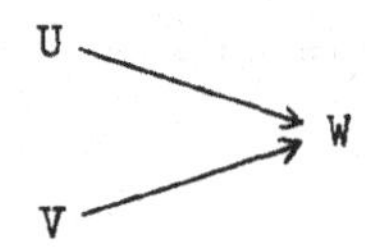

in $\underline{F}$.

(c) Given $U \overset{\alpha}{\underset{\beta}{\rightrightarrows}} V$ in $\underline{F}$, there exists $V \xrightarrow{\gamma} W$ having equal composites with α and β.

Now to any internal presheaf F on a category $\underline{A}$, we can associate a discrete fibration $\underline{F} \xrightarrow{\gamma} \underline{A}$ in $\underset{\sim}{cat}(\mathcal{E})$. (Specifically, F_0 is $\Sigma_{\underset{\sim}{a}} F(\underset{\sim}{a})$, and $\underline{F}(\underset{\sim}{f}_1, \underset{\sim}{f}_2)$ is the subobject of $\underline{A}(\gamma_0 \underset{\sim}{f}_1, \gamma_0 \underset{\sim}{f}_2)$ defined by a certain equalizer.) Now we have

<u>1.3 Proposition</u> A presheaf F on $\underline{A}$ is flat iff the total category $\underline{F}$ of the corresponding discrete fibration $\underline{F} \xrightarrow{\gamma} \underline{A}$ is filtered. Explicitly, we may express this condition in terms of the presheaf F in the following "elementary" form:

(a) $F_0 \longrightarrow 1$ is epi, where $F_0 = \Sigma_{\underset{\sim}{a}} F(\underset{\sim}{a})$.

(b) Given a pair of I-elements $I \overset{f_1}{\underset{f_2}{\rightrightarrows}} F_0$, there exists an epi $J \overset{\varepsilon}{\twoheadrightarrow} I$, a J-element $J \xrightarrow{f_3} F_0$ and elements

$$J \xrightarrow{b_i} \underline{A}(\gamma_0 f_i \varepsilon, \gamma_0 f_3)$$

$(i = 1,2)$ such that $\alpha(b_i, f_3) = f_i \varepsilon : J \longrightarrow F_0$.

(c) Given I-elements $I \overset{f_1}{\underset{f_2}{\rightrightarrows}} F_0$ and $I \overset{b_1}{\underset{b_2}{\rightrightarrows}} \underline{A}(\gamma_0 f_1, \gamma_0 f_2)$

such that $\alpha(b_1,f_2) = f_1 = \alpha(b_2,f_2)$, there exists $J \xrightarrow{\varepsilon} I$,
$J \xrightarrow{f_3} F_0$ and $J \xrightarrow{b_3} \underline{A}(\gamma_0 f_2 \varepsilon, \gamma_0 f_3)$ such that
$\alpha(b_3,f_3) = f_2\varepsilon$ and $c(b_1\varepsilon, b_3) = c(b_2\varepsilon, b_3)$.
(Here α denotes the action of $\underline{A}$ on $F(\underset{\sim}{a})$, and c denotes the composition map of the category $\underline{A}$.) □

CHAPTER II : NATURAL NUMBER OBJECTS

We recall that a natural number object in a topos $\mathcal{E}$ is defined to be an object N equipped with morphisms

$$1 \xrightarrow{o} N \xrightarrow{s} N$$

such that, given any diagram $1 \xrightarrow{x} X \xrightarrow{t} X$ in $\mathcal{E}$, there exists a unique $f\colon N \longrightarrow X$ satisfying $fo = x$ and $fs = tf$. More succinctly, we can say that a natural number object is an initial object in the category of $\mathbb{T}$-models in $\mathcal{E}$, where $\mathbb{T}$ is the free algebraic theory generated by one unary and one nullary operation.

The definition means that morphisms whose domain is N (or, using the exponential adjunction, an object of the form $N \times X$) may be defined "by recursion". In particular, we use recursion to define the arithmetic operations of addition, multiplication and exponentiation as morphisms $N \times N \longrightarrow N$, and (using the uniqueness clause) to prove that they satisfy the usual laws of arithmetic. For example, $N \times N \xrightarrow{+} N$ is defined as the morphism whose exponential transpose satisfies

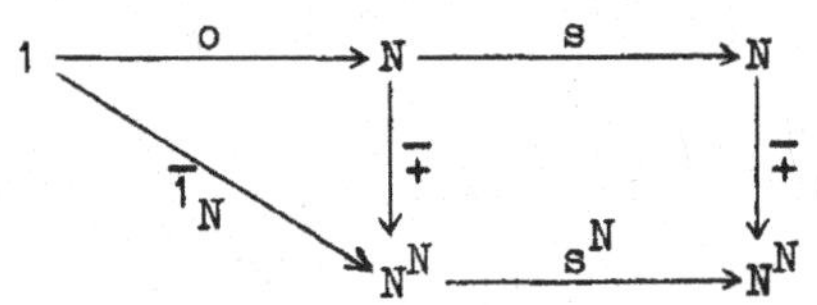

However, in mathematics there is another way in which we frequently use the notion of recursion, namely the construction of N-indexed families of objects - i.e. objects of $\mathbb{C}^N$, where $\mathbb{C}$ is some indexed category over $\mathcal{E}$. For example, suppose we are given an object X of $\mathcal{E}$ and a "process" T for constructing new objects out of old. (We shall loosely refer to this information as "recursion data in $\mathcal{E}$".) Then we wish to find an object $F(\underset{\sim}{n})$ of $\mathcal{E}/N$ which internalizes the notion of the sequence $(X, TX, T^2X, \ldots)$; i.e. such that

$$F(o) = X \quad \text{and} \quad F(s\underset{\sim}{n}) = T^N(F(\underset{\sim}{n})) \quad ,$$

where T^N denotes the process T "applied fibrewise" to objects of $\mathcal{E}/N$.

Now in the topos $\mathcal{S}$, which has no internal structure, we can clearly solve this problem when T is any function; but for a general topos $\mathcal{E}$, it is essential that T should "respect the internal structure of $\mathcal{E}$". To take a very simple example, let $\mathcal{E}$ be the topos $\mathcal{S} \times \mathcal{S}$ of pairs of sets and functions, and let T be the operation defined by $T(X,Y) = (X \sqcup Y, X \sqcup Y)$. A solution to this recursion problem would have to consist of a pair of sequences of sets $((X_p),(Y_q))$ satisfying $X_{p+1} \cong X_p \sqcup Y_q$ for all q, and $Y_{q+1} \cong X_p \sqcup Y_q$ for all p, which is clearly impossible unless $X_o = Y_o = \emptyset$.

Later in this paper, we shall prove an existence theorem (4.8) which tells us that any recursion problem in a suitably-defined class has a solution. For the present, we give a uniqueness theorem which we shall use frequently in the arguments which follow.

__2.1 Lemma__ ("Section Criterion") Suppose given an object $X(\underset{\sim}{n})$ of $\mathcal{E}/N$, together with

(i) an element $x_o: 1 \longrightarrow X(o)$ in $\mathcal{E}$, and

(ii) a morphism $t: X(\underset{\sim}{n}) \longrightarrow X(s\underset{\sim}{n})$ in $\mathcal{E}/N$.

Then there exists a unique element x of $X(\underset{\sim}{n})$ in $\mathcal{E}/N$ such that

$$o^*(x) = x_o \quad \text{and} \quad s^*(x) = tx \ .$$

__Proof__ Write A for $\Sigma_{\underset{\sim}{n}} X(\underset{\sim}{n})$; then an element of $X(\underset{\sim}{n})$ is simply a section of $A \longrightarrow N$ in $\mathcal{E}$. Now consider the diagram

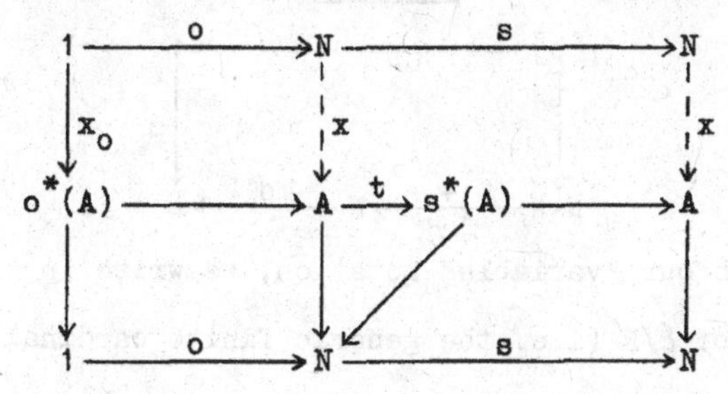

.

Here x exists uniquely by the definition of N, and the composite $N \xrightarrow{x} A \longrightarrow N$ is the identity by the uniqueness of recursively defined morphisms. □

2.2 Theorem ("Uniqueness Theorem") Let $\mathcal{C}$ be a locally internal category over a topos $\mathcal{E}$, and suppose we are given an object X of $\mathcal{C}^1$ and a locally internal functor $T: \mathcal{C} \longrightarrow \mathcal{C}$. Then if there exists an object $F(\underset{\sim}{n})$ of $\mathcal{C}^N$ satisfying

$$F(o) \cong X \quad \text{and} \quad F(s\underset{\sim}{n}) \cong T^N F(\underset{\sim}{n}) \quad ,$$

it is unique up to canonical isomorphism.

Proof Let F, F' be two solutions of the given recursion problem. Since $\mathcal{C}^N$ is enriched over $\mathcal{E}/N$, we can form the object of isomorphisms $\mathrm{Iso}(F(\underset{\sim}{n}),F'(\underset{\sim}{n}))$ in $\mathcal{E}/N$. Now the strength of the functor T^N induces a morphism

$$\mathrm{Iso}(F(\underset{\sim}{n}),F'(\underset{\sim}{n})) \longrightarrow \mathrm{Iso}(T^N F(\underset{\sim}{n}),T^N F'(\underset{\sim}{n})) \cong s^*\mathrm{Iso}(F(\underset{\sim}{n}),F'(\underset{\sim}{n})) \quad .$$

And the identity morphism on X defines an element

$$1 \longrightarrow \mathrm{Iso}(X,X) \cong \mathrm{Iso}(F(o),F'(o)) \cong o^*\mathrm{Iso}(F(\underset{\sim}{n}),F'(\underset{\sim}{n})) \quad .$$

Applying Lemma 2.1 to this data, we obtain a canonical element of $\mathrm{Iso}(F(\underset{\sim}{n}),F'(\underset{\sim}{n}))$, i.e. an isomorphism $F(\underset{\sim}{n}) \xrightarrow{\cong} F'(\underset{\sim}{n})$ in $\mathcal{C}^N$. □

An important application of natural number objects, due originally to J. Bénabou [4], is the definition of a finite cardinal, which provides a notion of "finite object in a topos" with many convenient properties. Given a natural number p in $\mathcal{E}$ (i.e. an element of N), we define its cardinal [p] by the pullback diagram

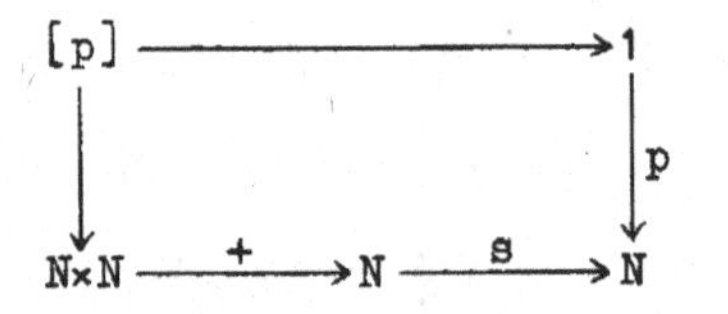

In keeping with our "variable" notation, we write $[\underset{\sim}{n}]$ for the object $(N \times N \xrightarrow{s+} N)$ of $\mathcal{E}/N$ (i.e. the generic finite cardinal). Using the

result of P. Freyd ([12], Proposition 5.11) that

$$1 \xrightarrow{\ o\ } N \xleftarrow{\ s\ } N$$

is a coproduct diagram in $\mathcal{E}$, it is not hard to show that $[\underset{\sim}{n}]$ satisfies the recursion data

$$[o] \cong 0 \quad , \quad [s\underset{\sim}{n}] \cong [\underset{\sim}{n}] \amalg 1 \ .$$

It is then an easy application of the uniqueness theorem to establish the isomorphisms

$$[p+q] \cong [p] \amalg [q] \quad , \quad [pq] \cong [p] \times [q] \quad , \quad [p^q] \cong [p]^{[q]}$$

where p, q are any two natural numbers in $\mathcal{E}$; so we are justified in thinking of [p] as a finite object having exactly p elements.

Further evidence of the "finiteness" of cardinals is provided by the following two results (whose proofs will be found in [17], § 6.2):

<u>2.3 Proposition</u> Let $\mathcal{F} \xrightarrow{f} \mathcal{E}$ be a geometric morphism. If $\mathcal{E}$ has a natural number object, then f^* preserves it. Moreover, we have canonical isomorphisms

$$f^*[p] \cong [f^*p] \quad \text{and} \quad f^*(X^{[p]}) \cong f^*X^{[f^*p]} \quad ,$$

where p is any natural number in $\mathcal{E}$ and X any object of $\mathcal{E}$. □

<u>2.4 Proposition</u> Let p be a natural number in a topos $\mathcal{E}$. Then the functor

$$(-)^{[p]}: \mathcal{E} \longrightarrow \mathcal{E}$$

preserves coequalizers of reflexive pairs. In particular, it preserves epimorphisms (since every epi in $\mathcal{E}$ is the coequalizer of its kernel-pair); we express this fact by saying that [p] is <u>internally projective</u> in $\mathcal{E}$. □

We shall require two further results concerning finite cardinals, both of which are proved in [17]. The first enables us to handle families of finite cardinals indexed by a cardinal; the second summarizes our information about the full subcategory $\mathcal{E}_{fc}$ of $\mathcal{E}$ whose objects are the finite cardinals.

2.5 Proposition Let $X \xrightarrow{f} [p]$ be a morphism of $\mathcal{E}$ whose codomain is a finite cardinal. Then X is isomorphic to a finite cardinal in $\mathcal{E}$ iff f is isomorphic to a finite cardinal in $\mathcal{E}/[p]$. □

2.6 Theorem Let $\mathcal{E}_{fc}$ be the full subcategory of $\mathcal{E}$ whose objects are finite cardinals. Then $\mathcal{E}_{fc}$ is a topos, and satisfies the axiom of choice; and the inclusion functor $\mathcal{E}_{fc} \longrightarrow \mathcal{E}$ preserves finite limits, finite colimits and exponentials. □

In chapter IV, we shall encounter an alternative concept of finiteness in a topos: that of "Kuratowski-finiteness", introduced by A. Kock, P. Lecouturier and C.J. Mikkelsen [21]. This has the advantage that its definition makes sense even in a topos without a natural number object, although we shall be concerned almost exclusively with toposes which do have a natural number object. Given an object X in a topos, we consider its power-object Ω^X as a semilattice under the operation of union of subobjects; and we define the Kuratowski semilattice K(X) to be the sub-semilattice of Ω^X generated by the subobject $X \xrightarrow{\{\}} \Omega^X$. X is said to be Kuratowski-finite if the maximal element $1 \xrightarrow{\ulcorner X \urcorner} \Omega^X$ factors through K(X).

Most of the important properties of Kuratowski-finiteness stem from the fact (first proved by C.J. Mikkelsen [30]) that K(X) is actually the free semilattice generated by X. From this it follows (see Lemma 5.2 below) that the functor K commutes up to isomorphism with the inverse image parts of geometric morphisms, and hence that inverse image functors preserve the property of being Kuratowski-finite. The same idea is at the heart of F.W. Lawvere's proof (cf. [23], p.116) of the following result:

2.7 Theorem Let $\mathcal{E}$ be a topos with a natural number object. Then an object X of $\mathcal{E}$ is Kuratowski-finite iff it is locally a quotient of a finite cardinal, i.e. iff there exists an object I of $\mathcal{E}$ having global support, a natural number p in $\mathcal{E}/I$ and an epimorphism

$$[p] \twoheadrightarrow I^*X \quad . \square$$

In Proposition 2.3 we observed that inverse image functors preserve exponentials when the exponent is a finite cardinal. We conclude this chapter with a pair of extensions of this result concerning objects of epimorphisms and of monomorphisms, which we shall need in chapter IV.

2.8 Proposition Let $\mathcal{F} \xrightarrow{f} \mathcal{E}$ be a geometric morphism, p a natural number in $\mathcal{E}$ and X an object of $\mathcal{E}$. Then

(i) if X is Kuratowski-finite, there is a canonical isomorphism

$$f^*(\mathrm{Epi}([p],X)) \cong \mathrm{Epi}([f^*p],f^*X) \quad .$$

(ii) if X is decidable (i.e. if the diagonal subobject $X \overset{\Delta}{\rightarrowtail} X \times X$ has a complement), there is a canonical isomorphism

$$f^*(\mathrm{Mono}([p],X)) \cong \mathrm{Mono}([f^*p],f^*X) \quad .$$

Proof (i) Normally, the object of epimorphisms is defined by the pullback

$$\begin{array}{ccc} \mathrm{Epi}([p],X) & \longrightarrow & X^{[p]} \\ \downarrow & & \downarrow \mathrm{im} \\ 1 & \xrightarrow{\ulcorner X \urcorner} & \Omega^X \end{array}$$

where im is the morphism which internalizes the operation of forming the image of a morphism $[p] \longrightarrow X$ (cf. [17], Example 5.46). However, the image of any such morphism is Kuratowski-finite by Theorem 2.7, so im factors through the subobject

$$K(X) \rightarrowtail \Omega^X \quad .$$

Thus if X itself is Kuratowski-finite, so that $\ulcorner X \urcorner$ factors through K(X), we may replace Ω^X by K(X) in the above square; and it is then clear that everything in the square is preserved by f^*.

(ii) Similarly, the object of monomorphisms is defined by the pullback

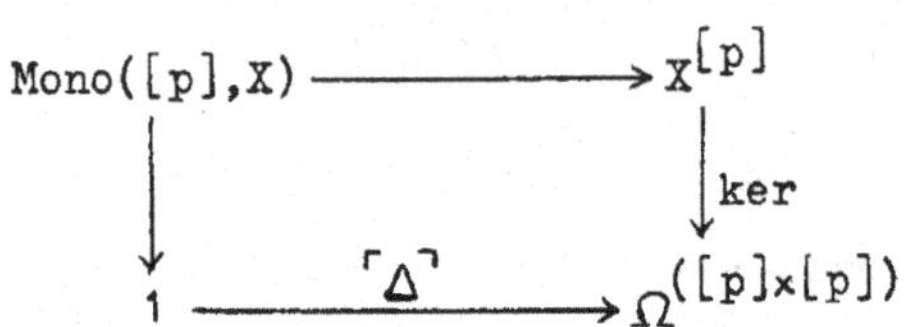

where ker sends a morphism $[p] \longrightarrow X$ to its kernel-pair. But if X is decidable, this kernel-pair is always a complemented subobject of $[p] \times [p] \cong [p^2]$; and since $[p]$ itself is decidable (which follows from the proof of Theorem 2.6), we may replace $\Omega^{([p]\times[p])}$ by the subobject $(1 \amalg 1)^{[p^2]}$ of complemented subobjects of $[p^2]$. The rest of the argument is similar to part (i). □

CHAPTER III : RECURSION FOR PROFUNCTORS

In this chapter our aim is to prove the following theorem, which was suggested to the second author by F.W. Lawvere in 1974:

3.1 Theorem Let $\mathcal{E}$ be a topos with a natural number object. Then the $\mathcal{E}$-topos $\mathcal{E}/N$ is a natural number object (in the sense appropriate to 2-categories) in the 2-category BTop/$\mathcal{E}$ of bounded $\mathcal{E}$-toposes ([17], 4.48); i.e., given a bounded $\mathcal{E}$-topos $\mathcal{F}$ and geometric morphisms $\mathcal{E} \xrightarrow{x} \mathcal{F} \xrightarrow{t} \mathcal{F}$ over $\mathcal{E}$, there exists (uniquely up to canonical isomorphism) a morphism $\mathcal{E}/N \xrightarrow{f} \mathcal{F}$ such that

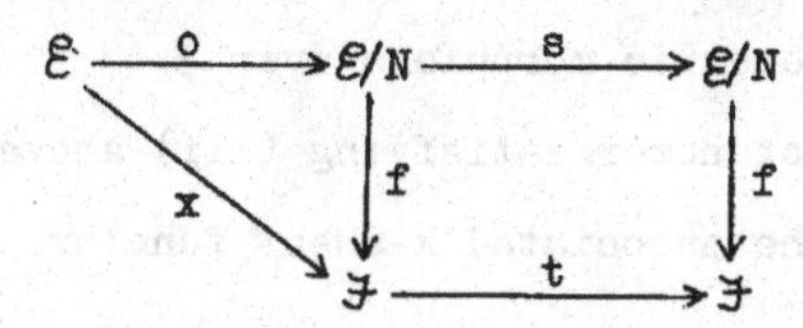

commutes up to isomorphism.

The key to proving this theorem is a result which enables us to describe geometric morphisms between bounded $\mathcal{E}$-toposes in terms of profunctors. Since the precise form of this result is rather complicated, we shall state it with what the reader may consider to be excessive care.

Let $\mathcal{F} = sh_j(\mathcal{E}^{\underline{A}})$ and $\mathcal{G} = sh_k(\mathcal{E}^{\underline{B}})$ be two bounded $\mathcal{E}$-toposes. We shall consider profunctors $\underline{A} \overset{T}{- - \rightarrow} \underline{B}$ which satisfy the following conditions:

(i) Given any finite diagram in $\mathcal{E}^{\underline{A}}$ with vertices H_α, the canonical map $(\varprojlim_\alpha H_\alpha)\otimes_{\underline{A}} T \longrightarrow \varprojlim_\alpha(H_\alpha \otimes_{\underline{A}} T)$ is k-bidense ("T is left k-almost flat").

(ii) If $H \longrightarrow K$ is a j-bidense map in $\mathcal{E}^{\underline{A}}$, then $H\otimes_{\underline{A}}T \longrightarrow K\otimes_{\underline{A}}T$ is k-bidense. ("T is k-almost j-continuous").

Since a composite of k-bidense morphisms is k-bidense, it is easily seen that (i) and (ii) may be combined into the single condition

(iii) If $H \longrightarrow \varprojlim_\alpha H_\alpha$ is a j-bidense map in $\mathcal{E}^{\underline{A}}$, then the induced map $H \otimes_{\underline{A}} T \longrightarrow \varprojlim_\alpha (H_\alpha \otimes_{\underline{A}} T)$ is k-bidense in $\mathcal{E}^{\underline{B}}$.

We shall write $\mathrm{Prof}_{j,k}(\underline{A},\underline{B})$ for the full subcategory of $\mathrm{Prof}(\underline{A},\underline{B})$ whose objects satisfy (iii).

<u>3.2 Proposition</u> Let $\mathcal{F}$ and $\mathcal{G}$ be as above. Then there exist functors

$$\underline{\underline{\mathrm{Top}}}/\mathcal{E}\ (\mathcal{G},\mathcal{F}) \xrightarrow{F} \mathrm{Prof}_{j,k}(\underline{A},\underline{B}) \xrightarrow{G} \underline{\underline{\mathrm{Top}}}/\mathcal{E}\ (\mathcal{G},\mathcal{F})$$

such that the composite GF is naturally isomorphic to the identity, and such that G sends tensor product of profunctors to composition of geometric morphisms over $\mathcal{E}$.

<u>Proof</u> Let T be a profunctor satisfying (iii) above, and let $L_k\colon \mathcal{E}^{\underline{B}} \longrightarrow \mathcal{G}$ be the associated k-sheaf functor. Then the composite

$$\mathcal{E}^{\underline{A}} \xrightarrow{(-)\otimes T} \mathcal{E}^{\underline{B}} \xrightarrow{L_k} \mathcal{G}$$

has a right adjoint and is left exact, so it defines a geometric morphism $\mathcal{G} \longrightarrow \mathcal{E}^{\underline{A}}$. But this morphism factors through the inclusion $\mathcal{F} \longrightarrow \mathcal{E}^{\underline{A}}$, since $L_k((-)\otimes T)$ inverts j-bidense morphisms; so we have a morphism $\mathcal{G} \xrightarrow{G(T)} \mathcal{F}$, which is clearly functorial in T.

Conversely, suppose given $\mathcal{G} \xrightarrow{g} \mathcal{F}$ over $\mathcal{E}$. Then the composite $\mathcal{G} \xrightarrow{g} \mathcal{F} \longrightarrow \mathcal{E}^{\underline{A}}$ corresponds (via Diaconescu's theorem) to a flat presheaf on $L_k\underline{B}^*(\underline{A})$ in $\mathcal{G} = \mathrm{sh}_k(\mathcal{E}^{\underline{B}})$. Pulling back this presheaf along the universal map $\underline{B}^*(\underline{A}) \longrightarrow L_k\underline{B}^*(\underline{A})$ in $\mathcal{E}^{\underline{B}}$, we obtain a presheaf on $\underline{B}^*(\underline{A})$, or equivalently a profunctor $\underline{A} \xrightarrow{F(g)} \underline{B}$ in $\mathcal{E}$. Since $X \longrightarrow L_k(X)$ is k-bidense for any X in $\mathcal{E}^{\underline{B}}$, it is now straightforward to verify that F(g) satisfies conditions (i) and (ii), and that it induces the morphism g in the manner described above, i.e. $GF(g) \cong g$.

The last part of the statement is immediate from the

associativity of tensor product. □

It is important to realise that if the topology k is nontrivial, then the functors F and G of Proposition 3.2 do <u>not</u> give an equivalence of categories. This may easily be seen by taking $\underline{A} = \underline{B} = \underline{1}$ and j = k: then there is (up to unique isomorphism) only one geometric morphism $\mathcal{G} \longrightarrow \mathcal{F}$ over $\mathcal{E}$, but $\mathrm{Prof}(\underline{1},\underline{1})$ is simply $\mathcal{E}$, and an object X of $\mathcal{E}$ is in $\mathrm{Prof}_{j,j}(\underline{1},\underline{1})$ iff $X \longrightarrow 1$ is j-bidense. This is the reason why we had to take particular care in stating Proposition 3.2. However, if k is the trivial topology (so that $\mathcal{G} = \mathcal{E}^{\underline{B}}$) then it follows from Diaconescu's theorem that F and G do define an equivalence.

Proposition 3.2 reduces the problem of constructing the "$\underset{\sim}{n}^{th}$ iterate" of a morphism $\mathcal{F} \xrightarrow{t} \mathcal{F}$ over $\mathcal{E}$, where $\mathcal{F} = \mathrm{sh}_j(\mathcal{E}^{\underline{A}})$, to that of constructing the "$\underset{\sim}{n}^{th}$ tensor power" of a profunctor $\underline{A} \overset{T}{-\!-\!\rightarrow} \underline{A}$. That is to say, we want to construct a profunctor $T^{\otimes\underset{\sim}{n}}$ from $\underline{A}$ to $\underline{A}$ in $\mathcal{E}/N$ such that $T^{\otimes o} \cong Y(\underline{A})$ and $T^{\otimes s\underset{\sim}{n}} \cong T^{\otimes\underset{\sim}{n}} \otimes_{\underline{A}} T \cong T \otimes_{\underline{A}} T^{\otimes\underset{\sim}{n}}$. As a first stage, we shall consider the case when $\underline{A}$ is a discrete category with object-of-objects I: in this case we may identify $\mathrm{Prof}(\underline{A},\underline{A})$ with $\mathcal{E}/I\times I$, and tensor product over $\underline{A}$ corresponds to the bifunctor $\wedge_I$ defined by

$$(X \wedge_I Y)(\underset{\sim}{i}_1, \underset{\sim}{i}_2) = \Sigma_{\underset{\sim}{i}_3} (X(\underset{\sim}{i}_1, \underset{\sim}{i}_3) \times Y(\underset{\sim}{i}_3, \underset{\sim}{i}_2)) \quad .$$

Similarly, the unit profunctor $Y(\underline{A})$ is identified with the diagonal map $I \xrightarrow{\Delta} I \times I$.

We shall now construct what is in effect a free-monoid functor for the monoidal category $(\mathcal{E}/I\times I, \wedge_I, \Delta)$. Specifically, we shall construct a functor

$$(-)^{\langle\underset{\sim}{n}\rangle} : \mathcal{E}/I\times I \longrightarrow \mathcal{E}/I\times I\times N$$

such that $X^{\langle o\rangle} \cong \Delta$ and $X^{\langle s\underset{\sim}{n}\rangle} \cong X^{\langle\underset{\sim}{n}\rangle} \wedge_I X = X \wedge_I X^{\langle\underset{\sim}{n}\rangle}$ for all X in $\mathcal{E}/I\times I$.

Let p be a natural number in $\mathcal{E}$. We saw in chapter II that there is a coproduct decomposition $[sp] \cong [p] \amalg 1$; but in fact we can form two different decompositions of this type, by regarding the factor 1 as either the "first" or the "last" element of $[sp]$. More explicitly, we have diagrams

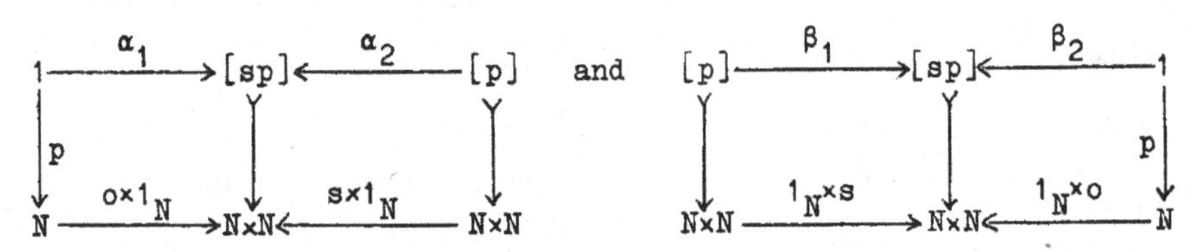

in which the squares are pullbacks and the bottom rows are coproducts by [12], Proposition 5.11, so that the top rows are also coproducts.

3.3 Lemma With the morphisms α_i and β_i defined as above, the following three diagrams commute:

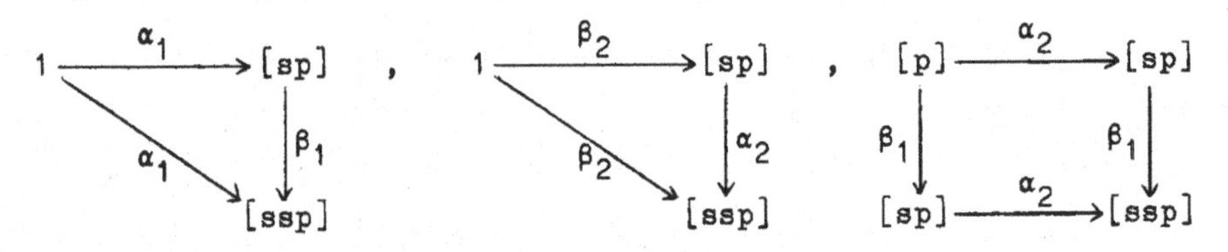

Proof In each case, the commutativity becomes trivial when we compose with the canonical monomorphism $[ssp] \rightarrowtail N \times N$ arising from the definition of finite cardinals. □

Since $1 \xrightarrow{o} N \xleftarrow{s} N$ is a coproduct diagram, it is clear that an object $X(\underset{\sim}{n})$ of $\mathcal{E}/N$ may be defined by specifying separately the two objects $X(o)$ and $X(s\underset{\sim}{n})$. We shall now make use of this fact in defining the object $X^{\langle \underset{\sim}{n} \rangle}$, where $X(\underset{\sim}{i}_1, \underset{\sim}{i}_2)$ is an object of $\mathcal{E}/I \times I$. First, we define $X^{\langle o \rangle}$ to be the unit object $I \xrightarrow{\Delta} I \times I$. Now let $S = \Sigma_{\underset{\sim}{i}_1, \underset{\sim}{i}_2} X(\underset{\sim}{i}_1, \underset{\sim}{i}_2)$, so that we can write X as

$$S \xrightarrow{(p_1, p_2)} I \times I \quad .$$

Then we define $\Sigma_{\underset{\sim}{i}_1, \underset{\sim}{i}_2} (X(\underset{\sim}{i}_1, \underset{\sim}{i}_2)^{\langle s\underset{\sim}{n} \rangle})$ to be the equalizer in $\mathcal{E}/N$ of

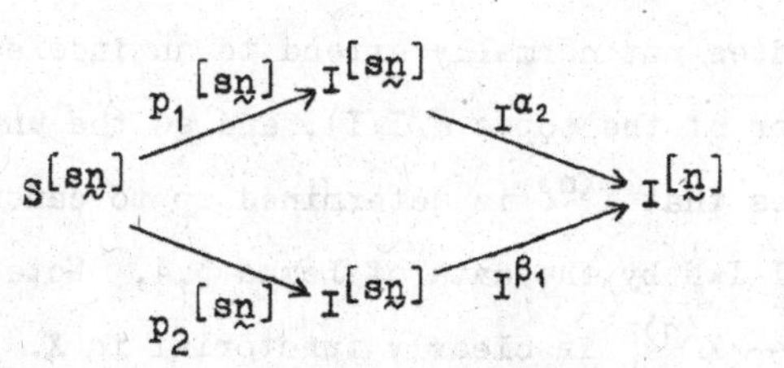

with projection to $I \times I$ induced by the maps

$$S^{[s\underset{\sim}{n}]} \xrightarrow{p_1^{[s\underset{\sim}{n}]}} I^{[s\underset{\sim}{n}]} \xrightarrow{I^{\alpha_1}} I \quad \text{and} \quad S^{[s\underset{\sim}{n}]} \xrightarrow{p_2^{[s\underset{\sim}{n}]}} I^{[s\underset{\sim}{n}]} \xrightarrow{I^{\beta_2}} I.$$

In terms of our informal language, we could equivalently define $X^{\langle so \rangle} = X$ and

$$X^{\langle ss\underset{\sim}{n} \rangle}(\underset{\sim}{i}_1, \underset{\sim}{i}_2) = \Sigma_{\underset{\sim}{u}}(X(\underset{\sim}{i}_1, \underset{\sim}{u}(o)) \times \Pi_{\underset{\sim}{j}} X(\underset{\sim}{u}(\underset{\sim}{j}), \underset{\sim}{u}(s\underset{\sim}{j})) \times X(\underset{\sim}{u}(\underset{\sim}{n}), \underset{\sim}{i}_2)) \quad ,$$

where $\underset{\sim}{u}$ is a variable of type $I^{[s\underset{\sim}{n}]}$, and $\underset{\sim}{j}$ is of type $[\underset{\sim}{n}]$.

<u>3.4 Lemma</u> The object $X^{\langle \underset{\sim}{n} \rangle}$ defined above satisfies the recursion data $X^{\langle o \rangle} = \Delta$ and $X^{\langle s\underset{\sim}{n} \rangle} \cong X^{\langle \underset{\sim}{n} \rangle} \wedge_I X \cong X \wedge_I X^{\langle \underset{\sim}{n} \rangle}$.

<u>Proof</u> The first isomorphism is part of the definition. To prove the second, observe first that $\Sigma_{\underset{\sim}{i}_1, \underset{\sim}{i}_2} X^{\langle so \rangle}(\underset{\sim}{i}_1, \underset{\sim}{i}_2)$ is the equalizer in $\mathcal{E}$ of a pair of maps $S \cong S^{[so]} \rightrightarrows I^{[o]} \cong 1$, and hence is isomorphic to S. From this it follows at once that $X^{\langle so \rangle} \cong X \cong \Delta \wedge_I X \cong X \wedge_I \Delta$ in $\mathcal{E}/I \times I$. Now we may use Lemma 3.3 and the remarks before it to express $\Sigma_{\underset{\sim}{i}_1, \underset{\sim}{i}_2} X^{\langle ss\underset{\sim}{n} \rangle}(\underset{\sim}{i}_1, \underset{\sim}{i}_2)$ as the equalizer of

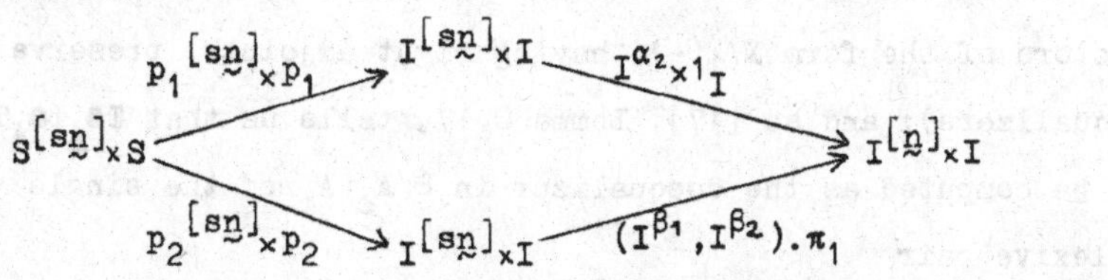

which is easily seen to be $\Sigma_{\underset{\sim}{i}_1, \underset{\sim}{i}_2}(X^{\langle s\underset{\sim}{n} \rangle} \wedge_I X)(\underset{\sim}{i}_1, \underset{\sim}{i}_2)$. Thus we obtain an isomorphism $X^{\langle ss\underset{\sim}{n} \rangle} \cong X^{\langle s\underset{\sim}{n} \rangle} \wedge_I X$ in $\mathcal{E}/I \times I \times N$; the proof that $X^{\langle ss\underset{\sim}{n} \rangle} \cong X \wedge_I X^{\langle s\underset{\sim}{n} \rangle}$ is similar. □

Note that the functor $(-)\wedge_I X$ is indexed over $\mathcal{E}$ in a natural way (although it does not normally extend to an indexed, or even a strong, endofunctor of the topos $\mathcal{E}/I\times I$), and so the uniqueness theorem 2.2 ensures that $X^{\langle \underset{\sim}{n} \rangle}$ is determined up to canonical isomorphism in $\mathcal{E}/I\times I\times N$ by the data of Lemma 3.4. Note also that the construction $X \longmapsto X^{\langle \underset{\sim}{n} \rangle}$ is clearly functorial in X.

Now let us return to the case of a general category $\underline{A}$ and a profunctor $\underline{A} \overset{T}{\dashrightarrow} \underline{A}$. By definition, $T\otimes_{\underline{A}}T(\underset{\sim}{a}_1,\underset{\sim}{a}_2)$ is the coequalizer in $\mathcal{E}/A_0\times A_0$ of the reflexive pair

$$T\wedge_{A_0}\underline{A}\wedge_{A_0}T\ (\underset{\sim}{a}_1,\underset{\sim}{a}_2) \underset{\beta\wedge 1}{\overset{1\wedge\alpha}{\rightrightarrows}} T\wedge_{A_0}T\ (\underset{\sim}{a}_1,\underset{\sim}{a}_2) \quad ,$$

where α and β denote the left and right actions of $\underline{A}$ on $T(\underset{\sim}{a}_1,\underset{\sim}{a}_2)$. Now consider the diagram

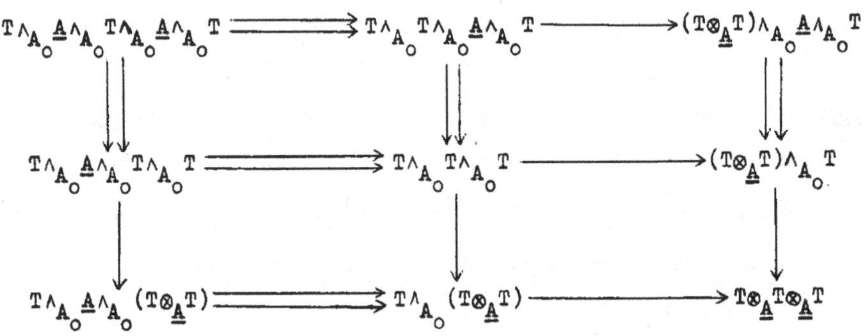

which is used in proving the associativity of tensor product. The rows and columns of this diagram are reflexive coequalizers (since functors of the form $X\wedge_I(-)$, having right adjoints, preserve coequalizers); and so [17], Lemma 0.17, tells us that $T\otimes_{\underline{A}}T\otimes_{\underline{A}}T\ (\underset{\sim}{a}_1,\underset{\sim}{a}_2)$ may be computed as the coequalizer in $\mathcal{E}/A_0\times A_0$ of the single reflexive pair

$$T\wedge_{A_0}\underline{A}\wedge_{A_0}T\wedge_{A_0}\underline{A}\wedge_{A_0}T\ (\underset{\sim}{a}_1,\underset{\sim}{a}_2) \underset{\beta\wedge\beta\wedge 1}{\overset{1\wedge\alpha\wedge\alpha}{\rightrightarrows}} T\wedge_{A_0}T\wedge_{A_0}T\ (\underset{\sim}{a}_1,\underset{\sim}{a}_2) \quad .$$

This provides the motivation for the construction used in the next proposition.

<u>3.5 Proposition</u> Let $\mathcal{E}$ be a topos with a natural number object, and

$\underline{A}$ an internal category in $\mathcal{E}$. Then there exists a functor

$$(-)^{\otimes \underset{\sim}{n}}: \text{Prof}(\underline{A},\underline{A}) \longrightarrow \text{Prof}(\underline{A},\underline{A})/N$$

satisfying the recursion data

$$T^{\otimes o} = Y(\underline{A}) \quad , \quad T^{\otimes s\underset{\sim}{n}} \cong T^{\otimes \underset{\sim}{n}} \otimes_{\underline{A}} T \cong T \otimes_{\underline{A}} T^{\otimes \underset{\sim}{n}}$$

for all profunctors T.

Proof   As before, we define $T^{\otimes \underset{\sim}{n}}$ in two stages. First we set $T^{\otimes o} = Y(\underline{A})$; then we define $T^{\otimes s\underset{\sim}{n}}(\underset{\sim}{a}_1, \underset{\sim}{a}_2)$ to be the coequalizer in $\mathcal{E}/A_o \times A_o \times N$ of

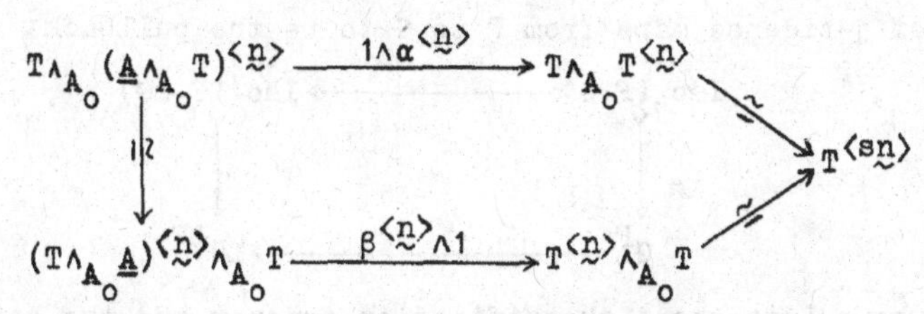

$$\begin{array}{ccc} T \wedge_{A_o} (\underline{A} \wedge_{A_o} T)^{\langle \underset{\sim}{n} \rangle} & \xrightarrow{1 \wedge \alpha^{\langle \underset{\sim}{n} \rangle}} & T \wedge_{A_o} T^{\langle \underset{\sim}{n} \rangle} \searrow^{\wr} \\ \downarrow \wr & & \qquad T^{\langle s\underset{\sim}{n} \rangle} \\ (T \wedge_{A_o} \underline{A})^{\langle \underset{\sim}{n} \rangle} \wedge_{A_o} T & \xrightarrow{\beta^{\langle \underset{\sim}{n} \rangle} \wedge 1} & T^{\langle \underset{\sim}{n} \rangle} \wedge_{A_o} T \nearrow_{\wr} \end{array}$$

where the vertical arrow on the left is the unique isomorphism arising from the fact that $T \wedge_{A_o} (\underline{A} \wedge_{A_o} T)^{\langle \underset{\sim}{n} \rangle}$ and $(T \wedge_{A_o} \underline{A})^{\langle \underset{\sim}{n} \rangle} \wedge_{A_o} T$ both satisfy the recursion data $F(o) \cong T$, $F(s\underset{\sim}{n}) \cong F(\underset{\sim}{n}) \wedge_{A_o} \underline{A} \wedge_{A_o} T$. The left and right actions of $\underline{A}$ on $T^{\otimes \underset{\sim}{n}}(\underset{\sim}{a}_1, \underset{\sim}{a}_2)$ are induced in the obvious way by α and β.

The proof that $T^{\otimes \underset{\sim}{n}}$ satisfies the given recursion data is similar to that of 3.4. First we note that

$$T^{\otimes so} \cong \text{coeq}(T \overset{1}{\underset{1}{\rightrightarrows}} T) \cong T \cong Y(\underline{A}) \otimes_{\underline{A}} T \cong T \otimes_{\underline{A}} Y(\underline{A}).$$

Then we use Lemma 3.4 to express $T^{\otimes ss\underset{\sim}{n}}(\underset{\sim}{a}_1, \underset{\sim}{a}_2)$ as the coequalizer of

$$T \wedge_{A_o} \underline{A} \wedge_{A_o} T \wedge_{A_o} (\underline{A} \wedge_{A_o} T)^{\langle \underset{\sim}{n} \rangle} \underset{\beta \wedge \beta^{\langle \underset{\sim}{n} \rangle} \wedge 1}{\overset{1 \wedge \alpha \wedge \alpha^{\langle \underset{\sim}{n} \rangle}}{\rightrightarrows}} T^{\langle ss\underset{\sim}{n} \rangle} \quad ,$$

and [17], Lemma 0.17, to show that this coequalizer is isomorphic to $T \otimes_{\underline{A}} T^{\otimes s\underset{\sim}{n}}(\underset{\sim}{a}_1, \underset{\sim}{a}_2)$. The remaining details are straightforward. □

Note that objects of $\text{Prof}(\underline{A},\underline{A})/N$ may be regarded equivalently as profunctors $\underline{A} - - \to \underline{A}$ in $\mathcal{E}/N$ or as profunctors $\underline{A} \times N - - \to \underline{A}$ or $\underline{A} - - \to \underline{A} \times N$ in $\mathcal{E}$, where we regard N as a discrete internal category.

Now let us consider a map of $\mathcal{E}$-toposes

$$t\colon \mathrm{sh}_j(\mathcal{E}^{\underline{A}}) \longrightarrow \mathrm{sh}_j(\mathcal{E}^{\underline{A}}) \quad .$$

By Proposition 3.2, we can regard this as being induced by a profunctor $\underline{A} \overset{T}{\dashrightarrow} \underline{A}$ satisfying condition (iii). And since $\mathcal{E}^{\underline{A}\times N} \simeq (\mathcal{E}^{\underline{A}})/N$, j defines a unique topology j' in $\mathcal{E}^{\underline{A}\times N}$ such that $\mathrm{sh}_{j'}(\mathcal{E}^{\underline{A}\times N}) \simeq (\mathrm{sh}_j(\mathcal{E}^{\underline{A}}))/N$.

3.6 Lemma If T is in $\mathrm{Prof}_{j,j}(\underline{A},\underline{A})$, then $T^{\otimes \underline{n}}$ is in $\mathrm{Prof}_{j,j'}(\underline{A},\underline{A}\times N)$.

Proof If F and G are internal diagrams on $\underline{A}$, we define the object of j-bidense maps from F to G to be the pullback

where L is the j-sheafification functor and the bottom arrow is the strength of L, i.e. the transpose of the composite

$$G^F \times LF \xrightarrow{\eta\times 1} L(G^F)\times LF \xrightarrow{\simeq} L(G^F\times F) \xrightarrow{L(ev)} LG \ .$$

If $a\colon F \longrightarrow G$ is named by the element $1 \xrightarrow{\ulcorner a \urcorner} G^F$, then a is j-bidense iff $\ulcorner a \urcorner$ factors through $\mathrm{Iso}_j(F,G)$.

Now $(-)\otimes_{\underline{A}}T\colon \mathcal{E}^{\underline{A}} \longrightarrow \mathcal{E}^{\underline{A}}$ is a locally internal functor over $\mathcal{E}$, and condition (ii) of Proposition 3.2 implies that its strength restricts to a morphism

$$\underline{A}_*\mathrm{Iso}_j(F,G) \longrightarrow \underline{A}_*\mathrm{Iso}_j(F\otimes_{\underline{A}}T, G\otimes_{\underline{A}}T)$$

where $\underline{A}_*$ is the direct image of the canonical geometric morphism $\mathcal{E}^{\underline{A}} \longrightarrow \mathcal{E}$, i.e. the functor $\varprojlim_{\underline{A}}$.

Suppose we are given a finite diagram in $\mathcal{E}^{\underline{A}}$ with vertices H_α, and a j-bidense map $H \longrightarrow \varprojlim_\alpha H_\alpha$. We wish to show that the induced map $H\otimes_{\underline{A}}T^{\otimes \underline{n}} \longrightarrow \lim_\alpha (H_\alpha \otimes_{\underline{A}} T^{\otimes \underline{n}})$ is j'-bidense, so we apply the section criterion 2.1 to the object

$$\underline{A}_*\mathrm{Iso}_{j'}(H\otimes_{\underline{A}}T^{\otimes \underline{n}}, \varprojlim_\alpha (H_\alpha \otimes_{\underline{A}} T^{\otimes \underline{n}}))$$

of $\mathcal{E}/N$. For n = o, we have the transpose

$$1 \longrightarrow \underline{A}_* \mathrm{Iso}_j(H, \varprojlim_\alpha H_\alpha)$$

of the given j-bidense map $H \longrightarrow \varprojlim_\alpha H_\alpha$. For the inductive step, we have a map

$$\begin{array}{c} \underline{A}_* \mathrm{Iso}_{j'}(H \otimes_{\underline{A}} T^{\otimes \underset{\sim}{n}}, \varprojlim_\alpha (H_\alpha \otimes_{\underline{A}} T^{\otimes \underset{\sim}{n}})) \\ \big\downarrow \\ \underline{A}_* \mathrm{Iso}_{j'}(H \otimes_{\underline{A}} T^{\otimes s \underset{\sim}{n}}, \varprojlim_\alpha (H_\alpha \otimes_{\underline{A}} T^{\otimes \underset{\sim}{n}}) \otimes_{\underline{A}} T) \\ \big\downarrow \\ \underline{A}_* \mathrm{Iso}_{j'}(H \otimes_{\underline{A}} T^{\otimes s \underset{\sim}{n}}, \varprojlim_\alpha (H_\alpha \otimes_{\underline{A}} T^{\otimes s \underset{\sim}{n}})) \end{array}$$

of which the first part is induced by the strength of $(-)\otimes_{\underline{A}} T$, and the second by composition with the j-bidense map

$$\varprojlim_\alpha (H_\alpha \otimes_{\underline{A}} T^{\otimes \underset{\sim}{n}}) \otimes_{\underline{A}} T \longrightarrow \lim_\alpha (H_\alpha \otimes_{\underline{A}} T^{\otimes \underset{\sim}{n}} \otimes_{\underline{A}} T)$$

whose existence is guaranteed by condition (i) of Proposition 3.2. Thus we obtain an element

$$1 \longrightarrow \underline{A}_* \mathrm{Iso}_{j'}(H \otimes_{\underline{A}} T^{\otimes \underset{\sim}{n}}, \varprojlim_\alpha (H_\alpha \otimes_{\underline{A}} T^{\otimes \underset{\sim}{n}}))$$

which is clearly the name of the canonical map

$$H \otimes_{\underline{A}} T^{\otimes \underset{\sim}{n}} \longrightarrow \varprojlim_\alpha (H_\alpha \otimes_{\underline{A}} T^{\otimes \underset{\sim}{n}}) .$$

So the latter map is j'-bidense. □

<u>Proof of Theorem 3.1</u> Suppose we are given a diagram

$$\mathcal{E} \xrightarrow{\ x\ } \mathcal{F} \xrightarrow{\ t\ } \mathcal{F}$$

in $\underline{\underline{\mathrm{BTop}}}/\mathcal{E}$, where $\mathcal{F}$ is the bounded $\mathcal{E}$-topos $\mathrm{sh}_j(\mathcal{E}^{\underline{A}})$. By Proposition 3.2, t is induced by a suitable profunctor $\underline{A} \overset{T}{- - \rightarrow} \underline{A}$, and then Lemma 3.6 tells us that $T^{\otimes \underset{\sim}{n}}$ induces a map of $\mathcal{E}$-toposes

$$\mathcal{F}/N \xrightarrow{\ t^{\underset{\sim}{n}}\ } \mathcal{F} .$$

Moreover, the isomorphism $T^{\otimes s \underset{\sim}{n}} \cong T \otimes_{\underline{A}} T^{\otimes \underset{\sim}{n}}$ implies that

$$\begin{array}{ccc} \mathcal{F}/N & \xrightarrow{\ s\ } & \mathcal{F}/N \\ \big\downarrow t^{\underset{\sim}{n}} & & \big\downarrow t^{\underset{\sim}{n}} \\ \mathcal{F} & \xrightarrow{\ t\ } & \mathcal{F} \end{array}$$

commutes up to isomorphism; and similarly the composite

$$\mathcal{F} \xrightarrow{\ o\ } \mathcal{F}/N \xrightarrow{\ t^{n}_{\sim}\ } \mathcal{F}$$

is isomorphic to the identity, since $T^{\otimes o} = Y(\underline{A})$.

Now since

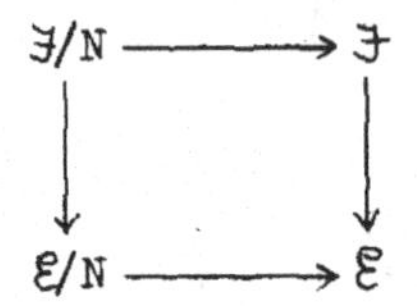

is a pullback in $\underline{\underline{\mathrm{BTop}}}/\mathcal{E}$, x induces a unique map of $(\mathcal{E}/N)$-toposes

$$\mathcal{E}/N \xrightarrow{\ x/N\ } \mathcal{F}/N \ .$$

It is now easy to verify that the composite

$$f:\ \mathcal{E}/N \xrightarrow{\ x/N\ } \mathcal{F}/N \xrightarrow{\ t^{n}_{\sim}\ } \mathcal{F}$$

makes the diagram

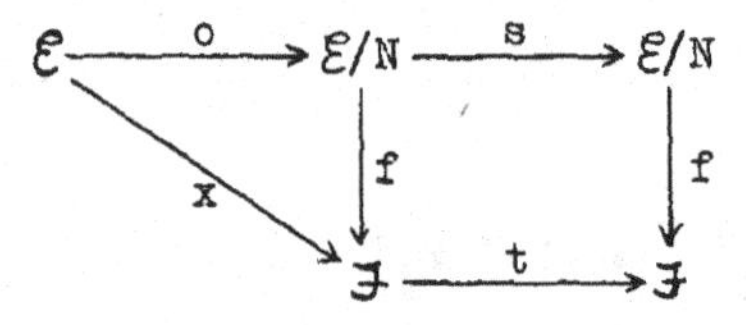

commute up to isomorphism, so the existence part of the theorem is proved.

The uniqueness part is once again an application of Theorem 2.2: we simply have to observe that the assignment

$$I \longmapsto \underline{\underline{\mathrm{Top}}}/\mathcal{E}\ (\mathcal{E}/I, \mathrm{sh}_j(\mathcal{E}^{\underline{A}}))$$

defines a locally internal category over $\mathcal{E}$ (in fact a full subcategory of the $\mathcal{E}$-topos $\mathcal{E}^{\underline{A}^{op}}$, by Diaconescu's theorem), and that the operation of composing with t induces a locally internal endofunctor of this category. □

CHAPTER IV : THE OBJECT CLASSIFIER

An object classifier for a topos $\mathcal{E}$ is defined to be an $\mathcal{E}$-topos $\mathcal{E}[U]$, together with an object U of $\mathcal{E}[U]$, such that for any $\mathcal{E}$-topos $\mathcal{F}$ the functor

$$\underline{\text{Top}}/\mathcal{E}\ (\mathcal{F},\mathcal{E}[U]) \longrightarrow \mathcal{F}\ ;\ f \longmapsto f^*(U)$$

is an equivalence of categories. In other words, each object X of $\mathcal{F}$ may be represented (uniquely up to canonical isomorphism) by a classifying map $\tilde{X}$: $\mathcal{F} \longrightarrow \mathcal{E}[U]$ over $\mathcal{E}$, and the assignment $X \longmapsto \tilde{X}$ is functorial. Equivalently again, an object classifier is a (pseudo-) representation of the forgetful 2-functor

$$(\underline{\text{Top}}/\mathcal{E})^{op} \longrightarrow \underline{\underline{\text{Cat}}}$$

which sends an $\mathcal{E}$-topos to its underlying category, and a map of $\mathcal{E}$-toposes to its inverse image.

It follows from the latter description that any two object classifiers for $\mathcal{E}$ are equivalent, and that any two such equivalences are uniquely naturally isomorphic. We shall indulge in the usual abuse of language, and talk of the object classifier $\mathcal{E}[U]$, when it exists. The object U will be referred to as the generic object.

We have purposely adopted a notation similar to that for polynomials, though it is unfortunate for this analogy that our arrows go in the wrong direction. That is to say, by common convention we think of a geometric morphism as pointing in the same direction as its direct image part, to emphasise the geometrical origins of topos theory; whereas for the algebraic aspects of the subject, the more important direction is that of the inverse image part.

The notion of object classifier makes precise an appealing but hitherto vague way of looking at things: if X is a topological space, one would like to say that a sheaf F on X is a continuous function

$x \longmapsto F_x$ from X to the "space of all sets". If one generalizes "space" to "$\mathcal{E}$-topos" and "continuous function" to "map of $\mathcal{E}$-toposes", then of course the "space of all sets" is simply the object classifier.

For every object T of $\mathcal{E}[U]$, we have its classifying map

$$\tilde{T}: \mathcal{E}[U] \longrightarrow \mathcal{E}[U]$$

and hence a natural endomorphism of the above mentioned forgetful 2-functor. That is to say, for each $\mathcal{E}$-topos $\mathcal{F}$ we have a functor

$$T_{\mathcal{F}}: \mathcal{F} \longrightarrow \mathcal{F}$$

commuting with inverse image parts of maps of $\mathcal{E}$-toposes up to coherent natural isomorphisms; i.e. for any geometric morphism $g: \mathcal{G} \longrightarrow \mathcal{F}$ over $\mathcal{E}$, we have $T_{\mathcal{G}}.g^* \cong g^*.T_{\mathcal{F}}$. Each of the three notions T, $\tilde{T}$ and $\{T_{\mathcal{F}}\}_{\mathcal{F}}$ determines the other two up to canonical isomorphism; we have $T \cong \tilde{T}^*(U) \cong T_{\mathcal{E}[U]}(U)$ and $T_{\mathcal{F}}(X) = \tilde{X}^*(T)$, so that $\widetilde{T_{\mathcal{F}}(X)} \cong \tilde{T}.\tilde{X}$.

Since pullback functors are a special case of inverse image functors, the functors $\{T_{\mathcal{F}/I}\}_I$ define an indexed endofunctor of any $\mathcal{E}$-topos $\mathcal{F}$. In particular, it follows from Lemma 1.1 that each $T_{\mathcal{F}}$ is an $\mathcal{F}$-functor.

If S, T are two objects of $\mathcal{E}[U]$, we define $S \otimes T$ by the formula

$$S \otimes T = \tilde{T}^*(S) \cong S_{\mathcal{E}[U]}(T).$$

It follows that $\widetilde{S \otimes T} \cong \tilde{S}.\tilde{T}$ and that $(S \otimes T)_{\mathcal{F}} \cong S_{\mathcal{F}}.T_{\mathcal{F}}$, from which we deduce that $\otimes$ is a bifunctor on $\mathcal{E}[U]$ which is associative up to coherent natural isomorphism, and for which U is a two-sided unit up to coherent natural isomorphisms. In this way we obtain a monoidal category $(\mathcal{E}[U], \otimes, U)$. This monoidal structure is not normally symmetric, but it is "closed on the right", i.e. for each T the functor $(-) \otimes T = \tilde{T}^*$ has a right adjoint $\tilde{T}_*$.

The above considerations are all consequences of the definition of an object classifier. Now we turn our attention to the question of existence. Recall that if $\mathcal{E}$ has a natural number object, then the generic finite cardinal $[\underset{\sim}{n}]$ in $\mathcal{E}/N$ is given by the map $N \times N \xrightarrow{s+} N$. We shall write $\underline{\underline{E}}_{fin}$ for the internal full subcategory $\mathrm{Full}_{\mathcal{E}}([\underset{\sim}{n}])$. Note that $\underline{\underline{E}}_{fin}$ "internalizes" the category $\mathcal{E}_{fc}$ of Theorem 2.6, in the sense that its (external) category of I-elements is equivalent to $(\mathcal{E}/I)_{fc}$; so we may deduce from 2.6 that $\underline{\underline{E}}_{fin}$ is an internal topos in $\mathcal{E}$.

We may now state the main theorem of this chapter:

<u>4.1 Theorem</u> If the topos $\mathcal{E}$ has a natural number object, then it has an object classifier $\mathcal{E}[U]$, which may be taken to be $\mathcal{E}^{(\underline{\underline{E}}_{fin})}$, with the "inclusion functor" $\underline{\underline{E}}_{fin} \longrightarrow \mathcal{E}$ playing the rôle of the generic object U.

In view of the fact that

$$\underline{\underline{E}}_{fin}(\underset{\sim}{n}_1, \underset{\sim}{n}_2) = [\underset{\sim}{n}_2]^{[\underset{\sim}{n}_1]} ,$$

we obtain from Proposition 2.3 and the remarks preceding it

<u>4.2 Lemma</u> If $\mathcal{F} \xrightarrow{f} \mathcal{E}$ is a geometric morphism and $\mathcal{E}$ has a natural number object, then

$$f^*(\underline{\underline{E}}_{fin}) \cong \underline{\underline{F}}_{fin} . \square$$

<u>4.3 Corollary</u> If $\mathcal{E}$ is a topos with a natural number object, there is an equivalence of categories

$$\underline{\underline{\mathrm{Top}}}/\mathcal{E}\ (\mathcal{F}, \mathcal{E}^{(\underline{\underline{E}}_{fin})}) \simeq \underline{\underline{\mathrm{Top}}}/\mathcal{F}\ (\mathcal{F}, \mathcal{F}^{(\underline{\underline{F}}_{fin})})$$

for any $\mathcal{E}$-topos $\mathcal{F}$.

<u>Proof</u> Apply Lemma 4.2 to the remark following Theorem 1.2. □

Corollary 4.3 reduces the problem of proving that $\mathcal{E}^{(\underline{\underline{E}}_{fin})}$ is an object classifier to that of proving that $\underline{\underline{\mathrm{Top}}}/\mathcal{E}\ (\mathcal{E}, \mathcal{E}^{(\underline{\underline{E}}_{fin})})$ is equivalent to $\mathcal{E}$. Diaconescu's theorem (1.2) in turn reduces this to the problem of proving that $\mathrm{Flat}(\underline{\underline{E}}_{fin}{}^{op}, \mathcal{E})$ is equivalent to $\mathcal{E}$.

For any object X of $\mathcal{E}$, we define a presheaf HX on $\underline{\underline{E}}_{fin}$ to be the object $X^{[\underset{\sim}{n}]}$ of $\mathcal{E}/N$, with $\underline{\underline{E}}_{fin}$-action defined by the composition map

$$[\underset{\sim}{n}_2]^{[\underset{\sim}{n}_1]} \times X^{[\underset{\sim}{n}_2]} \longrightarrow X^{[\underset{\sim}{n}_1]}$$

in $\mathcal{E}/N\times N$. For any map $u: X \longrightarrow Y$, we have a map of presheaves $H(u): HX \longrightarrow HY$ induced by $u^{[\underset{\sim}{n}]}: X^{[\underset{\sim}{n}]} \longrightarrow Y^{[\underset{\sim}{n}]}$. In this way we obtain a functor

$$H: \mathcal{E} \longrightarrow \mathcal{E}^{(\underline{\underline{E}}_{fin})^{op}}.$$

4.4 Lemma For any X, HX is a flat presheaf on $\underline{\underline{E}}_{fin}$.

Proof We must verify the conditions of Proposition 1.3 for the discrete fibration $\underline{HX} \longrightarrow \underline{\underline{E}}_{fin}$ corresponding to HX. But $HX_o = \Sigma_{\underset{\sim}{n}} X^{[\underset{\sim}{n}]}$; and since $X^{[o]} \cong X^0 \cong 1$, the map $HX_o \longrightarrow 1$ has a section and is therefore epimorphic. Thus condition (a) is satisfied.

Now suppose we are given a diagram of the form

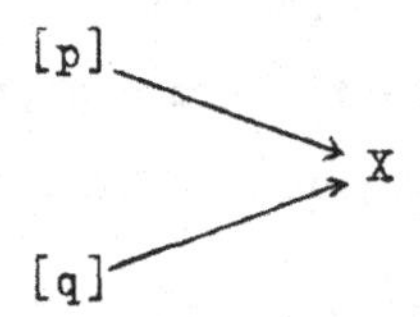

in $\mathcal{E}/I$ for some I. We may extend it to a diagram

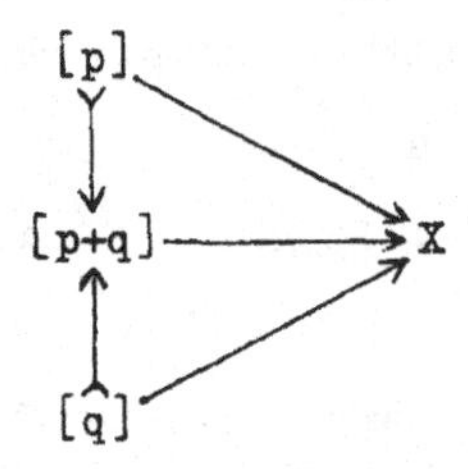

in $\mathcal{E}/I$, since $[p+q] \cong [p] \amalg [q]$. So condition (b) is satisfied.

Finally, let

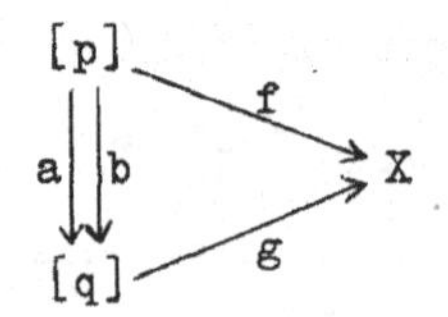

be a diagram in $\mathcal{E}/I$ with $ga = f = gb$. By Theorem 2.6, there exists a natural number r in $\mathcal{E}/I$ and a coequalizer diagram

$$[p] \underset{b}{\overset{a}{\rightrightarrows}} [q] \xrightarrow{c} [r] ,$$

giving us a diagram

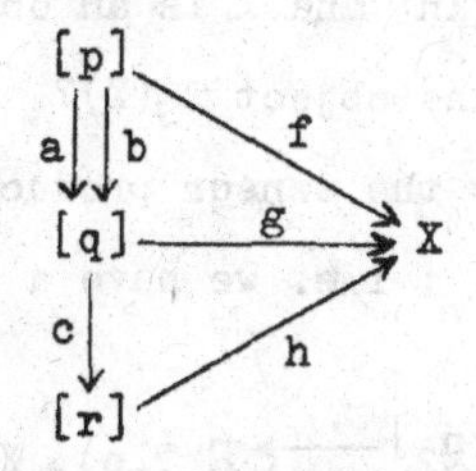

in $\mathcal{E}/I$ which verifies condition (c). □

In view of Lemma 4.4 we have a functor

$$H: \mathcal{E} \longrightarrow \mathrm{Flat}(\underline{E}_{fin}{}^{op}, \mathcal{E}) .$$

4.5 Lemma The functor H defined above is an equivalence of categories.

Proof Let V denote the functor

$$\text{"evaluate at 1"}: \mathcal{E}^{(\underline{E}_{fin})^{op}} \longrightarrow \mathcal{E} \quad ;$$

i.e. if F is a presheaf on $\underline{E}_{fin}$, then V(F) is the pullback of $F(\underset{\sim}{n})$ along $1 \xrightarrow{so} N$. Now it is clear that

$$V(HX) = X^{[so]} \cong X^1 = X \quad ;$$

we have to show that if F is a flat presheaf then $HV(F) \cong F$.

But if $\mathcal{F} \xrightarrow{f} \mathcal{E}$ is a geometric morphism, it is clear that $f^*(V(F)) = V(f^*F)$; and H similarly commutes with inverse image functors, by Proposition 2.3. So it is sufficient to establish the isomorphism $HV(F) \cong F$ in the case when F is the generic flat presheaf on $\underline{E}_{fin}$, i.e. the Yoneda profunctor $Y(\underline{E}_{fin})$. But the latter is precisely HU, where U is the object of $\mathcal{E}^{(\underline{E}_{fin})}$ defined by $U(\underset{\sim}{n}) = [\underset{\sim}{n}]$, with $\underline{E}_{fin}$ acting by the evaluation map

$$[\underset{\sim}{n}_1] \times [\underset{\sim}{n}_2]^{[\underset{\sim}{n}_1]} \longrightarrow [\underset{\sim}{n}_2]$$

(i.e. the "inclusion functor $\underline{E}_{fin} \longrightarrow \mathcal{E}$" defined in chapter I).

And it is also clear that $U \cong V(Y(\underline{E}_{fin}))$; so the result is established. □

This completes the proof of Theorem 4.1. □

If T is an object of $\mathcal{E}^{(\underline{E}_{fin})}$ and X is an object of an $\mathcal{E}$-topos $(\mathcal{F} \xrightarrow{f} \mathcal{E})$, how do we describe the object $T_{\mathcal{F}}(X)$? Disentangling the definitions, we find that it is the tensor product of the profunctors $1 \overset{f^*T}{\dashrightarrow} \underline{E}_{fin}$ and $\underline{E}_{fin} \overset{HX}{\dashrightarrow} 1$; i.e. we have a reflexive coequalizer diagram in $\mathcal{F}$

$$\Sigma_{\underset{\sim}{n}_1,\underset{\sim}{n}_2} T(\underset{\sim}{n}_1) \times [\underset{\sim}{n}_2]^{[\underset{\sim}{n}_1]} \times X^{[\underset{\sim}{n}_2]} \rightrightarrows \Sigma_{\underset{\sim}{n}} T(\underset{\sim}{n}) \times X^{[\underset{\sim}{n}]} \longrightarrow T_{\mathcal{F}}(X).$$

It is suggestive to think of this as a "polynomial" expression in X with coefficients determined by T. Indeed, T itself should be thought of as a "polynomial" in the "indeterminate" U, since we have $T \cong T_{\mathcal{E}[U]}(U)$. The polynomial analogy becomes even more striking when T is the free object (relative to the monadic adjunction $\mathcal{E}/N \rightleftarrows \mathcal{E}^{(\underline{E}_{fin})}$) generated by an object $A(\underset{\sim}{n})$ of $\mathcal{E}/N$: in this case we have $T(\underset{\sim}{n}) = \Sigma_{\underset{\sim}{n}'} A(\underset{\sim}{n}') \times [\underset{\sim}{n}]^{[\underset{\sim}{n}']}$, and the above coequalizer diagram splits to give an isomorphism $T_{\mathcal{F}}(X) \cong \Sigma_{\underset{\sim}{n}} A(\underset{\sim}{n}) \times X^{[\underset{\sim}{n}]}$. The reason why the general case requires the more complicated formula given above is that we have to interpret "addition" as corresponding to arbitrary $\mathcal{E}$-indexed colimits, and not just to coproducts.

<u>4.6 Proposition</u> Let T be an object of $\mathcal{E}^{(\underline{E}_{fin})}$. Then for any $\mathcal{E}$-topos $\mathcal{F}$, the functor $T_{\mathcal{F}}: \mathcal{F} \longrightarrow \mathcal{F}$ preserves coequalizers of reflexive pairs.

<u>Proof</u> Consider first the functor $\Sigma_{\underset{\sim}{n}} T(\underset{\sim}{n}) \times (-)^{[\underset{\sim}{n}]}: \mathcal{F} \longrightarrow \mathcal{F}$, i.e. the composite

$$\mathcal{F} \xrightarrow{N^*} \mathcal{F}/N \xrightarrow{(-)^{[\underset{\sim}{n}]}} \mathcal{F}/N \xrightarrow{T(\underset{\sim}{n}) \times (-)} \mathcal{F}/N \xrightarrow{\Sigma_N} \mathcal{F} .$$

Now all the functors in this composite, except the second, have right adjoints and therefore preserve all colimits; and the functor $(-)^{[\underset{\sim}{n}]}$ preserves reflexive coequalizers by Proposition

2.4. Similarly, we may show that the functor

$$\Sigma_{\underset{\sim}{n}_1,\underset{\sim}{n}_2} T(\underset{\sim}{n}_1) \times [\underset{\sim}{n}_2]^{[\underset{\sim}{n}_1]} \times (-)^{[\underset{\sim}{n}_2]}$$

preserves reflexive coequalizers.

Now suppose we have a reflexive coequalizer diagram

$$X \rightrightarrows Y \longrightarrow Z$$

in $\mathfrak{F}$. Then we have a diagram

$$\begin{array}{ccccc}
\Sigma_{\underset{\sim}{n}_1,\underset{\sim}{n}_2} T(\underset{\sim}{n}_1) \times [\underset{\sim}{n}_2]^{[\underset{\sim}{n}_1]} \times X^{[\underset{\sim}{n}_2]} & \rightrightarrows & \Sigma_{\underset{\sim}{n}} T(\underset{\sim}{n}) \times X^{[\underset{\sim}{n}]} & \longrightarrow & T_{\mathfrak{F}}(X) \\
\downdownarrows & & \downdownarrows & & \downdownarrows \\
\Sigma_{\underset{\sim}{n}_1,\underset{\sim}{n}_2} T(\underset{\sim}{n}_1) \times [\underset{\sim}{n}_2]^{[\underset{\sim}{n}_1]} \times Y^{[\underset{\sim}{n}_2]} & \rightrightarrows & \Sigma_{\underset{\sim}{n}} T(\underset{\sim}{n}) \times Y^{[\underset{\sim}{n}]} & \longrightarrow & T_{\mathfrak{F}}(Y) \\
\downarrow & & \downarrow & & \downarrow \\
\Sigma_{\underset{\sim}{n}_1,\underset{\sim}{n}_2} T(\underset{\sim}{n}_1) \times [\underset{\sim}{n}_2]^{[\underset{\sim}{n}_1]} \times Z^{[\underset{\sim}{n}_2]} & \rightrightarrows & \Sigma_{\underset{\sim}{n}} T(\underset{\sim}{n}) \times Z^{[\underset{\sim}{n}]} & \longrightarrow & T_{\mathfrak{F}}(Z)
\end{array}$$

in which the rows are coequalizers by the definition of $T_{\mathfrak{F}}$, and the first two columns are coequalizers by the remarks above. An easy diagram-chase (cf. [37], Lemma 2.10) now shows that the third column is a coequalizer. □

The object classifier enables us to give a solution to the problem of recursive definitions in a topos, which we mentioned in chapter II. Recall that we proved there (Theorem 2.2) that if X is an object of a topos $\mathcal{E}$ and T: $\mathcal{E} \longrightarrow \mathcal{E}$ a locally internal functor, there can be at most one object $F(\underset{\sim}{n})$ of $\mathcal{E}/N$ (up to canonical isomorphism) satisfying $F(o) \cong X$ and $F(s\underset{\sim}{n}) \cong T^N(F(\underset{\sim}{n}))$. However, even the requirement that T be locally internal is not sufficient to ensure the existence of such an $F(\underset{\sim}{n})$ in a general topos $\mathcal{E}$, as the following example indicates:

<u>4.7 Example</u> For convenience of notation, we assume the Generalized Continuum Hypothesis, although the example can easily be constructed without it. Let $\mathcal{E}$ denote the category of sets of

cardinality less than $\aleph_\omega$, and all functions between them; then $\mathcal{E}$ is closed under finite limits and formation of power-sets, and so forms a (logical) sub-topos of $\mathcal{S}$. Let $T: \mathcal{E} \longrightarrow \mathcal{E}$ be the covariant power-set functor, and let X be the set of natural numbers. Then any solution $F(\underset{\sim}{n})$ of the recursion problem determined by X and T must have $\text{card}(F(p)) = \aleph_p$ for all p, and hence $\text{card}(\Sigma_{\underset{\sim}{n}} F(\underset{\sim}{n})) = \aleph_\omega$. So the problem cannot be solved in $\mathcal{E}$.

The reason why Example 4.7 occurs is clearly that the axioms of elementary topos theory do not include any analogue of the set-theoretic axiom of Replacement; so we cannot deduce the existence of arbitrary countable coproducts from the existence of N in a topos. (See also [17], Example 4.49(ii).) Thus if we wish to prove an existence theorem for recursively defined objects in an arbitrary topos, we must impose some sort of "boundedness condition" on the functor T, which ensures that the terms of the sequence $(X, TX, T^2X, \ldots)$ do not grow in size too rapidly.

The "polynomial" description of the functors which arise from objects of $\mathcal{E}[U]$, given above, suggests that the "boundedness condition" we require should be satisfied if we impose the condition that T should extend naturally to an endofunctor $T_{\mathcal{F}}$ of every $\mathcal{E}$-topos $\mathcal{F}$ (not just to $\mathcal{E}$-toposes of the form $\mathcal{E}/I$). Indeed this condition is sufficient:

<u>4.8 Theorem</u> Let $\mathcal{E}$ be a topos with a natural number object, X an object of $\mathcal{E}$ and T an object of $\mathcal{E}[U]$. Then there exists (uniquely up to canonical isomorphism) an object $F(\underset{\sim}{n})$ of $\mathcal{E}/N$ such that

$F(o) \cong X$ and $F(s\underset{\sim}{n}) \cong T_{\mathcal{E}/N}(F(\underset{\sim}{n}))$.

<u>Proof</u> Taking $\mathcal{F} = \mathcal{E}[U]$ in Theorem 3.1, we obtain a diagram

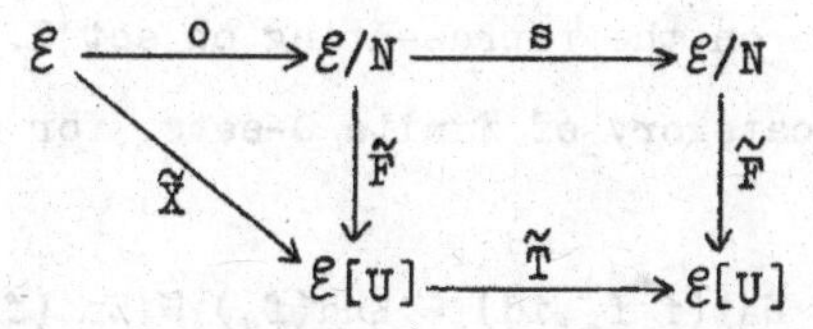

in BTop/$\mathcal{E}$; and $F(\underset{\sim}{n})$ is clearly the object classified by $\tilde{F}$. □

At least in the case when T is freely generated by an object $A(\underset{\sim}{n})$ of $\mathcal{E}/N$, it is possible to give a more explicit description of the object $F(\underset{\sim}{n})$ constructed in Theorem 4.8 in terms of X and $A(\underset{\sim}{n})$. Specifically, $F(\underset{\sim}{n})$ can be described as an "object of trees" whose nodes are labelled with elements of $\Sigma_{\underset{\sim}{n}} A(\underset{\sim}{n})$ or of X, in a manner similar to that used for constructing free algebraic theories in [38], Appendix A. A detailed account of this method of constructing $F(\underset{\sim}{n})$ is given in [15], Chapter VIII.

It is an open question whether the converse of Theorem 4.1 is true: i.e. does the existence of an object classifier for $\mathcal{E}$ imply that $\mathcal{E}$ has a natural number object? Corroborative evidence of a weak kind is provided by the following:

<u>4.9 Proposition</u> The topos $\mathcal{S}_f$ of finite sets does not have an object classifier.

<u>Proof</u> Suppose $\mathcal{S}_f[U]$ exists, and consider the image factorization

$$\mathcal{S}_f \xrightarrow{g} \mathcal{E} \xrightarrow{h} \mathcal{S}_f[U]$$

of a geometric morphism $f\colon \mathcal{S}_f \longrightarrow \mathcal{S}_f[U]$. Using the two alternative constructions of the image topos, we may describe $\mathcal{E}$ either as the category of sheaves for a Lawvere-Tierney topology in $\mathcal{S}_f[U]$, or as the category of coalgebras for a left exact comonad $\mathbb{C}$ on $\mathcal{S}_f$. But the Special Adjoint Functor Theorem (with "small" interpreted as "finite" throughout) implies that <u>any</u> left exact functor $C\colon \mathcal{S}_f \longrightarrow \mathcal{S}_f$ has a left adjoint and is therefore representable; and a comonad structure on C corresponds to a monoid structure on the representing object G. Thus $\mathcal{E}$ can be described as the category of finite G-sets, for some finite monoid G. But

$$G \cong \mathrm{Nat}((-)^G,\mathrm{id}) \cong \mathrm{Nat}(f^*f_*,\mathrm{id}) \cong \mathrm{End}(f_*) \cong \mathrm{End}(f^*)^{op}$$
$$\cong \mathrm{End}(f^*(U))^{op}.$$

Thus as $f^*(U)$ ranges over all finite sets, we get an infinite number of inequivalent categories $\mathcal{E}$. But $\mathcal{S}_f[U]$ is an $\mathcal{S}_f$-topos, and so has finite hom-sets; in particular it has only a finite number of Lawvere-Tierney topologies, and hence a finite number of sheaf subtoposes $\mathcal{E}$. So we have obtained a contradiction. □

We conclude this chapter with a slight digression, in which we consider a couple of other classifying toposes closely related to the object classifier. Let $\underline{\underline{E}}_{fine}$ denote the internal category of finite cardinals and epimorphisms in $\mathcal{E}$: i.e. $(E_{fine})_0 = N$, but we take $\underline{\underline{E}}_{fine}(\underset{\sim}{n}_1,\underset{\sim}{n}_2)$ to be the object of epimorphisms $\mathrm{Epi}([\underset{\sim}{n}_1],[\underset{\sim}{n}_2])$ in $\mathcal{E}/N{\times}N$, rather than the exponential used in constructing $\underline{\underline{E}}_{fin}$. Similarly, we may define $\underline{\underline{E}}_{finm}$ using monomorphisms instead of epimorphisms. Since finite cardinals are both decidable and Kuratowski-finite, it follows from Proposition 2.8 that both these categories are preserved by inverse image functors.

Now let X be an object of $\mathcal{E}$. The presheaf HX which we constructed on $\underline{\underline{E}}_{fin}$ restricts to presheaves on the subcategories $\underline{\underline{E}}_{fine}$ and $\underline{\underline{E}}_{finm}$; but we also have presheaves EX, MX on $\underline{\underline{E}}_{fine}$ and $\underline{\underline{E}}_{finm}$ respectively, defined by

$$EX(\underset{\sim}{n}) = \mathrm{Epi}([\underset{\sim}{n}],X) \quad \text{and} \quad MX(\underset{\sim}{n}) = \mathrm{Mono}([\underset{\sim}{n}],X)$$

with the appropriate category acting by composition.

<u>4.10 Lemma</u> (i) The presheaf EX is flat iff X is Kuratowski-finite.

(ii) MX is flat iff X is decidable.

<u>Proof</u> (i) Condition (a) of 1.3 says that $\Sigma_{\underset{\sim}{n}}\mathrm{Epi}([\underset{\sim}{n}],X)$ has global

support; but this is precisely equivalent to Kuratowski-finiteness of X, by Theorem 2.7.

To verify condition (b), suppose we are given a diagram

in $\mathcal{E}/I$ for some I. We must complete it locally (i.e. in $\mathcal{E}/J$ for some $J \twoheadrightarrow I$) to a diagram

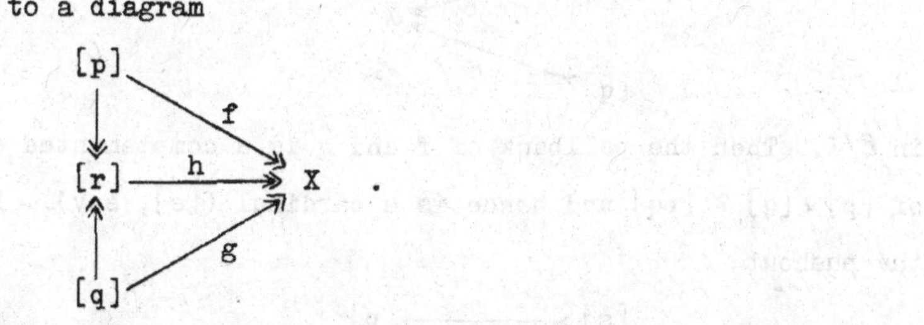

But since [p] is internally projective in $\mathcal{E}/I$ (2.4), there exists (locally) a map d: $[p] \longrightarrow [q]$ with gd = f; let [r] be the image of d in $(\mathcal{E}/J)_{fc}$, and let h be the composite $[r] \rightarrowtail [q] \xrightarrow{g} X$ (which is clearly epi, since its composite with $[p] \twoheadrightarrow [r]$ is). Since $(\mathcal{E}/J)_{fc}$ is Boolean, the inclusion $[r] \rightarrowtail [q]$ has a complement $[s] \rightarrowtail [q]$ say; then by internal projectivity of [s] we can locally construct a map $[s] \longrightarrow [r]$ such that

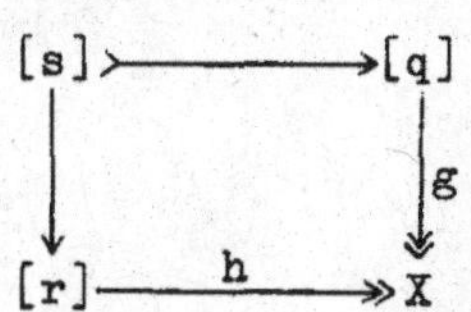

commutes. Combining this map with the identity on [r], we obtain a splitting for the inclusion $[r] \rightarrowtail [q]$, such that

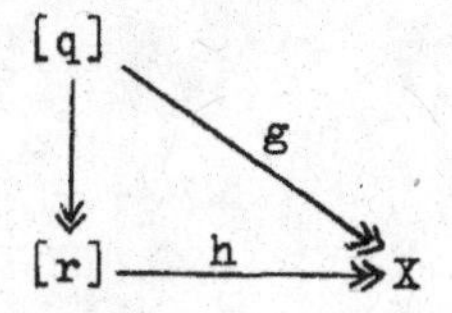

commutes. Thus we have constructed the required diagram.

Condition (c) is verified exactly as in Lemma 4.4, since coequalizers are automatically epi in $\mathcal{E}/I$.

(ii) Here condition (a) is trivial, since $\Sigma_{\underset{\sim}{n}}\mathrm{Mono}([\underset{\sim}{n}],X)$ always has a global element corresponding to $[o] \rightarrowtail X$.

For condition (b), suppose X is decidable and we are given

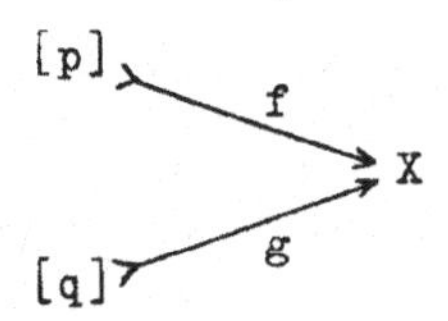

in $\mathcal{E}/I$. Then the pullback of f and g is a complemented subobject of $[p] \times [q] \cong [pq]$ and hence is a cardinal ([s], say). Now form the pushout

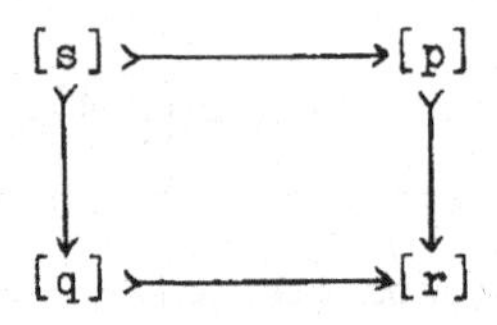

in $(\mathcal{E}/I)_{fc}$; then [r] is the union of [p] and [q] in the lattice of subobjects of X, and so we have a commutative diagram

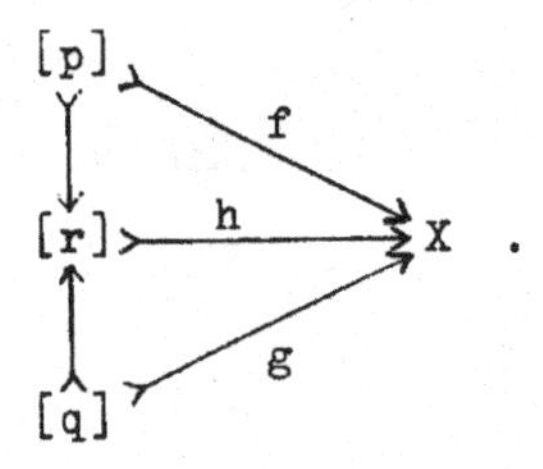

Condition (c) is again trivial, this time because the commutativity of

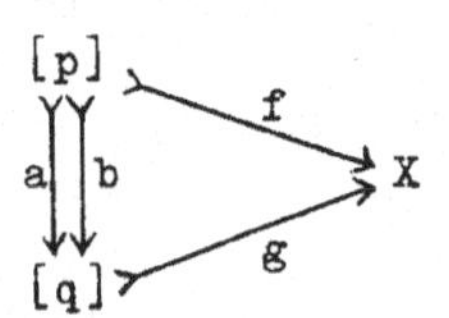

forces a to be equal to b.

Conversely, suppose MX is flat. Taking $I = X\times X$, we have a generic pair of maps $1 \overset{f}{\underset{g}{\rightrightarrows}} X$ in $\mathcal{E}/I$. Now there exists (locally) a diagram

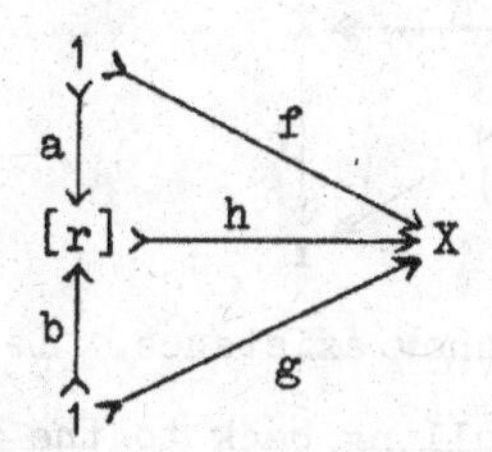

which implies that the equalizer of f and g is also the equalizer of a and b, and in particular that it is (locally, and hence globally) a complemented subobject of 1. But this equalizer is just the diagonal $X \overset{\Delta}{\rightarrowtail} X\times X$, regarded as a subobject of 1 in $\mathcal{E}/I$; so X is decidable. □

Now if $f\colon X \twoheadrightarrow Y$ is an epimorphism, the operation of composing with f induces a map of presheaves $EX \longrightarrow EY$; so we have a functor

$$E\colon \mathcal{E}_{kfe} \longrightarrow \mathrm{Flat}(\underline{E}_{fine}{}^{op}, \mathcal{E}) \quad ,$$

where $\mathcal{E}_{kfe}$ denotes the category of Kuratowski-finite objects and epimorphisms in $\mathcal{E}$. Moreover, Proposition 2.8(i) ensures that E commutes up to isomorphism with inverse image functors. But the generic flat presheaf on $\underline{E}_{fine}$ has the form EK for some object K of $\mathcal{E}^{(\underline{E}_{fine})}$; specifically, $K(\underset{\sim}{n}) = [\underset{\sim}{n}]$, with $\underline{E}_{fine}$ acting by the evaluation map $[\underset{\sim}{n}_1] \times \mathrm{Epi}([\underset{\sim}{n}_1],[\underset{\sim}{n}_2]) \longrightarrow [\underset{\sim}{n}_2]$. Thus we obtain a functor

$$S\colon \underline{\underline{\mathrm{Top}}}/\mathcal{E}\ (\mathcal{F}, \mathcal{E}^{(\underline{E}_{fine})}) \longrightarrow \mathcal{F}_{kf};\ g \longmapsto g^*(K) \quad .$$

To show that E and S give rise to an equivalence of categories

$$\mathcal{F}_{kfe} \simeq \mathrm{Flat}(\underline{E}_{fine}{}^{op}, \mathcal{F}) \simeq \underline{\underline{\mathrm{Top}}}/\mathcal{E}\ (\mathcal{F}, \mathcal{E}^{(\underline{E}_{fine})}) \quad ,$$

it is sufficient to show that any map of presheaves $h\colon EX \longrightarrow EY$ is induced by a unique epimorphism $e\colon X \twoheadrightarrow Y$.

But since X is Kuratowski-finite, there exists $I \twoheadrightarrow 1$ in $\mathcal{E}$ and $[p] \xrightarrow{g} X$ in $\mathcal{E}/I$. By regarding g as an element of EX(p), we can apply h to it to obtain $[p] \xrightarrow{h(g)} Y$ in $\mathcal{E}/I$. Uniqueness of e is now clear from the fact that

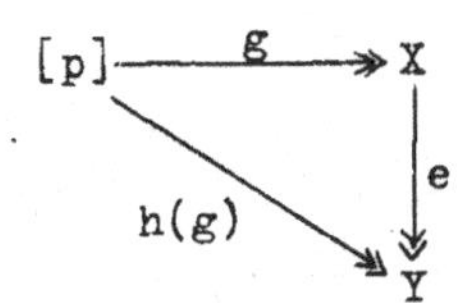

must commute, so it remains to show existence. Let $R \overset{a}{\underset{b}{\rightrightarrows}} [p]$ be the kernel-pair of g; then on pulling back to the $(\mathcal{E}/I)$-topos $\mathcal{E}/R$, we have a generic pair of elements of [p] coequalized by g. Replacing these by the corresponding epimorphisms

$$[sp] \cong [p] \amalg 1 \rightrightarrows [p] \quad ,$$

we see from the naturality of h that h(g) must coequalize a and b; so we obtain a factorization $X \xrightarrow{e} Y$ of h(g) through g in $\mathcal{E}/I$. A similar argument now shows that e satisfies the "descent condition" for the epi $I \twoheadrightarrow 1$, so that it is actually a morphism $X \twoheadrightarrow Y$ in $\mathcal{E}$. We have thus proved

<u>4.11 Theorem</u> The topos $\mathcal{E}^{(\underline{E}_{fine})}$ is a classifying topos for the theory of Kuratowski-finite objects and epimorphisms between them; i.e. for any $\mathcal{E}$-topos $\mathcal{F}$ we have

$$\underline{\underline{Top}}/\mathcal{E}\ (\mathcal{F}, \mathcal{E}^{(\underline{E}_{fine})}) \simeq \mathcal{F}_{kfe} \ . \ \square$$

We leave it to the reader to verify the remaining details of

<u>4.12 Theorem</u> The topos $\mathcal{E}^{(\underline{E}_{finm})}$ is a classifying topos for the theory of decidable objects and monomorphisms between them. □

It should be said that slicker proofs of the "classifiability" of these two theories can be given using the general machinery of [36]; but the explicit descriptions of the classifying toposes which we have obtained here have a number of advantages. For example, it follows from [17], Exercise 7.5, that $\mathcal{S}^{(\underline{\underline{S}}_{fine})}$ is an étendue (= locally spatial topos); thus Theorem 4.11 not only answers an outstanding question of F.W. Lawvère ([23], p.115), but also provides further evidence in support of Lawvere's contention that the étendues are an interesting class of toposes for deeper investigation. (However, Lawvere's conjecture ([23], p.130) that the étendues are precisely the "internalizable" $\mathcal{S}$-toposes is false: the topos of sets-with-an-idempotent-endomorphism is not an étendue, but is internalizable in every $\mathcal{S}$-topos.)

It is also of interest that, if we define $\underline{\underline{E}}_{fini}$ to be the category of finite cardinals and isomorphisms (i.e. the intersection of $\underline{\underline{E}}_{fine}$ and $\underline{\underline{E}}_{finm}$), then flat presheaves on $\underline{\underline{E}}_{fini}$ correspond to objects which are locally isomorphic to finite cardinals. But by [17], Exercise 9.5, these are precisely the decidable Kuratowski-finite objects; so the theory classified by $\mathcal{E}^{(\underline{\underline{E}}_{fini})}$ is the intersection of those defined in 4.11 and 4.12. In other words, the pullback diagram

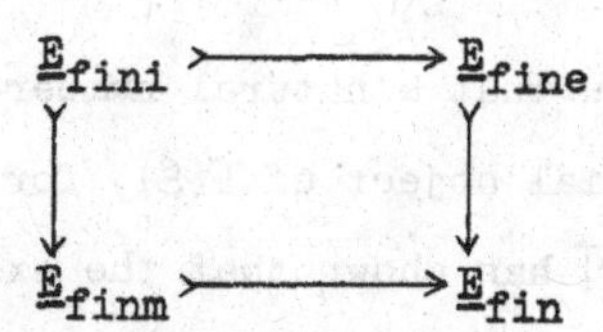

in $\underset{\sim}{cat}(\mathcal{E})$ is preserved by the functor

$$\mathcal{E}^{(-)}: \underset{\sim}{cat}(\mathcal{E}) \longrightarrow \underline{\underline{Top}}/\mathcal{E}$$

(although this functor does not always preserve finite limits: see [17], Exercise 6.7).

CHAPTER V : FINITARY ALGEBRAIC THEORIES

In this chapter we reach the principal goal of the present work: the formulation of a definition of "finitary algebraic theory" which is internal to an arbitrary topos with a natural number object. Before giving this definition, however, and to provide motivation for it, we shall briefly review the model theory of "external" algebraic theories in an elementary topos. For this purpose, we shall initially find it convenient to adopt the original "universal-algebra" definition of an algebraic theory [6,9]: i.e. an algebraic theory $\mathbb{T} = (\Omega, E)$ is determined by a set Ω of finitary operations, each equipped with an "arity" which is a natural number, and a set E of finitary equations between (well-formed) words in the operations. A $\mathbb{T}$-model in a category $\mathcal{E}$ with finite products consists of an object A, together with a morphism $\omega_A\colon A^m \longrightarrow A$ for each m-ary operation ω of $\mathbb{T}$, such that each equation of $\mathbb{T}$ gives rise to a commutative diagram in $\mathcal{E}$ in an obvious way. We write $\mathbb{T}(\mathcal{E})$ for the category of $\mathbb{T}$-models in $\mathcal{E}$.

In chapter II, we saw that a natural number object in a topos $\mathcal{E}$ may be regarded as an initial object of $\mathbb{T}(\mathcal{E})$, for a suitable $\mathbb{T}$. More generally, B. Lesaffre [25] has shown that the existence of a natural number object implies the existence, not merely of initial models, but of arbitrary free models for any finitely-presented theory. Explicitly, she proved

5.1 Theorem Let $\mathcal{E}$ be a topos with a natural number object, and $\mathbb{T}$ a finitely-presented, finitary algebraic theory (i.e. one such that the sets Ω and E are finite). Then

(i) There exists a free functor $F\colon \mathcal{E} \longrightarrow \mathbb{T}(\mathcal{E})$ which is left adjoint to the forgetful functor.

(ii) $\mathbb{T}(\mathcal{E})$ is monadic over $\mathcal{E}$.

(iii) $\mathbb{T}(\mathcal{E})$ has finite colimits. □

Now let $\mathcal{F} \xrightarrow{f} \mathcal{E}$ be a geometric morphism, and $\mathbb{T}$ a finitary algebraic theory. Since the functors f_* and f^* both preserve finite products, they can clearly be lifted to a pair of adjoint functors (also denoted f_* and f^*) between $\mathbb{T}(\mathcal{F})$ and $\mathbb{T}(\mathcal{E})$.

<u>5.2 Lemma</u> Let f and $\mathbb{T}$ be as above, and suppose that free $\mathbb{T}$-model functors exist in $\mathcal{E}$ and $\mathcal{F}$. Then the square

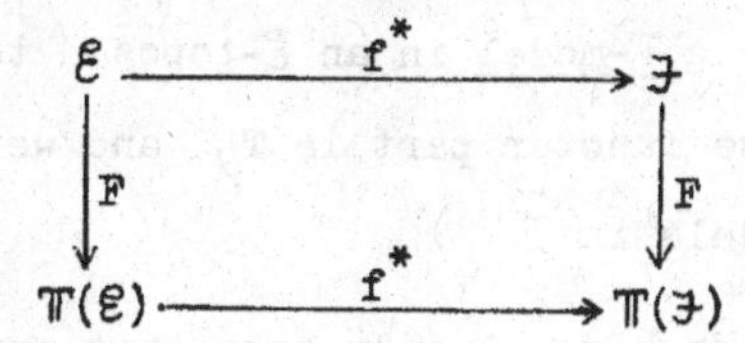

commutes up to natural isomorphism.

<u>Proof</u> The corresponding diagram

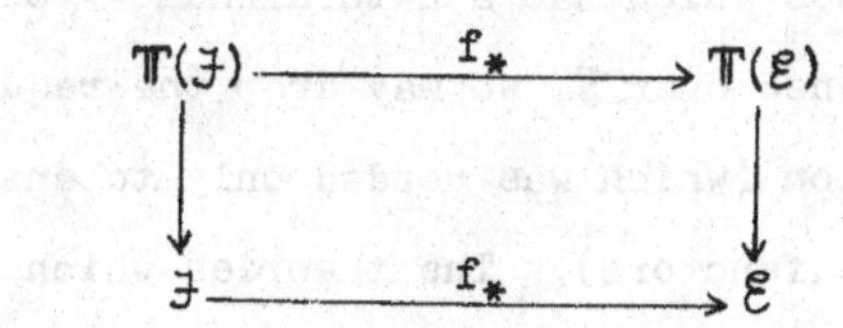

of right adjoints clearly commutes, so this is immediate from the uniqueness of adjoints. □

Thus if we define $T_{\mathcal{F}}$ to be the composite functor

$$\mathcal{F} \xrightarrow{F} \mathbb{T}(\mathcal{F}) \longrightarrow \mathcal{F} ,$$

we have (for any topos $\mathcal{E}$ with a natural number object) a "natural endomorphism of $\mathcal{E}$-toposes" of the type considered in chapter IV. Moreover, the functor $T_{\mathcal{F}}$ has a monad structure which is natural in $\mathcal{F}$; this clearly corresponds to a monoid structure (for the product ⊗) on the corresponding object T of $\mathcal{E}[U]$. We recall also ([26], Proposition 6) that a monad $\mathbb{T} = (T,\mu,\eta)$ on $\mathcal{S}$ corresponds to a finitary algebraic theory iff the functor T is isomorphic to the left Kan extension along the inclusion $\underline{S}_{fin} \longrightarrow \mathcal{S}$ of a functor $\check{T}: \underline{S}_{fin} \longrightarrow \mathcal{S}$; in view of the coequalizer formula for $T_{\mathcal{F}}$ before

Proposition 4.6, this is equivalent to saying that T has the form $(\check{T})_{\mathcal{S}}$ for some object $\check{T}$ of $\mathcal{S}[U]$.

We are therefore led to make the following definition:

5.3 Definition Let $\mathcal{E}$ be a topos with a natural number object. We define an (internal) finitary algebraic theory in $\mathcal{E}$ to be a monoid in the monoidal category $(\mathcal{E}[U],\otimes,U)$. We write $\underset{\sim}{\text{alg}}(\mathcal{E})$ for the category of such monoids. If $\mathbb{T} = (T,m,e)$ is an object of $\underset{\sim}{\text{alg}}(\mathcal{E})$, we define a $\mathbb{T}$-model in an $\mathcal{E}$-topos $\mathcal{F}$ to be an algebra for the monad $\mathbb{T}_{\mathcal{F}}$ whose functor part is $T_{\mathcal{F}}$, and we write $\mathbb{T}(\mathcal{F})$ for the category of $\mathbb{T}$-models in $\mathcal{F}$.

5.4 Examples (a) We have already seen that any finitely-presented external finitary algebraic theory may be regarded as an internal theory in any topos which has a natural number object. And if our topos is defined over $\mathcal{S}$, we may drop the requirement of finite presentation (which was needed only to ensure the existence of free functors). The theories which arise in this way should be thought of as "constant", in that their operations and equations do not depend on the "domain of variation" represented by the base topos.

(b) Let R be an internal ring in a topos $\mathcal{E}$. For any $\mathcal{E}$-topos $\mathcal{F}$, we may construct the free left R-module functor

$${}_R F\colon \mathcal{F} \longrightarrow R\text{-}\underset{\sim}{\text{mod}}(\mathcal{F})$$

by ${}_R F(X) = R \otimes F_{\underset{\sim}{ab}}(X)$, where $F_{\underset{\sim}{ab}}$ denotes the free abelian group functor and tensor product of abelian groups in $\mathcal{F}$ is defined in the usual way. It is easy to see that ${}_R F$ commutes up to isomorphism with inverse image parts of maps of $\mathcal{E}$-toposes, and so we get an internal theory $R\text{-}\underset{\sim}{\text{mod}}$ in $\mathcal{E}$, whose models are left R-modules. The "object of unary operations" of $R\text{-}\underset{\sim}{\text{mod}}$ is clearly isomorphic to R; thus the operations vary "from point to point" within $\mathcal{E}$.

(c) Let I be a subobject of 1 in $\mathcal{E}$, and let $\mathbb{S} \longrightarrow \mathbb{T}$ be a quotient map of (external) theories (i.e. $\mathbb{T}$ has the same operations as $\mathbb{S}$, but may have additional equations). Then we may construct "the theory of $\mathbb{S}$-models which become $\mathbb{T}$-models when restricted to I", as follows:

We shall see shortly that the pullback functor $I^*: \mathbb{S}(\mathcal{E}) \longrightarrow \mathbb{S}(\mathcal{E}/I)$ has a left adjoint $I_\#$. Now if X is an object of an $\mathcal{E}$-topos $\mathcal{F}$, we may form a pushout diagram

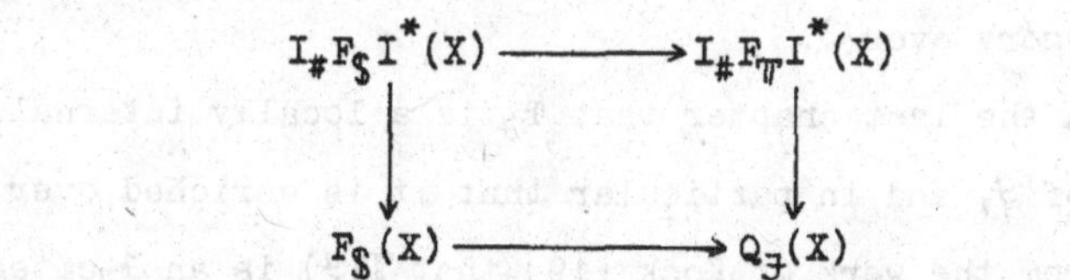

in $\mathbb{S}(\mathcal{F})$. Then it is easy to verify that $I^* Q_{\mathcal{F}}(X)$ is a $\mathbb{T}$-model in $\mathcal{F}/I$, and the composite $X \longrightarrow F_{\mathbb{S}}(X) \longrightarrow Q_{\mathcal{F}}(X)$ is universal among maps from X into $\mathbb{S}$-models with this property. Moreover, it is clear from the construction that $Q_{\mathcal{F}}$ commutes up to isomorphism with inverse images of maps of $\mathcal{E}$-toposes, and so it defines an internal algebraic theory $\mathbb{Q}$, in which the operations are constant but the equations vary.

(d) Let $\mathcal{F}$ be an $\mathcal{E}$-topos and Y an object of $\mathcal{F}$. Consider the object $\tilde{Y}_*(Y)$ of $\mathcal{E}[U]$. Since morphisms $T \longrightarrow \tilde{Y}_*(Y)$ in $\mathcal{E}[U]$ correspond to morphisms $T_{\mathcal{F}}(Y) = \tilde{Y}^*(T) \longrightarrow Y$ in $\mathcal{F}$, we have morphisms $U \xrightarrow{e} \tilde{Y}_*(Y)$ corresponding to $Y \xrightarrow{1} Y$ and

$$\tilde{Y}_*(Y) \otimes \tilde{Y}_*(Y) \xrightarrow{m} \tilde{Y}_*(Y)$$

corresponding to

$$(\tilde{Y}_*(Y)_{\mathcal{F}})^2(Y) \xrightarrow{\tilde{Y}_*(Y)_{\mathcal{F}}(\varepsilon_Y)} \tilde{Y}_*(Y)_{\mathcal{F}}(Y) \xrightarrow{\varepsilon_Y} Y$$

where ε is the counit of $(\tilde{Y}^* \dashv \tilde{Y}_*)$. It is easily verified that these morphisms make $\mathcal{Y} = (\tilde{Y}_*(Y), m, e)$ into a $\otimes$-monoid in $\mathcal{E}[U]$, i.e. an internal finitary algebraic theory.

Now if $\mathbb{T}$ is any other $\otimes$-monoid in $\mathcal{E}[U]$, it is again easy to check that a morphism $T \longrightarrow \tilde{Y}_*(Y)$ is a monoid homomorphism $\mathbb{T} \longrightarrow \mathbb{Y}$ iff its transpose $T_{\mathcal{F}}(Y) \longrightarrow Y$ is a $\mathbb{T}$-model structure for Y. So we can describe $\mathbb{Y}$ as the generic finitary algebraic theory of which Y is a model.

5.5 Lemma Let $\mathcal{E}$ be a topos with a natural number object, $\mathbb{T}$ a finitary algebraic theory in $\mathcal{E}$ and $\mathcal{F}$ an $\mathcal{E}$-topos. Then the category $\mathbb{T}(\mathcal{F})$ of $\mathbb{T}$-models in $\mathcal{F}$ has the structure of a locally internal category over $\mathcal{F}$.

Proof We saw in the last chapter that $T_{\mathcal{F}}$ is a locally internal endofunctor of $\mathcal{F}$, and in particular that it is enriched over $\mathcal{F}$. It follows from the work of Kock [19] that $\mathbb{T}(\mathcal{F})$ is an $\mathcal{F}$-category. Explicitly, if (A,α) and (B,β) are $\mathbb{T}$-models in $\mathcal{F}$, we may define the object of $\mathbb{T}$-model homomorphisms from (A,α) to (B,β) by the equalizer diagram

$$\mathbb{T}(\mathcal{F})((A,\alpha),(B,\beta)) \rightarrowtail B^A \underset{T_{A,B}\ \searrow\ T_{\mathcal{F}}(B)^{T_{\mathcal{F}}(A)}\ \nearrow\ \beta^{T_{\mathcal{F}}(A)}}{\overset{B^\alpha}{\longrightarrow}} B^{T_{\mathcal{F}}(A)}$$

in $\mathcal{F}$, where $T_{A,B}$ denotes the strength of $T_{\mathcal{F}}$.

But pullback functors preserve exponentials and equalizers, and hence commute up to isomorphism with the above construction; so the assignment

$$I \longmapsto \mathbb{T}(\mathcal{F}/I)$$

defines a locally internal category over $\mathcal{F}$. □

5.6 Lemma With the same hypotheses as Lemma 5.5, the forgetful functor $\mathbb{T}(\mathcal{F}) \longrightarrow \mathcal{F}$ creates reflexive coequalizers.

Proof By Proposition 4.6, the functor $T_{\mathcal{F}}$ preserves reflexive coequalizers. Thus if we have a reflexive pair

$$(A,\alpha) \underset{v}{\overset{u}{\rightrightarrows}} (B,\beta)$$

in $\mathbb{T}(\mathcal{F})$, and a coequalizer diagram

$$A \underset{v}{\overset{u}{\rightrightarrows}} B \xrightarrow{\ w\ } C$$

in $\mathcal{F}$, we obtain a diagram

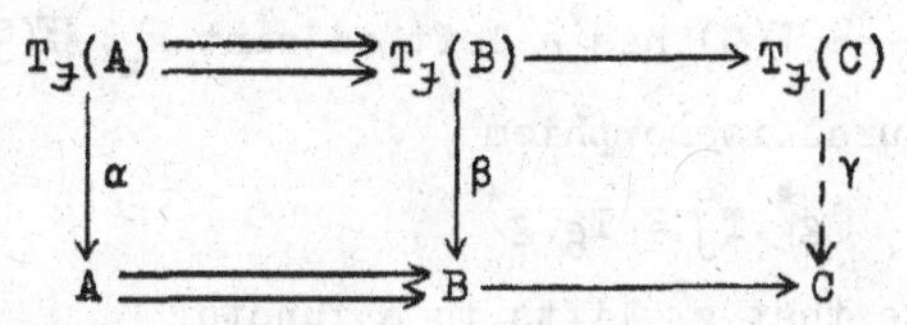

in which both the rows are coequalizers. Hence there is a unique map γ making the right-hand square commute; it is now easy to check that (C,γ) is a $\mathbb{T}$-model, and that it is the coequalizer in $\mathbb{T}(\mathcal{F})$ of the pair (u,v). □

<u>5.7 Corollary</u> $\mathbb{T}(\mathcal{F})$ is an exact category in the sense of Barr [2], and has all finite limits and colimits.

<u>Proof</u> Existence of finite limits is trivial since the forgetful functor, being monadic, creates them. Since it also creates coequalizers of equivalence relations, by a special case of Lemma 5.6, we can simply "lift" each of the axioms defining an exact category from $\mathcal{F}$ to $\mathbb{T}(\mathcal{F})$. Note in particular that a morphism in $\mathbb{T}(\mathcal{F})$ is regular epi (i.e. occurs as a coequalizer) iff its underlying morphism in $\mathcal{F}$ is epi.

Lemma 5.6 tells us that $\mathbb{T}(\mathcal{F})$ has coequalizers of reflexive pairs. To obtain more general finite colimits, we use a well-known result of Linton ([27], Corollary 2) and the fact that $\mathcal{F}$ has finite colimits [32]. □

Of course, we cannot in general hope to form infinite external limits or colimits in $\mathbb{T}(\mathcal{F})$; but we should be able to form limits and colimits indexed by internal categories in $\mathcal{F}$. The next lemma enables us to do so.

<u>5.8 Lemma</u> Let $\mathbb{T}$ be an internal finitary algebraic theory in $\mathcal{E}$,

and let $\mathcal{G} \xrightarrow{g} \mathcal{F}$ be a map of $\mathcal{E}$-toposes. Then g_* and g^* can be lifted to an adjoint pair of functors (which we denote by the same symbols) between $\mathbb{T}(\mathcal{G})$ and $\mathbb{T}(\mathcal{F})$. Furthermore, if g is essential, i.e. g^* has a left adjoint $g_!: \mathcal{G} \longrightarrow \mathcal{F}$, then its lifting $g^*: \mathbb{T}(\mathcal{F}) \longrightarrow \mathbb{T}(\mathcal{G})$ has a left adjoint $g_\#: \mathbb{T}(\mathcal{G}) \longrightarrow \mathbb{T}(\mathcal{F})$.

<u>Proof</u> We have a natural isomorphism

$$g^*.T_{\mathcal{F}} \cong T_{\mathcal{G}}.g^* \quad ,$$

so it is immediate that g^* lifts to a functor

$$g^*: \mathbb{T}(\mathcal{F}) \longrightarrow \mathbb{T}(\mathcal{G}) \quad .$$

From the counit of $(g^* \dashv g_*)$, we get a natural map

$$g^* T_{\mathcal{F}} g_* \cong T_{\mathcal{G}} g^* g_* \longrightarrow T_{\mathcal{G}}$$

which transposes to give a map $T_{\mathcal{F}} g_* \xrightarrow{\theta} g_* T_{\mathcal{G}}$. We may now define $g_*: \mathbb{T}(\mathcal{G}) \longrightarrow \mathbb{T}(\mathcal{F})$ by

$$g_*(A,\alpha) = (g_*A, g_*(\alpha).\theta_A) \quad .$$

For the proof that this does define a functor right adjoint to g^*, see [16], Theorem 4.

Since $\mathbb{T}(\mathcal{F})$ has reflexive coequalizers by Lemma 5.6, the construction of $g_\#$ when g is essential is a straightforward application of [16], Theorem 2. □

<u>5.9 Corollary</u> The locally internal category $\mathbb{T}(\mathcal{F})$ is $\mathcal{F}$-complete and cocomplete, in the sense of [17], Theorem A.11.

<u>Proof</u> We have to check that $\mathbb{T}(\mathcal{F})$ has locally internal equalizers and coequalizers, that pullback functors $u^*: \mathbb{T}(\mathcal{F}/J) \longrightarrow \mathbb{T}(\mathcal{F}/I)$ induced by morphisms $I \xrightarrow{u} J$ in $\mathcal{F}$ have left and right adjoints $u_\#$, u_*, and the "Beck condition" that

$$\begin{array}{ccc} \mathbb{T}(\mathcal{F}/J) & \xrightarrow{u^*} & \mathbb{T}(\mathcal{F}/I) \\ \downarrow{\scriptstyle w_*} & & \downarrow{\scriptstyle v_*} \\ \mathbb{T}(\mathcal{F}/L) & \xrightarrow{x^*} & \mathbb{T}(\mathcal{F}/K) \end{array}$$

commutes up to isomorphism whenever

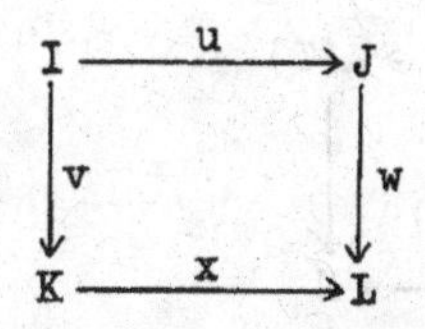

is a pullback square in $\mathcal{J}$. But the first condition is trivial from Corollary 5.7, the second is a special case of Lemma 5.8, and the Beck condition follows from the corresponding condition for $\mathcal{J}$ itself. □

We now turn from investigating the category of models of an individual algebraic theory to investigating the category $\underset{\sim}{\text{alg}}(\mathcal{E})$ itself. Our first objective is to construct a free ⊗-monoid functor $\mathcal{E}[U] \longrightarrow \underset{\sim}{\text{alg}}(\mathcal{E})$; it turns out that the "iterated profunctor composition" which we described in chapter III holds the key to this construction.

Lemma 4.5 tells us that $\mathcal{E}[U]$ is equivalent to the category of left flat profunctors $\underline{E}_{fin} \dashrightarrow \underline{E}_{fin}$. So it follows from Lemma 3.6 that the functor

$$(-)^{\otimes \underset{\sim}{n}}: \text{Prof}(\underline{E}_{fin}, \underline{E}_{fin}) \longrightarrow \text{Prof}(\underline{E}_{fin}, \underline{E}_{fin})/N$$

induces a functor (which we shall also denote by $(-)^{\otimes \underset{\sim}{n}}$)

$$\mathcal{E}[U] \longrightarrow \mathcal{E}[U]/N \quad .$$

This functor clearly satisfies the recursion data

$$T^{\otimes o} \cong U \quad , \quad T^{\otimes s\underset{\sim}{n}} \cong T^{\otimes \underset{\sim}{n}} \otimes T \cong T \otimes T^{\otimes \underset{\sim}{n}}$$

for all objects T of $\mathcal{E}[U]$. We write A(T) for $\sum_{\underset{\sim}{n}} T^{\otimes \underset{\sim}{n}}$.

<u>5.10 Proposition</u> A(T) has the structure of a ⊗-monoid $\mathbb{A}(T)$, which is the free ⊗-monoid generated by T.

<u>Proof</u> The unit of the monoid structure on A(T) is defined by the pullback

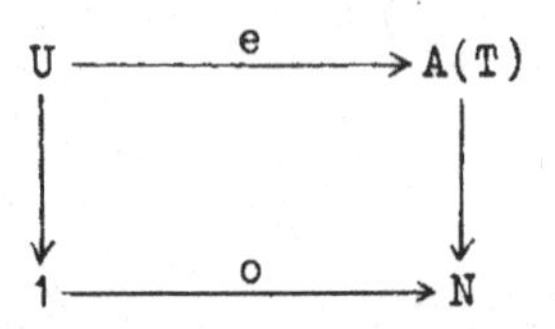

arising from the isomorphism $T^{\otimes o} \cong U$. By an application of the uniqueness theorem 2.2, we obtain an isomorphism

$$T^{\otimes(\underset{\sim}{n}_1+\underset{\sim}{n}_2)} \cong T^{\otimes \underset{\sim}{n}_1} \otimes T^{\otimes \underset{\sim}{n}_2}$$

in $\mathcal{E}[U]/N\times N$, giving rise to a pullback diagram

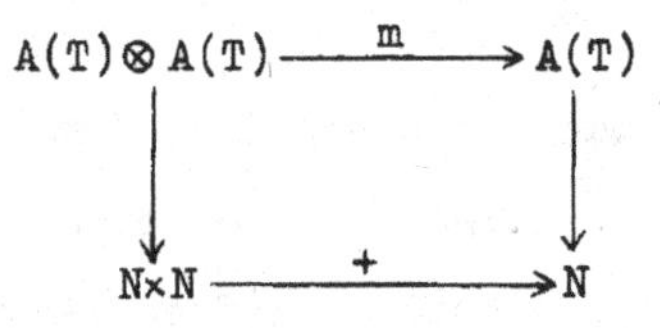

which defines the multiplication on A(T). The fact that the above squares are pullbacks makes it easy to verify that (A(T),m,e) is indeed a monoid.

The unit of the adjunction (A ⊣ forgetful functor) similarly derives from the pullback

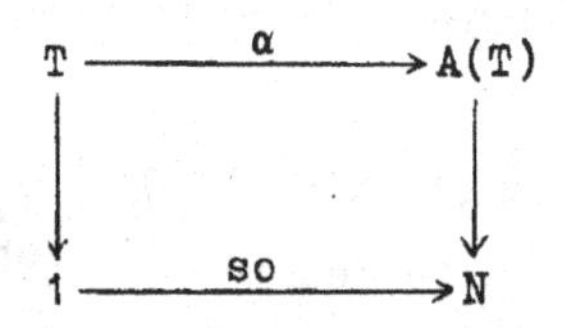

To construct the counit, suppose T has a $\otimes$-monoid structure (T,μ,η). We want to define a map $A(T) \xrightarrow{\beta} T$ in $\mathcal{E}[U]$, or equivalently a map $T^{\otimes \underset{\sim}{n}} \longrightarrow T$ in $\mathcal{E}[U]/N$, or equivalently again a U-element

$$U \longrightarrow T^{\otimes \underset{\sim}{n}} \pitchfork T$$

in $\mathcal{E}[U]/N$ (where we are temporarily using the "internal hom" notation $S \pitchfork (-)$ for the right adjoint of $(-)\otimes S = \tilde{S}^*$). We construct the latter by (a slight generalization of) the section criterion 2.1, using the data

$$U \xrightarrow{\eta} T \cong U \pitchfork T \cong T^{\otimes o} \pitchfork T \quad \text{and} \quad T^{\otimes \underset{\sim}{n}} \pitchfork T \xrightarrow{\theta} T^{\otimes s\underset{\sim}{n}} \pitchfork T ,$$

where the transpose of θ is the composite

$$(T^{\otimes \underline{n}} \pitchfork T) \otimes T^{\otimes s\underline{n}} \cong (T^{\otimes \underline{n}} \pitchfork T) \otimes T^{\otimes \underline{n}} \otimes T \xrightarrow{ev \otimes 1} T \otimes T \xrightarrow{\mu} T .$$

Straightforward inductive arguments can now be used to show that β is in fact a monoid homomorphism $\mathbb{A}(T) \longrightarrow \mathbb{T}$, and that α and β satisfy the "triangular identities". □

<u>5.11 Lemma</u> The forgetful functor $\underset{\sim}{\text{alg}}(\mathcal{E}) \longrightarrow \mathcal{E}[U]$ creates reflexive coequalizers.

<u>Proof</u> Let

$$R \underset{g}{\overset{f}{\rightrightarrows}} S \xrightarrow{h} T$$

be a reflexive coequalizer diagram in $\mathcal{E}[U]$. Consider the diagram

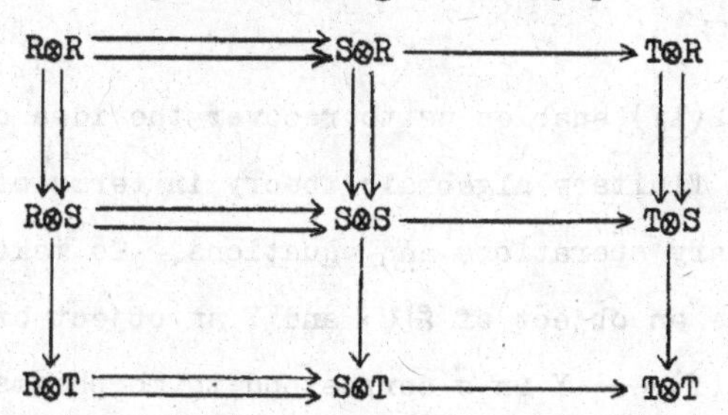

Here the rows are reflexive coequalizers since $(-)\otimes R = \tilde{R}^*$ preserves all colimits, and the columns are reflexive coequalizers by Proposition 4.6, since $R\otimes(-) = R_{\mathcal{E}[U]}$. So by [17], Lemma 0.17, the diagonal

$$R\otimes R \underset{g\otimes g}{\overset{f\otimes f}{\rightrightarrows}} S\otimes S \xrightarrow{h\otimes h} T\otimes T$$

is a coequalizer. It now follows exactly as in Lemma 5.6 that if R and S have $\otimes$-monoid structures and f and g are monoid homomorphisms, then there is a unique $\otimes$-monoid structure on T making h a monoid homomorphism; and h is then the coequalizer of f and g in $\underset{\sim}{\text{alg}}(\mathcal{E})$. □

<u>5.12 Theorem</u> (i) $\underset{\sim}{\text{alg}}(\mathcal{E})$ is monadic over $\mathcal{E}[U]$.

(ii) $\underset{\sim}{\text{alg}}(\mathcal{E})$ is monadic over $\mathcal{E}/N$.

(iii) $\underset{\sim}{\text{alg}}(\mathcal{E})$ has finite limits and colimits.

(iv) $\underset{\sim}{\mathrm{alg}}(\mathcal{E})$ has the structure of a locally internal category over $\mathcal{E}$, and is $\mathcal{E}$-complete and cocomplete.

Proof Proposition 5.10 and Lemma 5.11 imply that the forgetful functor $\underset{\sim}{\mathrm{alg}}(\mathcal{E}) \longrightarrow \mathcal{E}[U]$ satisfies all the hypotheses of the Crude Tripleability Theorem (in its "reflexive-coequalizer" form, see [17], Theorem 0.13), so (i) is immediate. (ii) follows from (i) and the fact that $\mathcal{E}[U]$ is monadic over $\mathcal{E}/N$, since the hypotheses of CTT are stable under composition. (iii) follows from (i) and Lemma 5.11, as in the proof of Corollary 5.7; and (iv) is verified by arguments similar to those used in proving 5.8 and 5.9. □

Theorem 5.12(ii) enables us to recover the idea of a presentation of a finitary algebraic theory in terms of two N-indexed families of finitary operations and equations. To make this more explicit, let T be an object of $\mathcal{E}[U]$ and Y an object of an $\mathcal{E}$-topos $\mathcal{F}$. Then morphisms $T_{\mathcal{F}}(Y) \longrightarrow Y$ in $\mathcal{F}$ correspond to morphisms $T \longrightarrow \tilde{Y}_*(Y)$ in $\mathcal{E}[U]$, and hence to monoid homomorphisms $\mathbb{A}(T) \longrightarrow \mathbb{Y}$, where $\mathbb{Y}$ is the theory of Example 5.4(d). But these in turn correspond to $\mathbb{A}(T)$-model structures on Y. In particular, if T is freely generated by an object $C(\underset{\sim}{n})$ of $\mathcal{E}/N$, then the remarks before Proposition 4.6 tell us that specifying an $\mathbb{A}(T)$-model structure on Y is equivalent to specifying a morphism

$$\Sigma_{\underset{\sim}{n}}\; C(\underset{\sim}{n}) \times Y^{[\underset{\sim}{n}]} \longrightarrow Y \quad ,$$

i.e. a "$C(\underset{\sim}{n})$-indexed family of finitary operations on Y".

We are now, therefore, in a position to "lift" most of the basic concepts and theorems of classical (i.e. $\mathcal{S}$-based) universal algebra to the internal case. By way of an example, we give the "internalization" of a familiar classical result:

5.13 Theorem Let $\mathbb{T}$ be an internal finitary algebraic theory in a topos $\mathcal{E}$. The following conditions are equivalent:

(i) $\mathbb{T}(\mathcal{E}/I)$ is abelian for all objects I of $\mathcal{E}$.

(ii) $\mathbb{T}(\mathcal{E})$ is enriched over $\underset{\sim}{ab}(\mathcal{E})$.

(iii) There exists a ring R in $\mathcal{E}$ such that $\mathbb{T} \cong R\text{-}\underset{\sim}{mod}$.

Proof Since $\mathbb{T}(\mathcal{E}/I)$ is exact and enriched over $\mathcal{E}/I$ (5.5 and 5.7), the equivalence of (i) and (ii) follows easily from Tierney's theorem ([2], Theorem I.3.11). The equivalence of (ii) and (iii) is proved as in [38], §4: we define the ring R to be T(so) (the ring structure arises from the fact that T(so) is the object of endomorphisms of the free $\mathbb{T}$-model on one generator), and then use induction to establish an isomorphism between the object of operations of $\mathbb{T}$ (i.e. the object $T(\underset{\sim}{n})$) and the object of operations of $R\text{-}\underset{\sim}{mod}$. □

We leave it to the reader to formulate the "internal" definitions of such concepts as Kronecker product and commutativity of theories, affine theories, Mal'cev theories [35], ... and to prove that they have their usual properties.

Next, we investigate the relationship between Definition 5.3 and the notion of finitary algebraic theory introduced in [22] by F.W. Lawvere. We recall that objects T of $\mathcal{E}[U]$ correspond to left flat profunctors $\underline{\underline{E}}_{fin} \overset{\check{T}}{- - \rightarrow} \underline{\underline{E}}_{fin}$ (explicitly, the correspondence is given by

$$\check{T}(\underset{\sim}{n}_1, \underset{\sim}{n}_2) = T(\underset{\sim}{n}_2)^{[\underset{\sim}{n}_1]}$$

with left action of $\underline{\underline{E}}_{fin}$ induced by composition and right action induced by that on $T(\underset{\sim}{n})$, and

$$T(\underset{\sim}{n}) = \check{T}(so, \underset{\sim}{n}) \qquad).$$

Moreover, a $\otimes$-monoid structure (T,m,e) on T clearly corresponds to a $\otimes_{\underline{\underline{E}}_{fin}}$-monoid structure $(\check{T}, \check{m}, \check{e})$ on $\check{T}$; an endoprofunctor equipped with such a structure is commonly called a profunctor monad.

Now according to M. Justersen [18], a profunctor monad $\check{\mathbb{T}} = (\check{T},\check{m},\check{e})$ on a category $\underline{A}$ corresponds uniquely to a functor

$$\underline{A} \xrightarrow{e} \underline{A}_{\check{\mathbb{T}}}$$

which is an isomorphism on objects. The "hom-sets" of $\underline{A}_{\check{\mathbb{T}}}$ are given by

$$\underline{A}_{\check{\mathbb{T}}}(\underset{\sim}{a}_1,\underset{\sim}{a}_2) = \check{T}(\underset{\sim}{a}_1,\underset{\sim}{a}_2)$$

with composition defined by the map

$$\Sigma_{\underset{\sim}{a}_2} \check{T}(\underset{\sim}{a}_1,\underset{\sim}{a}_2) \times \check{T}(\underset{\sim}{a}_2,\underset{\sim}{a}_3) \longrightarrow\!\!\!\!\!\rightarrow \check{T}\otimes_{\underline{A}}\check{T}\ (\underset{\sim}{a}_1,\underset{\sim}{a}_3) \xrightarrow{\check{m}} \check{T}(\underset{\sim}{a}_1,\underset{\sim}{a}_3)\ ,$$

and the functor e is simply $\check{e}$. Conversely, if we are given a functor $\underline{A} \xrightarrow{e} \underline{B}$ which is the identity on objects, we can define left and right actions of $\underline{A}$ on $\underline{B}(\underset{\sim}{a}_1,\underset{\sim}{a}_2)$ by

$$\Sigma_{\underset{\sim}{a}_2}\underline{A}(\underset{\sim}{a}_1,\underset{\sim}{a}_2) \times \underline{B}(\underset{\sim}{a}_2,\underset{\sim}{a}_3) \xrightarrow{e\times 1} \Sigma_{\underset{\sim}{a}_2}\underline{B}(\underset{\sim}{a}_1,\underset{\sim}{a}_2) \times \underline{B}(\underset{\sim}{a}_2,\underset{\sim}{a}_3) \xrightarrow{c} \underline{B}(\underset{\sim}{a}_1,\underset{\sim}{a}_3)\ ,$$

etc., where c is the composition map for $\underline{B}$; and it is then clear that $\underline{B}(\underset{\sim}{a}_1,\underset{\sim}{a}_2)$ has the structure of a profunctor monad on $\underline{A}$. We might well call $\underline{A}_{\check{\mathbb{T}}}$ the Kleisli category of $\check{\mathbb{T}}$; as frequently happens, the construction for profunctors is smoother (because more symmetric) than that for functors.

In the case when $\underline{A} = \underline{E}_{fin}$, we shall normally write $\underline{\underline{FFT}}$ for $(\underline{E}_{fin})_{\check{\mathbb{T}}}$; the reason for this notation will become clear shortly. Now it is easily verified that the left Kan extension functor

$$\varinjlim_e \colon \mathcal{E}^{(\underline{E}_{fin})} \longrightarrow \mathcal{E}^{\underline{\underline{FFT}}}$$

(i.e. the left adjoint of pullback along e) is given by

$$(\varinjlim_e(S))(\underset{\sim}{n}) = (S\otimes_{\underline{E}_{fin}}\check{T})(\underset{\sim}{n})\ ,$$

with right action of $\underline{\underline{FFT}}$ induced by the multiplication $\check{m}$. Thus the left flatness of $\check{T}$ may be expressed by saying that $\varinjlim_e$ is left exact. In general, a functor $\underline{A} \xrightarrow{f} \underline{B}$ such that $\varinjlim_f \colon \mathcal{E}^{\underline{A}} \longrightarrow \mathcal{E}^{\underline{B}}$ is left exact is said to be <u>internally right exact</u>; if $\mathcal{E} = \mathcal{S}$, this condition is equivalent to saying that f preserves all finite colimits, provided $\underline{A}$ has them.

We have thus proved

5.14 Theorem Let $\mathcal{E}$ be a topos with a natural number object. Define an internal Lawvere theory in $\mathcal{E}$ to be an internal functor

$$\underline{\underline{E}}_{fin} \xrightarrow{e} \underline{\underline{T}}$$

which is internally right exact and isomorphic on objects, and a map of internal Lawvere theories to be a commutative triangle

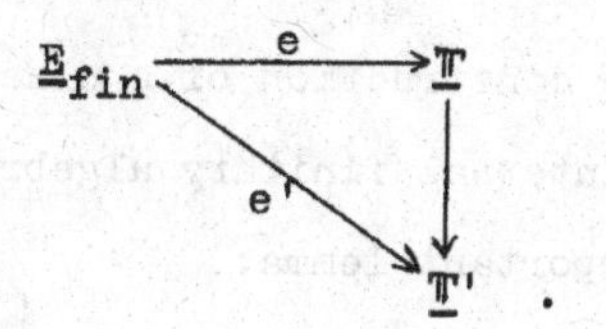

Then the category of internal Lawvere theories is equivalent to the category $\underset{\sim\sim}{alg}(\mathcal{E})$ defined in 5.3. □

The internal category $\underline{\underline{FFT}}$ provides the answer to another naturally arising question. If T is an object of $\mathcal{E}[U]$, specifying a $\otimes$-monoid structure on T is clearly equivalent to specifying a monad structure on the functor $(-)\otimes T = \tilde{T}^*$, or to specifying a comonad structure on its right adjoint $\tilde{T}_*$. But $\tilde{T}_*$ is left exact, and so by a well-known result ([17], Theorem 2.32) the category of $\tilde{\mathbb{T}}_*$-coalgebras (which is isomorphic to the category of $\tilde{\mathbb{T}}^*$-algebras) is a topos. How can we give an explicit site of definition for this topos?

5.15 Proposition Let $\mathbb{T}$ be an internal algebraic theory in a topos $\mathcal{E}$, and let $\tilde{\mathbb{T}}_*$ denote the corresponding comonad on $\mathcal{E}[U]$. Then the category of $\tilde{\mathbb{T}}_*$-coalgebras is equivalent to $\mathcal{E}^{\underline{\underline{FFT}}}$.

Proof Since the internal functor $\underline{\underline{E}}_{fin} \xrightarrow{e} \underline{\underline{FFT}}$ is epimorphic on objects, it follows from [17], Exercise 4.2, that the induced geometric morphism

$$\mathcal{E}^{(\underline{\underline{E}}_{fin})} \xrightarrow{e} \mathcal{E}^{\underline{\underline{FFT}}}$$

is a surjection, i.e. that $\mathcal{E}^{\underline{\underline{FFT}}}$ is equivalent to the category of coalgebras for the comonad induced by $(e^* \dashv \varprojlim_e)$, or of algebras for the monad induced by $(\varinjlim_e \dashv e^*)$. But it follows from the remarks before Theorem 5.14 that the composite

$$\mathcal{E}^{(\underline{\mathbb{E}}_{fin})} \xrightarrow{\varinjlim_e} \mathcal{E}^{\underline{FFT}} \xrightarrow{e^*} \mathcal{E}^{(\underline{\mathbb{E}}_{fin})}$$

is simply the functor

$$(-)\otimes_{\underline{\mathbb{E}}_{fin}} \check{\mathbb{T}}: \mathcal{E}^{(\underline{\mathbb{E}}_{fin})} \longrightarrow \mathcal{E}^{(\underline{\mathbb{E}}_{fin})} \quad ,$$

i.e. the functor $\tilde{\mathbb{T}}^*$. The remaining details are straightforward.□

We now embark on the construction of a classifying topos for $\mathbb{T}$-models, where $\mathbb{T}$ is any internal finitary algebraic theory. We begin with a simple but important lemma:

<u>5.16 Lemma</u> Let $\mathbb{T}$ be an internal theory in $\mathcal{E}$, let $\mathcal{F}$ be an $\mathcal{E}$-topos, and let

$$\mathcal{F} \xrightarrow{F} \mathbb{T}(\mathcal{F}) \quad , \quad \mathbb{T}(\mathcal{F}) \xrightarrow{G} \mathcal{F}$$

denote the free and forgetful functors. Then the adjunction $(F \dashv G)$ is enriched over $\mathcal{F}$; i.e. we have a natural isomorphism

$$\mathbb{T}(\mathcal{F})(FX,A) \cong GA^X$$

for objects X, A of $\mathcal{F}$ and $\mathbb{T}(\mathcal{F})$ respectively.

<u>Proof</u> The argument is similar to that of Lemma 1.1. The "external" adjunction between F and G establishes a bijection between the global elements of these two objects of $\mathcal{F}$. But F and G are both indexed functors between locally internal categories; hence this bijection extends naturally to the I-elements of the two objects for any I, and so by the principle of the Yoneda lemma they must be isomorphic as objects of $\mathcal{F}$. □

<u>5.17 Corollary</u> Let p be a natural number in $\mathcal{F}$. Then the free $\mathbb{T}$-model F[p] is internally projective in $\mathbb{T}(\mathcal{F})$, in the sense that the functor

$$\mathbb{T}(\mathcal{F})(F[p],-): \mathbb{T}(\mathcal{F}) \longrightarrow \mathcal{F}$$

preserves regular epimorphisms.

<u>Proof</u> This is immediate from combining Lemma 5.16 and Proposition

2.4, together with the fact (which we observed in Corollary 5.7) that regular epis in $\mathbb{T}(\mathcal{F})$ are epi in $\mathcal{F}$. □

Since $\mathbb{T}(\mathcal{E})$ is a locally internal category over $\mathcal{E}$, we can construct internal full subcategories of $\mathbb{T}(\mathcal{E})$ generated by indexed families of $\mathbb{T}$-models in $\mathcal{E}$. In particular, we can consider the internal category $\mathrm{Full}_{\mathbb{T}(\mathcal{E})}(F[\underset{\sim}{n}])$ of finitely-generated free $\mathbb{T}$-models in $\mathcal{E}$. By Lemma 5.16, the hom-sets of this category are given by

$$\mathbb{T}(\mathcal{E}/N\times N)(F[\underset{\sim}{n}_1],F[\underset{\sim}{n}_2]) \cong T(\underset{\sim}{n}_2)^{[\underset{\sim}{n}_1]} \cong \check{T}(\underset{\sim}{n}_1,\underset{\sim}{n}_2) \quad ,$$

where $\check{T}$ is the profunctor corresponding to T. It is thus not difficult to show that we have an isomorphism

$$\mathrm{Full}_{\mathbb{T}(\mathcal{E})}(F[\underset{\sim}{n}]) \cong (\underline{E}_{fin})_{\check{\mathbb{T}}}$$

where the right-hand side is the Kleisli category which we considered before Theorem 5.14. This explains the name of $\underline{FF\mathbb{T}}$ which we gave to the latter category, and which we shall henceforth use for $\mathrm{Full}_{\mathbb{T}(\mathcal{E})}(F[\underset{\sim}{n}])$.

We shall say that a $\mathbb{T}$-model A is <u>finitely-presented</u> if there exist natural numbers p, q in $\mathcal{E}$ and a coequalizer diagram

$$F[q] \rightrightarrows F[p] \longrightarrow A$$

in $\mathbb{T}(\mathcal{E})$.

<u>5.18 Lemma</u> Let A and B be $\mathbb{T}$-models in $\mathcal{E}$ such that A is finitely-presented. Then the object $\mathbb{T}(\mathcal{E})(A,B)$ is preserved by inverse image functors, in the sense that for any $\mathcal{F} \xrightarrow{f} \mathcal{E}$ we have

$$f^*(\mathbb{T}(\mathcal{E})(A,B)) \cong \mathbb{T}(\mathcal{F})(f^*A,f^*B) \ .$$

<u>Proof</u> If A is a finitely-generated free model, this is immediate from Lemma 5.16 and Proposition 2.3, since we have

$$\mathbb{T}(\mathcal{E})(F[p],B) \cong GB^{[p]} \ .$$

In general, a finite presentation of A as above gives rise to an equalizer diagram

$$\mathbb{T}(\mathcal{E})(A,B) \longrightarrow GB^{[p]} \rightrightarrows GB^{[q]}$$

and the result follows from the fact that f^* preserves equalizers. □

Now consider the object

$$P = \Sigma_{\underset{\sim}{n}_1,\underset{\sim}{n}_2}(\underline{\mathrm{FFT}}(\underset{\sim}{n}_1,\underset{\sim}{n}_2) \times \underline{\mathrm{FFT}}(\underset{\sim}{n}_1,\underset{\sim}{n}_2)) \quad .$$

We think of P as the object of finite presentations of $\mathbb{T}$-models in $\mathcal{E}$; if we write d_1, d_2 for the two projections $P \rightrightarrows N$, we have a generic parallel pair of maps

$$F[d_1\underset{\sim}{p}] \rightrightarrows F[d_2\underset{\sim}{p}]$$

in $\mathbb{T}(\mathcal{E}/P)$. If we denote the coequalizer of this pair by

$$F[d_2\underset{\sim}{p}] \twoheadrightarrow M(\underset{\sim}{p}) \quad ,$$

then $M(\underset{\sim}{p})$ is the generic finitely-presented $\mathbb{T}$-model. We write $\underline{\mathrm{FPT}}$ for $\mathrm{Full}_{\mathbb{T}(\mathcal{E})}(M(\underset{\sim}{p}))$, the internal category of finitely-presented $\mathbb{T}$-models in $\mathcal{E}$.

<u>5.19 Lemma</u> The internal categories $\underline{\mathrm{FFT}}$ and $\underline{\mathrm{FPT}}$ are preserved by inverse image functors.

<u>Proof</u> Proposition 2.3 and Lemma 5.18 together imply that inverse image functors preserve all the machinery involved in the definition of these two categories. □

Now let A be a $\mathbb{T}$-model in $\mathcal{E}$. We define a presheaf HA on $\underline{\mathrm{FPT}}$ by $HA(\underset{\sim}{p}) = \mathbb{T}(\mathcal{E}/P)(M(\underset{\sim}{p}),A)$, with action of $\underline{\mathrm{FPT}}$ given by composition as usual.

<u>5.20 Lemma</u> For any $\mathbb{T}$-model A, HA is a flat presheaf on $\underline{\mathrm{FPT}}$.

<u>Proof</u> Once again, we have to verify the conditions of Proposition 1.3. But condition (a) is trivial, since the initial object F(0) of $\mathbb{T}(\mathcal{E})$ is finitely-presented, and so $\Sigma_{\underset{\sim}{p}}HA(\underset{\sim}{p})$ has a global element corresponding to $F(0) \longrightarrow A$.

For condition (b), suppose we are given morphisms

$$M_1 \longrightarrow A \longleftarrow M_2$$

in $\mathbb{T}(\mathcal{E}/I)$ for some I, where each M_i has a finite presentation

$$F[q_i] \rightrightarrows F[p_i] \longrightarrow M_i \ .$$

Since F preserves coproducts, we can give a finite presentation

$$F[q_1+q_2] \rightrightarrows F[p_1+p_2] \longrightarrow M_1 * M_2$$

of the coproduct of M_1 and M_2 in $\mathbb{T}(\mathcal{E}/I)$, which enables us to complete the required diagram.

For condition (c), suppose we have a diagram

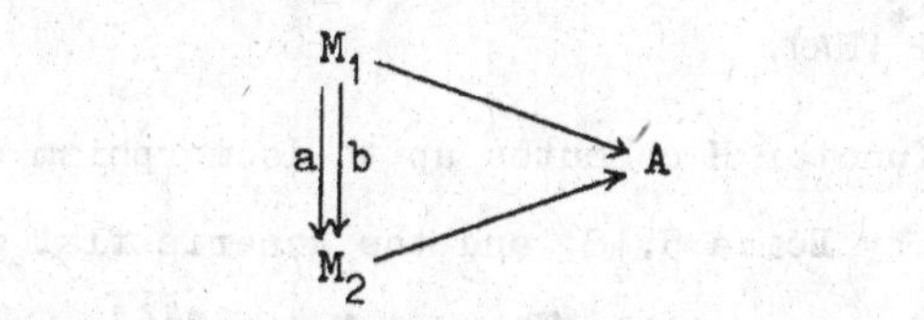

in $\mathbb{T}(\mathcal{E}/I)$, where the M_i are finitely-presented as above. In order to give a finite presentation

$$F[q_2+p_1] \rightrightarrows F[p_2] \longrightarrow M_3$$

of the coequalizer of a and b, we need to construct liftings of the composites

$$F[p_1] \twoheadrightarrow M_1 \overset{a}{\underset{b}{\rightrightarrows}} M_2$$

across the epimorphism $F[p_2] \twoheadrightarrow M_2$. But since $F[p_1]$ is internally projective by Corollary 5.17, the liftings we require exist locally, i.e. in $\mathbb{T}(\mathcal{E}/J)$ for some $J \twoheadrightarrow I$; and so we can complete the above diagram to one of the required form in $\mathbb{T}(\mathcal{E}/J)$.□

Let i_o: $N \longrightarrow P$ be the N-element of P corresponding to the presentation

$$F[o] \rightrightarrows F[\underset{\sim}{n}] \longrightarrow F[\underset{\sim}{n}]$$

of the generic finitely-generated free model. (In fact it is easily seen that i_o is the object-map of a full embedding of internal categories

$$\underline{\text{FFT}} \xrightarrow{\ i\ } \underline{\text{FPT}} \ .)$$

Now $HA(i_o\underset{\sim}{n}) \cong \mathbb{T}(\mathcal{E}/N)(F[\underset{\sim}{n}],A) \cong GA^{[\underset{\sim}{n}]}$, and in particular

$$HA(i_o so) \cong GA ,$$

i.e. we can recover the underlying object of A from the presheaf HA. But in fact we can recover the $\mathbb{T}$-model structure as well, since the map

$$\Sigma_{\underset{\sim}{n}} \; T(\underset{\sim}{n}) \times GA^{[\underset{\sim}{n}]} \longrightarrow GA$$

which induces it is simply the action map

$$\Sigma_{\underset{\sim}{n}} \; \underline{FFT}(so,\underset{\sim}{n}) \times HA(i_o\underset{\sim}{n}) \longrightarrow HA(i_o so)$$

of $\underline{FFT}$ on the presheaf $i^*(HA)$.

Furthermore, the functor H commutes up to isomorphism with inverse image functors, by Lemma 5.18; and the generic flat presheaf $Y(\underline{FPT})$ has the form HM for a suitable $\mathbb{T}$-model M in $\mathcal{E}^{\underline{FPT}}$. (The underlying object of M is simply the object $GM(\underset{\sim}{p})$ of $\mathcal{E}/P$, with $\underline{FPT}$ acting by the evaluation map - it is easy to check that the $\mathbb{T}$-model structure on $M(\underset{\sim}{p})$ extends to one on M.) We may thus deduce as in Lemma 4.5 that the functor

$$H\colon \mathbb{T}(\mathcal{E}) \longrightarrow \mathrm{Flat}(\underline{FPT}^{op},\mathcal{E})$$

is an equivalence of categories. So we have

<u>5.21 Theorem</u> The topos $\mathcal{E}^{\underline{FPT}}$ is a classifying topos for $\mathbb{T}$-models in $\mathcal{E}$-toposes; i.e. for any $\mathcal{E}$-topos $\mathcal{F}$ the functor

$$\underline{Top}/\mathcal{E}\;(\mathcal{F},\mathcal{E}^{\underline{FPT}}) \longrightarrow \mathbb{T}(\mathcal{F});\; g \longmapsto g^*M$$

is an equivalence of categories, where M is the particular $\mathbb{T}$-model in $\mathcal{E}^{\underline{FPT}}$ defined above. □

Since the internal functor $\underline{FFT} \xrightarrow{\;i\;} \underline{FPT}$ is a full embedding, it follows from [17], Exercise 4.2, that the geometric morphism

$$\mathcal{E}^{\underline{FFT}} \xrightarrow{\;\;i\;\;} \mathcal{E}^{\underline{FPT}}$$

which it induces is an inclusion. In fact $\mathcal{E}^{\underline{FFT}}$ may be characterized as the image of the classifying map

$$\mathcal{E}[U] \xrightarrow{\overline{FU}} \mathcal{E}^{\underline{FPT}}$$

of the free $\mathbb{T}$-model generated by the generic object U; for the latter is easily seen to be induced by the composite internal functor

$$\underline{E}_{fin} \xrightarrow{e} \underline{FFT} \xrightarrow{i} \underline{FPT} \quad .$$

It is clearly a matter of interest to characterize those $\mathbb{T}$-models A whose classifying maps factor through the inclusion $\mathcal{E}^{\underline{FFT}} \longrightarrow \mathcal{E}^{\underline{FPT}}$ - i.e. those for which HA restricts to a flat presheaf on $\underline{FFT}$.

We shall say that a $\mathbb{T}$-model A in $\mathcal{F}$ is <u>flat</u> if it is expressible as a filtered colimit of free models, i.e. if there exists a filtered category $\underline{C}$ in $\mathcal{F}$ and a $\mathbb{T}$-model B in $\mathcal{F}^{\underline{C}}$ such that $B(\underline{c})$ is a free model in $\mathcal{F}/C_o$ and $A \cong \varinjlim_{\underline{C}}(B)$. A well-known theorem of D. Lazard [24] asserts that for the theory of modules over a ring (in $\mathcal{S}$), this definition of flatness is equivalent to the usual one in terms of tensor product. We note also that the property of flatness, as defined above, is preserved by inverse image functors.

<u>5.22 Theorem</u> The topos $\mathcal{E}^{\underline{FFT}}$ is a classifying topos for the theory of flat $\mathbb{T}$-models in $\mathcal{E}$-toposes; i.e. a map of $\mathcal{E}$-toposes

$$\mathcal{F} \longrightarrow \mathcal{E}^{\underline{FPT}}$$

factors through $\mathcal{E}^{\underline{FFT}}$ iff the $\mathbb{T}$-model which it classifies is flat.

<u>Proof</u> First we observe that if B is a free $\mathbb{T}$-model in $\mathcal{F}$, then its classifying map factors through $\mathcal{E}[U] \xrightarrow{\overline{FU}} \mathcal{E}^{\underline{FPT}}$ and therefore through $\mathcal{E}^{\underline{FFT}}$. Also, if $\mathcal{G} \xrightarrow{p} \mathcal{F}$ is a surjection and B is a $\mathbb{T}$-model in $\mathcal{F}$ such that p^*B is free, then we have a diagram

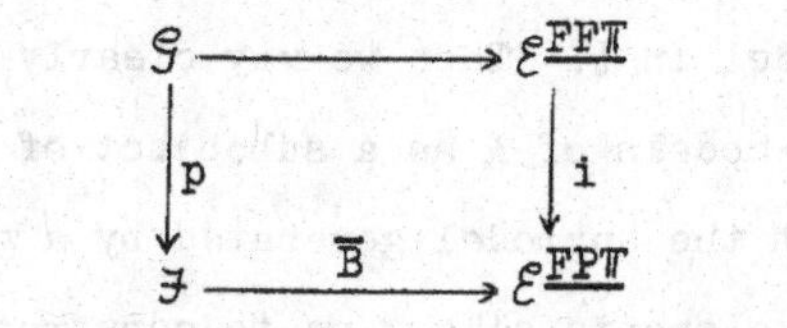

in <u>Top</u>/$\mathcal{E}$ where p is a surjection and i is an inclusion, from which we deduce that $\overline{B}$ factors through $\mathcal{E}^{\underline{FFT}}$. Hence if A is a flat $\mathbb{T}$-model in $\mathcal{F}$, we can factor its classifying map as

$$\mathcal{F} \longrightarrow \mathcal{F}^{\underline{C}} \longrightarrow \mathcal{E}^{\underline{FFT}} \xrightarrow{\;i\;} \mathcal{E}^{\underline{FPT}} ,$$

where the first factor is the geometric morphism whose inverse image is $\varinjlim_{\underline{C}}$, and the remainder is the classifying map of B. So $\bar{A}$ factors through $\mathcal{E}^{\underline{FFT}}$.

Conversely, suppose the classifying map of A factors through $\mathcal{E}^{\underline{FFT}}$. Then $i^*(HA)$ is a flat presheaf on $\underline{FFT}$ in $\mathcal{F}$; let

$$\underline{C} \xrightarrow{\;\gamma\;} \underline{FFT}$$

be the corresponding discrete fibration, so that $\underline{C}$ is a filtered category in $\mathcal{F}$ by Proposition 1.3. Now if M is the generic $\mathbb{T}$-model, we have an isomorphism

$$A \cong M \otimes_{\underline{FPT}} HA \cong i^* M \otimes_{\underline{FFT}} i^* HA \quad ,$$

which we can re-interpret as an isomorphism

$$A \cong \varinjlim_{\underline{C}} (\gamma^* i^* M) \quad .$$

But $\gamma^* i^* M\,(\underset{\sim}{c}) = M(i_o \gamma_o \underset{\sim}{c}) \cong F[\gamma_o \underset{\sim}{c}]$ is a free $\mathbb{T}$-model in $\mathcal{F}/C_o$; so A is flat. □

Two further subcategories of $\underline{FPT}$ which we may study are the categories $\underline{FPT}_e$ and $\underline{FPT}_m$, defined analogously to the categories $\underline{E}_{fine}$ and $\underline{E}_{finm}$ which we considered at the end of chapter IV. (Throughout the following discussion, we shall interpret "epi" to mean "epi in $\mathcal{E}$" - by Corollary 5.7, this coincides with "regular epi in $\mathbb{T}(\mathcal{E})$" - and we shall write $\mathrm{Epi}_{\mathbb{T}}(A,B)$ for the object of $\mathbb{T}$-model epimorphisms from A to B.)

Let A be a $\mathbb{T}$-model in $\mathcal{E}$. Then we may clearly construct the object $\Omega_{\mathbb{T}}(A)$ of sub-$\mathbb{T}$-models of A as a subobject of Ω^A. Moreover, since we can construct the submodel generated by a given subobject, the method of generic elements allows us to construct a left adjoint left inverse

$$\Omega^A \xrightarrow{\quad g \quad} \Omega_{\mathbb{T}}(A)$$

for the inclusion $\Omega_{\mathbb{T}}(A) \rightarrowtail \Omega^A$. $\Omega_{\mathbb{T}}(A)$ has a lattice structure, and g is a $\vee$-semilattice homomorphism. Let $K_{\mathbb{T}}(A) \rightarrowtail \Omega_{\mathbb{T}}(A)$ denote the image of the semilattice homomorphism

$$K(A) \rightarrowtail \Omega^A \xrightarrow{\quad g \quad} \Omega_{\mathbb{T}}(A);$$

we shall say that A is Kuratowski-finitely-generated if the maximal element $1 \xrightarrow{\ulcorner A \urcorner} \Omega_{\mathbb{T}}(A)$ factors through $K_{\mathbb{T}}(A)$. (Note that a Kuratowski-finite $\mathbb{T}$-model is automatically Kuratowski-finitely-generated.)

Adapting the argument of Theorem 2.7, it is not hard to show that A is Kuratowski-finitely-generated iff it is locally a quotient of a (cardinal-)finitely-generated free model, i.e. iff there exists $F[p] \twoheadrightarrow A$ in $\mathbb{T}(\mathcal{E}/I)$ for some $I \twoheadrightarrow 1$ and some p. To prove the analogue of Proposition 2.8(i), we need to show that $K_{\mathbb{T}}(A)$ is preserved by inverse image functors. But we observe first that the "membership relation" $E(A) \rightarrowtail A \times K(A)$ is preserved, since it may be defined by the pullback diagram

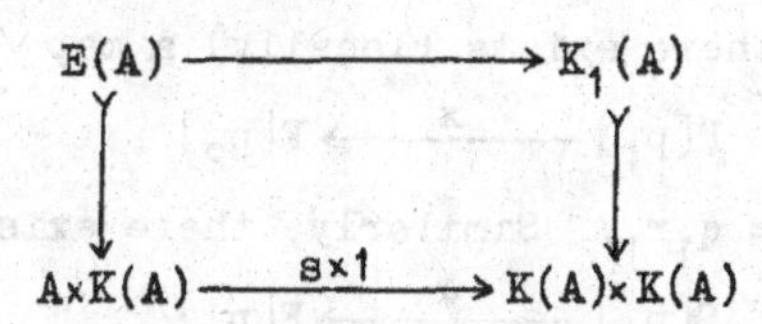

where $K_1(A)$ is the order-relation on $K(A)$ and s is the factorization of the singleton map $A \longrightarrow \Omega^A$ through $K(A)$. Now regard $E(A)$ as a subobject of A in $\mathcal{E}/K(A)$, and let $R(A) \rightarrowtail A \times K(A)$ be the sub-$\mathbb{T}$-model which it generates. (Thus $R(A)$ is the object of pairs $(\underset{\sim}{a}, \underset{\sim}{k})$ such that $\underset{\sim}{a}$ is in the sub-$\mathbb{T}$-model generated by $\underset{\sim}{k}$.) Now consider the map

$$R(A) \rightarrowtail A \times K(A) \xrightarrow{(\pi_2, \vee(s \times 1))} K(A) \times K(A) \quad .$$

Since $K(A)$ is generated as a semilattice by $A \overset{s}{\rightarrowtail} K(A)$, it is easy to see that the semilattice congruence on $K(A)$ generated by the image of this map is precisely the kernel-pair of $K(A) \twoheadrightarrow K_{\mathbb{T}}(A)$.

So the diagram

$$F_{\underset{\sim}{slat}}R(A) \rightrightarrows K(A) \twoheadrightarrow K_{\mathbb{T}}(A)$$

is a coequalizer in the category $\underset{\sim}{slat}(\mathcal{E})$ of semilattices in $\mathcal{E}$; hence $K_{\mathbb{T}}(A)$ is preserved by inverse image functors.

Now if A is any $\mathbb{T}$-model, the object

$$Epi_{\mathbb{T}}(M(\underset{\sim}{p}),A)$$

has the structure of a presheaf on $\underline{FP\mathbb{T}}_e$; and as in Lemma 4.10(i) we can show that this presheaf is flat iff A is Kuratowski-finitely-generated. The argument for conditions (a) and (c) is as before; for condition (b), we proceed as follows:

Suppose given a diagram

$$\begin{array}{ccccc} F[q_1] \overset{a_1}{\underset{b_1}{\rightrightarrows}} F[p_1] \overset{r_1}{\twoheadrightarrow} M_1 & & \\ & \searrow^{q_1} & \\ & & A \\ & \nearrow_{q_2} & \\ F[q_2] \overset{a_2}{\underset{b_2}{\rightrightarrows}} F[p_2] \overset{r_2}{\twoheadrightarrow} M_2 & & \end{array}$$

in $\mathbb{T}(\mathcal{E}/I)$ for some I, where the rows are finite presentations of M_1 and M_2. Since $F[p_1]$ is internally projective in $\mathbb{T}(\mathcal{E}/I)$ by Corollary 5.17, there exists (locally) a map

$$F[p_1] \xrightarrow{x} F[p_2]$$

such that $q_2r_2x = q_1r_1$. Similarly, there exists

$$F[p_2] \xrightarrow{y} F[p_1]$$

such that $q_1r_1y = q_2r_2$.

Let $F[p_1] \overset{s_1}{\twoheadrightarrow} Q$ be the joint coequalizer of the pairs (a_1,b_1), (ya_2,yb_2) and $(1_{F[p_1]},yx)$. Clearly, Q has a finite presentation of the form

$$F[q_1+q_2+p_1] \rightrightarrows F[p_1] \quad .$$

From the universal property of coequalizers, we obtain morphisms

$$M_1 \overset{t_1}{\twoheadrightarrow} Q \quad \text{and} \quad Q \overset{u}{\twoheadrightarrow} A$$

such that $t_1r_1 = s_1$ and $us_1 = q_1r_1$ (whence $ut_1 = q_1$, since r_1

is epi).

Let s_2 be the composite $s_1y\colon F[p_2] \longrightarrow Q$, and consider a map $F[p_2] \xrightarrow{z} T$ such that $za_2 = zb_2$, $zxa_1 = zxb_1$ and $z = zxy$. Then $zx = zxyx$, whence zx factors uniquely through s_1 (say by $Q \xrightarrow{w} T$), and $ws_2 = ws_1y = zxy = z$. But any w satisfying $ws_2 = z$ satisfies $ws_1 = ws_1yx = ws_2x = zx$; so s_2 is the joint coequalizer of (a_2,b_2), (xa_1,xb_1) and $(1,xy)$. Hence by symmetry we have a map

$$M_2 \xrightarrow{t_2} Q$$

with $t_2r_2 = s_2$ and $ut_2 = q_2$. So we have constructed a diagram

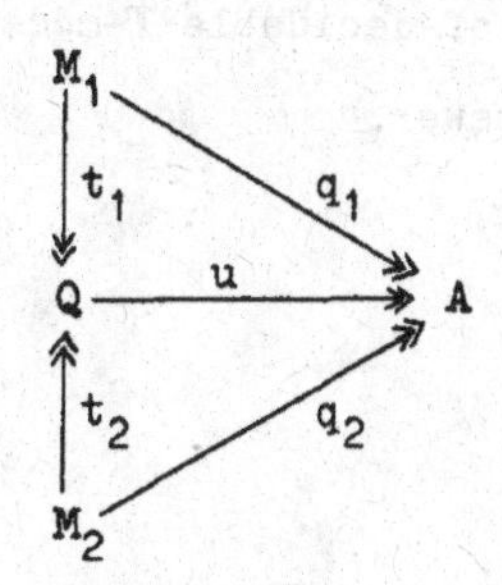

as required.

We may now proceed to prove

5.23 Theorem The topos $\mathcal{E}^{\mathbf{FPT}_e}$ is a classifying topos for the theory of Kuratowski-finitely-generated $\mathbb{T}$-models and (regular) epimorphisms between them. □

Unfortunately, however, the corresponding argument in the "mono" case fails to generalize. In the first place, it is not true even for a decidable $\mathbb{T}$-model A that the object $\Sigma_{\underset{\sim}{p}} \mathrm{Mono}_{\mathbb{T}}(M(\underset{\sim}{p}),A)$ need have global support: for example if $\mathcal{E} = \mathcal{S}$ and $\mathbb{T}$ is a free theory having infinitely many nullary operations, then the one-element $\mathbb{T}$-model has no finitely-presented submodel. Similarly, condition (b) of flatness need not be satisfied even in $\mathcal{S}$: for if we take $\mathbb{T}$ to be the theory of (non-abelian) groups, M_1 and M_2 to be finitely-generated free groups, and A to be the free product of M_1 and M_2 with amalgamation of a common non-finitely-generated subgroup, then there is no finitely-presented subgroup of A containing both M_1 and M_2.

From these examples, it would appear that the correct condition for a $\mathbb{T}$-model A to yield a flat presheaf on $\underline{\mathrm{FP}\mathbb{T}}_m$ is that the poset of (cardinal-)finitely-presented submodels of A should be cofinal in the semilattice $K_{\mathbb{T}}(A)$. But this does not seem to be a particularly interesting or natural condition to impose, even for a familiar theory such as that of groups. Equally, the question "What is the classifying topos for the theory of decidable $\mathbb{T}$-models?" does not appear to have any very simple answer.

CHAPTER VI : MANY-SORTED THEORIES

In this chapter, our aim is to investigate theories such as the theory of categories, which are algebraic in the sense of being defined by finitary operations and equations, but which are not single-sorted - i.e. their models in a category $\mathcal{E}$ cannot be considered as single objects of $\mathcal{E}$ equipped with algebraic operations, but their description involves several different objects whose elements are of different "sorts". We follow R.B. Coates [8] in adopting the viewpoint that the sorts of a many-sorted theory should be described by a category rather than a set; the level of generality which we consider is thus intermediate between that of the many-sorted theories of J. Bénabou [3] and that of the essentially algebraic theories of P. Freyd [12]. Much of what we do can be generalized fairly easily to the essentially algebraic case; we leave the details to the reader.

We consider first the special case of a "diagram theory", i.e. a theory having multiple sorts but no operations. If $\underline{D}$ is an internal category in a topos $\mathcal{E}$, we shall write $\mathbb{D}$ for the "theory of diagrams of type $\underline{D}$", defined by $\mathbb{D}(\mathcal{F}) = \mathcal{F}^{\underline{D}}$ for any $\mathcal{E}$-topos $\mathcal{F}$. We use the same notation when $\underline{D}$ is a finite external category (which may, of course, be identified with an internal category in any topos $\mathcal{E}$ by identifying a p-element set with the p-fold copower of 1 in $\mathcal{E}$).

Clearly $\mathbb{D}(\mathcal{F})$, being an $\mathcal{F}$-topos, has the structure of a locally internal category over $\mathcal{F}$. Thus if we are given a $\mathbb{D}$-model $M(\underset{\sim}{i})$ in $\mathcal{F}/I$ for some I, we may construct the internal full subcategory $\mathrm{Full}_{\mathbb{D}(\mathcal{F})}(M)$ in the usual way; and we may regard the inclusion functor

$$\mathrm{Full}_{\mathbb{D}(\mathcal{F})}(M) \longrightarrow \mathbb{D}(\mathcal{F})$$

as a $\mathbb{D}$-model in the internal diagram topos $\mathcal{F}^{(\mathrm{Full}_{\mathbb{D}(\mathcal{F})}(M))}$.

We shall say that a $\mathbb{D}$-model F is <u>essentially finite</u> if its classifying map

$$\mathcal{E}^{\underline{D}} \xrightarrow{\tilde{F}} \mathcal{E}[U]$$

is induced by an internal functor $\underline{D} \longrightarrow \underline{E}_{fin}$ in $\mathcal{E}$, i.e. if F is isomorphic to the pullback of the generic diagram U along such an internal functor. (Note: this is not equivalent to the condition that F is isomorphic to a finite cardinal in $\mathcal{E}^{\underline{D}}$. The latter implies additionally that F is "constant on connected components of $\underline{D}$", i.e. that the internal functor $\underline{D} \longrightarrow \underline{E}_{fin}$ factors through the discrete subcategory $N \rightarrowtail \underline{E}_{fin}$ with the same objects.)

Now let $G = \underset{\sim}{cat}(\mathcal{E})(\underline{D}, \underline{E}_{fin})$. Then in the topos $\mathcal{E}/G$, we have a generic essentially finite $\mathbb{D}$-model $M(\underset{\sim}{g})$, obtained by pulling back U along the generic internal functor $\underline{D} \longrightarrow \underline{E}_{fin}$. We write $\underline{\mathbb{FD}}$ for $\mathrm{Full}_{\mathbb{D}(\mathcal{E})}(M)$, the internal category of essentially finite $\mathbb{D}$-models in $\mathcal{E}$.

<u>6.1 Lemma</u> $\underline{\mathbb{FD}}$ is isomorphic to the exponential $(\underline{E}_{fin})^{\underline{D}}$ in $\underset{\sim}{cat}(\mathcal{E})$.

<u>Proof</u> From the definition of exponentials in $\underset{\sim}{cat}(\mathcal{E})$, it is clear that they have the same object of objects, so we need to show that their hom-sets are isomorphic. In terms of generalized elements, this amounts to showing that if we are given two internal functors

$$\underline{D} \overset{g_1}{\underset{g_2}{\rightrightarrows}} \underline{E}_{fin} ,$$

then internal natural transformations $g_1 \longrightarrow g_2$ are in natural 1-1 correspondence with maps $g_1{}^*U \longrightarrow g_2{}^*U$ in $\mathcal{E}^{\underline{D}}$. But the latter correspond to morphisms $[g_1\underset{\sim}{d}] \longrightarrow [g_2\underset{\sim}{d}]$ in $\mathcal{E}/D_0$ satisfying a certain naturality condition, and the former to elements of $\underline{E}_{fin}(g_1\underset{\sim}{d}, g_2\underset{\sim}{d}) = [g_2\underset{\sim}{d}]^{[g_1\underset{\sim}{d}]}$ satisfying the same condition. □

We shall say that a category $\underline{D}$ is <u>Yoneda-finite</u> if there exists an internal functor

$$\underline{D}^{op} \xrightarrow{\ h\ } \mathbf{F}\mathbb{D}$$

such that the Yoneda profunctor $Y(\underline{D})$ (regarded as a $\mathbb{D}$-model in $\mathcal{E}^{\underline{D}^{op}}$) is isomorphic to h^*M, where M is the $\mathbb{D}$-model in $\mathcal{E}^{\mathbf{F}\mathbb{D}}$ corresponding to the inclusion functor $\mathbf{F}\mathbb{D} \longrightarrow \mathbb{D}(\mathcal{E})$. (If $\mathcal{E} = \mathcal{S}$, this is equivalent to saying that $\underline{D}$ has finite hom-sets; we avoid the common term "locally finite", since the word "locally" has other uses in topos theory.)

6.2 Lemma (i) A $\mathbb{D}$-model F is essentially finite iff $F(\underset{\sim}{d})$ is isomorphic to a finite cardinal in $\mathcal{E}/D_o$.

(ii) $\underline{D}$ is Yoneda-finite iff $\underline{D}(\underset{\sim}{d}_1, \underset{\sim}{d}_2)$ is isomorphic to a finite cardinal in $\mathcal{E}/D_o \times D_o$.

Proof (i) Let $p_o\colon D_o \longrightarrow N$ be a natural number in $\mathcal{E}/D_o$ whose cardinal is isomorphic to $F(\underset{\sim}{d})$. Now the action of $\underline{D}$ on $F(\underset{\sim}{d})$ is an associative, unitary morphism

$$[p_o \underset{\sim}{d}_1] \times \underline{D}(\underset{\sim}{d}_1, \underset{\sim}{d}_2) \longrightarrow [p_o \underset{\sim}{d}_2]$$

in $\mathcal{E}/D_o \times D_o$, or equivalently

$$\underline{D}(\underset{\sim}{d}_1, \underset{\sim}{d}_2) \longrightarrow [p_o \underset{\sim}{d}_2]^{[p_o \underset{\sim}{d}_1]} = \underline{E}_{fin}(p_o \underset{\sim}{d}_1, p_o \underset{\sim}{d}_2) \quad .$$

So p_o is the object-map of an internal functor $p\colon \underline{D} \longrightarrow \underline{E}_{fin}$; and clearly $p^*U \cong F$ as an object of $\mathcal{E}^{\underline{D}}$.

The converse is immediate, since $U(\underset{\sim}{n}) = [\underset{\sim}{n}]$ is a finite cardinal in $\mathcal{E}/N$.

(ii) Suppose $\underline{D}(\underset{\sim}{d}_1, \underset{\sim}{d}_2)$ is isomorphic to a finite cardinal. Then by part (i), the $\mathbb{D}$-model in $\mathcal{E}/D_o$ defined by $\underline{D}(\underset{\sim}{d}_1, \underset{\sim}{d}_2)$, with right action of $\underline{D}$ by composition, is essentially finite; so it determines a D_o-element h_o of G. But now h_o extends to an internal functor

$$h\colon \underline{D}^{op} \longrightarrow \mathbf{F}\mathbb{D}$$

in the same way that p_o was extended in part (i), using the fact that $\mathbf{F}\mathbb{D}$ is an internal full subcategory. And the converse is again immediate, since $M(\underset{\sim}{g})$ is essentially finite. □

6.3 Theorem Let $\underline{D}$ be a Yoneda-finite category. Then there is an equivalence of categories

$$\mathbb{D}(\mathcal{E}) \simeq \mathrm{Flat}(\underline{F\mathbb{D}}^{op},\mathcal{E}) .$$

Proof As usual, we define a functor $H\colon \mathbb{D}(\mathcal{E}) \longrightarrow \mathcal{E}^{\underline{F\mathbb{D}}^{op}}$ by

$$HX(\underset{\sim}{g}) = \mathbb{D}(\mathcal{E}/G)(M(\underset{\sim}{g}),X)$$

with action of $\underline{F\mathbb{D}}$ given by composition. Using Lemma 6.2(i), it is easy to show that the category of essentially finite $\mathbb{D}$-models has finite colimits, and hence (as in Lemma 4.4) that HX is flat for any X.

Now the Yoneda lemma gives us a natural isomorphism

$$\mathbb{D}(\mathcal{E}/D_0)(M(h_0\underset{\sim}{d}),X) \cong X(\underset{\sim}{d})$$

in $\mathcal{E}/D_0$, and it follows easily that the composite

$$\mathcal{E}^{\underline{D}} \xrightarrow{\quad H \quad} \mathcal{E}^{\underline{F\mathbb{D}}^{op}} \xrightarrow{\quad h^* \quad} \mathcal{E}^{\underline{D}}$$

is isomorphic to the identity. But the generic flat presheaf on $\underline{F\mathbb{D}}$ is simply H(M), so we deduce as in Lemma 4.5 that H is an equivalence of categories. □

To ensure that the category $\underline{F\mathbb{D}}$ is preserved by inverse image functors, we have to assume a stronger condition on $\underline{D}$: namely that it is actually finite in the sense that D_0 and D_1 are finite cardinals (i.e. $\underline{D}$ is an internal category in $\mathcal{E}_{fc}$). It follows at once from Proposition 2.5 and Lemma 6.2(ii) that a finite category is automatically Yoneda-finite.

6.4 Lemma Let $\underline{C}$ and $\underline{D}$ be internal categories in $\mathcal{E}$, and suppose $\underline{D}$ is finite. Then for any geometric morphism $\mathcal{F} \xrightarrow{f} \mathcal{E}$, we have

$$f^*(\underset{\sim}{cat}(\mathcal{E})(\underline{D},\underline{C})) \cong \underset{\sim}{cat}(\mathcal{F})(f^*\underline{D},f^*\underline{C}) \quad .$$

Proof The argument is similar to that of Lemma 5.18, using the fact that $\underset{\sim}{cat}(\mathcal{E})(\underline{D},\underline{C})$ may be described as a subobject of $C_0{}^{D_0} \times C_1{}^{D_1}$

obtained by intersecting certain equalizers. □

<u>6.5 Theorem</u> If $\underline{D}$ is a finite category in $\mathcal{E}$, then the $\mathcal{E}$-topos $\mathcal{E}^{\underline{F\mathbb{D}}}$ is a classifying topos for the theory of diagrams of type $\underline{D}$; i.e. for any $\mathcal{E}$-topos $\mathcal{F}$ we have

$$\underline{\text{Top}}/\mathcal{E}\ (\mathcal{F},\mathcal{E}^{\underline{F\mathbb{D}}}) \simeq \mathbb{D}(\mathcal{F}) \quad .$$

<u>Proof</u> Lemma 6.4 ensures that the object G, and hence the category $\underline{F\mathbb{D}}$, are preserved by inverse image functors. Hence we may reduce to the case $\mathcal{F} = \mathcal{E}$, which follows at once from Theorem 6.3. □

J. Bénabou [5] has pointed out that, at least if we are interested only in <u>external</u> finite categories $\underline{D}$, it is sufficient to construct the $\mathbb{D}$-model classifier only in the special cases $\underline{D} = \underline{1}$ and $\underline{D} = \underline{2}$. For any finite category can be expressed as a finite colimit of copies of $\underline{1}$, $\underline{2}$ and the "commutative triangle" category $\underline{3}$; and $\underline{3}$ itself is the pushout of two copies of $\underline{2}$ under a copy of $\underline{1}$. But any colimit diagram in $\text{cat}(\mathcal{S}_{fc})$ clearly gives rise to a corresponding <u>limit</u> diagram of classifying toposes in $\underline{\text{BTop}}/\mathcal{E}$; and Diaconescu's theorem enables us to construct finite limits in this 2-category.

It is also true that the general techniques of Bénabou and M. Tierney [36] can be used to prove the existence of a classifying topos for $\mathbb{D}$-models when $\underline{D}$ is <u>any</u> internal category, even without the assumption of Yoneda-finiteness. (See also [17], Example 6.60(i).) However, it seems that the finiteness assumptions of Theorems 6.3 and 6.5 are necessary if we are to give an explicit description of this classifying topos in terms of diagrams on some internal category.

Now let $\underline{D}$ be a particular finite category in $\mathcal{E}$, and write $\mathcal{D}$ for the $\mathbb{D}$-model classifier $\mathcal{E}^{\underline{F\mathbb{D}}}$. It follows from Theorem 6.5 that objects of the topos $\mathcal{D}^{\underline{D}}$ are equivalent to geometric endomorphisms of $\mathcal{D}$ over $\mathcal{E}$ (or to left flat profunctors $\underline{F\mathbb{D}}$- - - →$\underline{F\mathbb{D}}$ in $\mathcal{E}$), and so this topos has a monoidal structure, similar to that on $\mathcal{E}[U]$. By a $\underline{D}$-sorted algebraic theory (or algebraic theory over $\mathbb{D}$), we mean a monoid in this monoidal category. We write $\text{alg}_{\underline{D}}(\mathcal{E})$ for the category of $\underline{D}$-sorted algebraic theories in $\mathcal{E}$. If $\mathbb{T} = (T,m,e)$ is an object of $\text{alg}_{\underline{D}}(\mathcal{E})$, we define a $\mathbb{T}$-model in an $\mathcal{E}$-topos $\mathcal{F}$ to be an algebra for the corresponding monad $\mathbb{T}_{\mathcal{F}}$ on $\mathcal{F}^{\underline{D}}$.

It is of interest to consider the relationship between $\underline{D}$-sorted algebraic theories in $\mathcal{E}$ and (single-sorted) algebraic theories in the topos $\mathcal{E}^{\underline{D}}$. For any $\mathcal{E}$-topos $(\mathcal{F} \xrightarrow{f} \mathcal{E})$, we have a pullback diagram

$$\begin{array}{ccc} \mathcal{F}^{\underline{D}} & \xrightarrow{f^{\underline{D}}} & \mathcal{E}^{\underline{D}} \\ \downarrow & & \downarrow \\ \mathcal{F} & \xrightarrow{f} & \mathcal{E} \end{array}$$

in $\underline{\underline{\text{Top}}}$; hence $\mathcal{F}^{\underline{D}}$ is an $(\mathcal{E}^{\underline{D}})$-topos, and any map of $\mathcal{E}$-toposes $\mathcal{G} \xrightarrow{g} \mathcal{F}$ induces a map of $(\mathcal{E}^{\underline{D}})$-toposes

$$\mathcal{G}^{\underline{D}} \xrightarrow{g^{\underline{D}}} \mathcal{F}^{\underline{D}} .$$

(Of course, not every $(\mathcal{E}^{\underline{D}})$-topos has the form $\mathcal{F}^{\underline{D}}$ for some $\mathcal{F}$; nor does every map of $(\mathcal{E}^{\underline{D}})$-toposes $\mathcal{G}^{\underline{D}} \longrightarrow \mathcal{F}^{\underline{D}}$ have the form $g^{\underline{D}}$ for some g.) It follows that we have a forgetful functor

$$\theta\colon \mathcal{E}^{\underline{D}}[U] \longrightarrow \mathcal{D}^{\underline{D}} \quad ,$$

obtained by regarding an object of $\mathcal{E}^{\underline{D}}[U]$ as a natural endofunctor of $(\mathcal{E}^{\underline{D}})$-toposes, and then forgetting its effect on toposes which are not of the form $\mathcal{F}^{\underline{D}}$. It is clear from the definition that θ is a monoidal functor, and so lifts to a functor

$$\theta\colon \text{alg}(\mathcal{E}^{\underline{D}}) \longrightarrow \text{alg}_{\underline{D}}(\mathcal{E}).$$

Note also that $\mathcal{D}^{\underline{D}}$ may be regarded as the topos of internal ($\mathcal{E}$-valued) diagrams on the product category $\underline{\mathbb{F}\mathbb{D}} \times \underline{D}$, and $\mathcal{E}^{\underline{D}}[U]$ may similarly be regarded as the topos of diagrams on $\underline{E}_{fin} \times \underline{D}$. So the identification of $\underline{\mathbb{F}\mathbb{D}}$ with the exponential $(\underline{E}_{fin})^{\underline{D}}$ (Lemma 6.1) gives us an internal functor

$$\underline{F\mathbb{D}} \times \underline{D} \xrightarrow{(ev,\pi_2)} \underline{E}_{fin} \times \underline{D}$$

over $\underline{D}$, and hence an essential geometric morphism

$$\mathcal{D}^{\underline{D}} \xrightarrow{v} \mathcal{E}^{\underline{D}}[U]$$

over $\mathcal{E}^{\underline{D}}$. It is not hard to see that v classifies the generic object of $\mathcal{D}^{\underline{D}}$, i.e. the object corresponding to the identity endomorphism of $\mathcal{D}$. Moreover, v is a split epimorphism in Top; its splitting

$$\mathcal{E}^{\underline{D}}[U] \xrightarrow{u} \mathcal{D}^{\underline{D}}$$

is induced by the functor

$$\underline{E}_{fin} \times \underline{D} \xrightarrow{c\times 1} \underline{F\mathbb{D}} \times \underline{D}$$

where c is the functor which sends a finite cardinal to the corresponding constant diagram of type $\underline{D}$, i.e. the transpose of

$$\underline{E}_{fin} \times \underline{D} \xrightarrow{\pi_1} \underline{E}_{fin}.$$

In particular, this shows that the functor v^* is faithful.

<u>6.6 Lemma</u> The inverse image of v is naturally isomorphic to the forgetful functor θ defined above.

<u>Proof</u> Let $\mathcal{F}$ be an $\mathcal{E}$-topos, X an object of $\mathcal{F}^{\underline{D}}$. We may classify X either by a morphism

$$\mathcal{F}^{\underline{D}} \xrightarrow{\tilde{X}} \mathcal{E}^{\underline{D}}[U]$$

over $\mathcal{E}^{\underline{D}}$, or by a morphism

$$\mathcal{F} \xrightarrow{\bar{X}} \mathcal{D}$$

over $\mathcal{E}$. Now the diagram

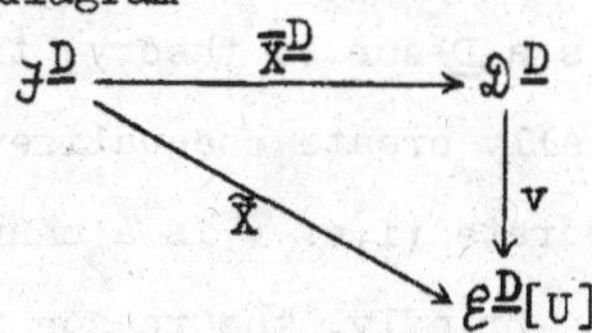

commutes, since both ways round map the generic object U to X. Similarly, if T is an object of $\mathcal{E}^{\underline{D}}[U]$ and $S = v^*T$, then the diagram

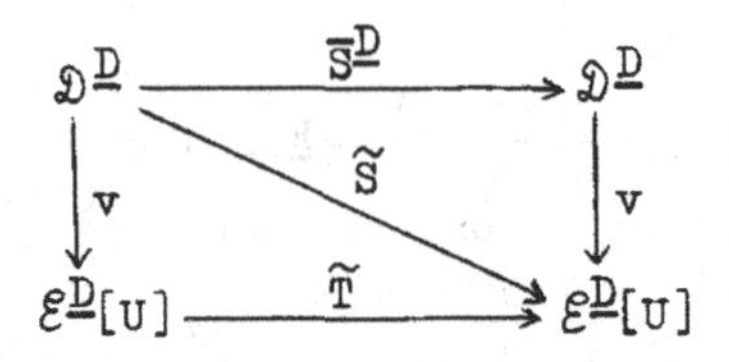

commutes. Combining the two diagrams, we obtain a natural isomorphism

$$S_{\mathfrak{Z}}(X) \cong T_{(\mathfrak{Z}^{\underline{D}})}(X)$$

in $\mathfrak{Z}^{\underline{D}}$. Thus v^* is identified with the forgetful functor. □

By a suitable modification of the argument of Proposition 5.10, we may construct a free functor

$$\mathfrak{D}^{\underline{D}} \longrightarrow \underset{\sim}{\mathrm{alg}}_{\underline{D}}(\mathcal{E}) \quad .$$

It is not true in general that the forgetful functor $\underset{\sim}{\mathrm{alg}}_{\underline{D}}(\mathcal{E}) \longrightarrow \mathfrak{D}^{\underline{D}}$ creates reflexive coequalizers (see below), but we may still prove that $\underset{\sim}{\mathrm{alg}}_{\underline{D}}(\mathcal{E})$ is monadic over $\mathfrak{D}^{\underline{D}}$ using the Precise Tripleability Theorem. We have already remarked that the functor v^*: $\mathcal{E}^{\underline{D}}[U] \longrightarrow \mathfrak{D}^{\underline{D}}$ lifts to a functor $\underset{\sim}{\mathrm{alg}}(\mathcal{E}^{\underline{D}}) \longrightarrow \underset{\sim}{\mathrm{alg}}_{\underline{D}}(\mathcal{E})$; and in fact standard techniques for lifting adjoints (see [16]) enable us to construct left and right adjoints for the latter functor, and hence to prove that $\underset{\sim}{\mathrm{alg}}(\mathcal{E}^{\underline{D}})$ is both monadic and comonadic over $\underset{\sim}{\mathrm{alg}}_{\underline{D}}(\mathcal{E})$.

One aspect of single-sorted theories which fails to generalize to the many-sorted case is the special role played by reflexive coequalizers (5.6): if $\mathbb{T}$ is a $\underline{D}$-sorted theory, the forgetful functor $\mathbb{T}(\mathcal{E}) \longrightarrow \mathcal{E}^{\underline{D}}$ does not normally create coequalizers of reflexive pairs, unless $\underline{D}$ happens to be discrete (i.e. $\mathbb{T}$ is a many-sorted theory in the sense of Bénabou [3]). Briefly, the reason for this is that reflexive coequalizers commute with finite products in a topos (cf. Proposition 2.4), but not with more general finite limits; we shall give an explicit instance of the failure in Example 6.8 below. It follows that we cannot simply appeal to Linton's theorem [27] to prove that $\mathbb{T}(\mathcal{E})$ has finite colimits, nor can we immediately mimic the arguments of the last chapter to construct the internal category $\underline{FP\mathbb{T}}$ and to prove that it yields a classifying topos for $\mathbb{T}$-models.

For constructing coequalizers in $\mathbb{T}(\mathcal{E})$, the General Adjoint Functor Theorem of Paré and Schumacher ([33], Theorem IV.1.5) would appear to provide us with a useful tool, and it is indeed possible to use this in a good many special cases. In the case $\mathcal{E} = \mathcal{S}$, R.B. Coates [8] has given a method of constructing colimits which works for any $\mathbb{T}$; but this method depends on a cardinality argument to establish a "solution-set condition", and it is not clear how one should extend this to a general base topos.

Rather than devote any further space to the pursuit of a general theorem of this type, however, we shall turn our attention for the rest of the chapter to a particular example of a many-sorted theory: the theory $\underset{\sim}{cat}$ of categories. We shall see that in this case we are able to circumvent the difficulties mentioned above by an <u>ad hoc</u> variant of our reflexive-coequalizer arguments; but first we need to give a description of $\underset{\sim}{cat}$ as an algebraic theory.

Let $\underline{D}$ be the (external) finite category represented diagrammatically by

we shall call a $\mathbb{D}$-model a <u>directed graph</u>, and write $\underset{\sim}{dgph}$ for $\mathbb{D}$.

<u>6.7 Proposition</u> Let $\mathcal{E}$ be a topos with a natural number object. Then the forgetful functor $\underset{\sim}{cat}(\mathcal{E}) \longrightarrow \underset{\sim}{dgph}(\mathcal{E})$ has a left adjoint.

<u>Proof</u> We may think of a directed graph $\underline{A}$ as consisting of an object A_o of $\mathcal{E}$ together with an object $\underline{A}(\underset{\sim}{a}_1,\underset{\sim}{a}_2)$ of $\mathcal{E}/A_o \times A_o$. Now a category structure on $\underline{A}$ is simply a monoid structure (for the product $\wedge_{A_o}$ of Lemma 3.4) on $\underline{A}(\underset{\sim}{a}_1,\underset{\sim}{a}_2)$. Hence if we define a

directed graph $\underline{FA}$ by $(FA)_o = A_o$ and

$$\underline{FA}\ (\underset{\sim}{a}_1, \underset{\sim}{a}_2) = \Sigma_{\underset{\sim}{n}}(\underline{A}^{\langle \underset{\sim}{n} \rangle}(\underset{\sim}{a}_1, \underset{\sim}{a}_2)) \quad ,$$

it is easy to adapt the argument of Proposition 5.10 to prove that $\underline{FA}$ has a category structure, and that it is the free category generated by $\underline{A}$. □

6.8 Example We now give an example to illustrate the failure of Lemma 5.6 in the case of many-sorted theories. Consider the coequalizer diagram

$$\underline{A} \underset{g}{\overset{f}{\rightrightarrows}} \underline{B} \xrightarrow{\quad h \quad} \underline{C}$$

in $\underset{\sim}{\text{dgph}}(\mathcal{S})$, where $\underline{A}$, $\underline{B}$ and $\underline{C}$ are represented diagrammatically by

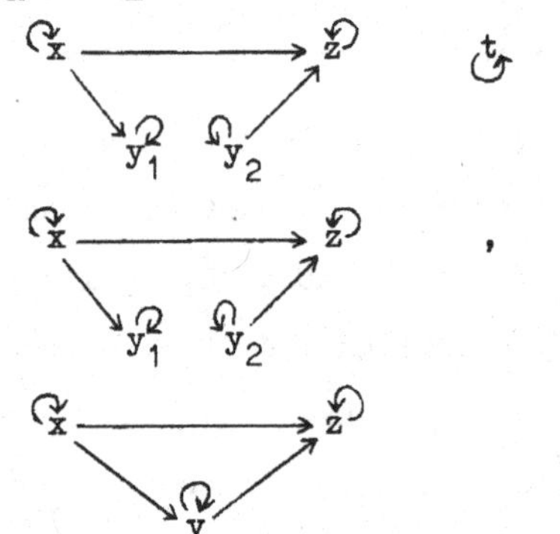

respectively, and f, g send the vertex t to y_1, y_2 respectively.

Then the pair (f,g) is clearly reflexive, with splitting given by the inclusion map $\underline{B} \longrightarrow \underline{A}$; but if we impose the unique category structures on $\underline{A}$, $\underline{B}$ and $\underline{C}$ (the endo-arrows being identities) then the diagram is no longer a coequalizer in $\underset{\sim}{\text{cat}}(\mathcal{S})$. In fact the coequalizer of f and g in $\underset{\sim}{\text{cat}}(\mathcal{S})$ has two morphisms from x to z, one being the image of the morphism $x \longrightarrow z$ in $\underline{B}$, and the other the composite of $x \longrightarrow y$ and $y \longrightarrow z$.

The reason why this difficulty arises is that if we consider the objects of "composable pairs" of arrows in $\underline{A}$, $\underline{B}$ and $\underline{C}$ (i.e. the objects

$$A_2 = \Sigma_{\underset{\sim}{a}_1,\underset{\sim}{a}_2,\underset{\sim}{a}_3} \underline{A}(\underset{\sim}{a}_1,\underset{\sim}{a}_2) \times \underline{A}(\underset{\sim}{a}_2,\underset{\sim}{a}_3) \quad ,$$

etc.), then the diagram

$$A_2 \underset{g_2}{\overset{f_2}{\rightrightarrows}} B_2 \xrightarrow{h_2} C_2$$

is not a coequalizer in $\mathcal{S}$ - in fact h_2 is not even epi.

However, for the particular case of the theory $\underset{\sim}{\text{cat}}$, we do have a direct method of overcoming this difficulty. Note that the free functor constructed in Proposition 6.7 is the identity on objects: accordingly, we define a parallel pair $\underline{A} \underset{g}{\overset{f}{\rightrightarrows}} \underline{B}$ in $\underset{\sim}{\text{cat}}(\mathcal{E})$ (or $\underset{\sim}{\text{dgph}}(\mathcal{E})$) to be <u>acceptable</u> if it is reflexive and, in addition, we have $A_o = B_o$ and $f_o = g_o = 1_{A_o}$.

<u>6.9 Lemma</u> The forgetful functor $\underset{\sim}{\text{cat}}(\mathcal{E}) \longrightarrow \underset{\sim}{\text{dgph}}(\mathcal{E})$ creates coequalizers of acceptable pairs.

<u>Proof</u> Let $\underline{A} \underset{g}{\overset{f}{\rightrightarrows}} \underline{B}$ be an acceptable pair in $\underset{\sim}{\text{cat}}(\mathcal{E})$, and $\underline{B} \xrightarrow{h} \underline{C}$ its coequalizer in $\underset{\sim}{\text{dgph}}(\mathcal{E})$. Then $C_o = A_o$, and we have a reflexive coequalizer diagram

$$\underline{A}(\underset{\sim}{a}_1,\underset{\sim}{a}_2) \rightrightarrows \underline{B}(\underset{\sim}{a}_1,\underset{\sim}{a}_2) \longrightarrow \underline{C}(\underset{\sim}{a}_1,\underset{\sim}{a}_2)$$

in $\mathcal{E}/A_o \times A_o$. Since finite products commute with reflexive coequalizers in a topos ([17], Exercise 0.1), the diagram

$$\underline{A}(\underset{\sim}{a}_1,\underset{\sim}{a}_2) \times \underline{A}(\underset{\sim}{a}_2,\underset{\sim}{a}_3) \rightrightarrows \underline{B}(\underset{\sim}{a}_1,\underset{\sim}{a}_2) \times \underline{B}(\underset{\sim}{a}_2,\underset{\sim}{a}_3) \longrightarrow \underline{C}(\underset{\sim}{a}_1,\underset{\sim}{a}_2) \times \underline{C}(\underset{\sim}{a}_2,\underset{\sim}{a}_3)$$

is a coequalizer in $\mathcal{E}/A_o \times A_o \times A_o$. It now follows exactly as in Lemma 5.6 that $\underline{C}$ has a unique category structure which makes h into a coequalizer in $\underset{\sim}{\text{cat}}(\mathcal{E})$. □

<u>6.10 Corollary</u> (i) The forgetful functor $\underset{\sim}{\text{cat}}(\mathcal{E}) \longrightarrow \underset{\sim}{\text{dgph}}(\mathcal{E})$ is monadic.

(ii) $\underset{\sim}{\text{cat}}(\mathcal{E})$ has finite colimits.

<u>Proof</u> Since the free functor $\underset{\sim}{\text{dgph}}(\mathcal{E}) \longrightarrow \underset{\sim}{\text{cat}}(\mathcal{E})$ is the identity on objects, it is clear that the "standard free presentation" of an object of $\underset{\sim}{\text{cat}}(\mathcal{E})$ is an acceptable pair. Now we simply have to

work through the proofs of the Crude Tripleability Theorem and of Linton's theorem [27], observing that all the coequalizers we require are in fact acceptable. □

The argument of 6.9 and 6.10 is of course peculiar to the theory of categories and certain closely-related theories (e.g. posets, groupoids); but it is worth noting that it can be applied in contexts other than the one which we are considering here. For example, it affords a considerable simplification of the proof by H. Wolff [37] of the corresponding results for categories and directed graphs enriched over a symmetric monoidal closed category $\mathcal{V}$.

Now it is clear that the free functor $\text{dgph}(\mathcal{E}) \longrightarrow \text{cat}(\mathcal{E})$ constructed in Proposition 6.7 commutes with inverse image functors, and so the monad on $\text{dgph}(\mathcal{E})$ which it defines can be extended to an algebraic theory over dgph. Since a coequalizer of maps between finite cardinals is a finite cardinal, it is not hard to show that any coequalizer diagram

$$\underline{FA} \rightrightarrows \underline{FB} \longrightarrow \underline{C}$$

where $\underline{A}$ and $\underline{B}$ are essentially finite directed graphs, can be replaced by an acceptable coequalizer

$$\underline{FA}' \rightrightarrows \underline{FB}' \longrightarrow \underline{C}$$

in which $\underline{A}'$ and $\underline{B}'$ are still essentially finite; so we define a category $\underline{C}$ to be <u>finitely-presented</u> if there exists a coequalizer of the latter type. Now we may construct the object of finite presentations and hence the internal category FPcat of finitely-presented categories; and it is then a straightforward extension of Theorem 5.21 to prove

<u>6.11 Theorem</u> Let $\mathcal{E}$ be a topos with a natural number object. Then the topos $\mathcal{E}^{\text{FPcat}}$ is a classifying topos for the theory of internal categories in $\mathcal{E}$-toposes. □

CHAPTER VII : FINITARY AND INFINITARY THEORIES

We have seen that the correct "semantical" interpretation of the notion of finitary algebraic theory in a topos $\mathcal{E}$ involves consideration, not only of the models of $\mathbb{T}$ in $\mathcal{E}$, but also of the models of $\mathbb{T}$ in toposes defined over $\mathcal{E}$. If we wish to drop the word "finitary", then we clearly cannot expect to have any particular relationship between $\mathbb{T}$-models in $\mathcal{E}$ and those in an arbitrary $\mathcal{E}$-topos; but we should still be able to talk about families of $\mathbb{T}$-models indexed by an object of $\mathcal{E}$ - i.e. we should at least consider $\mathbb{T}$-models in $\mathcal{E}$-toposes of the form $\mathcal{E}/I$. It therefore seems reasonable to make the following provisional definition:

7.1 Definition Let $\mathcal{E}$ be a topos. An (infinitary) internal algebraic theory in $\mathcal{E}$ is a locally internal monad on $\mathcal{E}$, i.e. a monad on $\mathcal{E}$ in the 2-category of locally internal categories over $\mathcal{E}$.

An internal algebraic theory $\mathbb{T}$ thus consists of a family of $(\mathcal{E}/I)$-functors

$$T^I : \mathcal{E}/I \longrightarrow \mathcal{E}/I$$

for each object I of $\mathcal{E}$, which commute up to coherent natural isomorphism with pullback functors, and are equipped with monad structures which are natural in I. (From Lemma 1.1, we know that the strength of the functors T^I is implied by the other hypotheses.) A $\mathbb{T}$-model in $\mathcal{E}/I$ is of course an algebra for the monad $\mathbb{T}^I$.

In this chapter we shall investigate some of the consequences of this definition, and its relationship to that of a finitary algebraic theory (5.3). We shall see that in some respects, these theories are perhaps not as well behaved as we would like, although in others they do appear to have the right properties. The question

"Is 7.1 the 'right' definition of an internal algebraic theory?" therefore remains open. We begin, however, with some examples.

7.2 Examples (a) Let $A(\underset{\sim}{i})$ be an object of $\mathcal{E}/I$, for some I. Then we may define "the free theory generated by an I-indexed family of $A(\underset{\sim}{i})$-ary operations" by specifying its models to be objects X equipped with a morphism

$$\Sigma_{\underset{\sim}{i}}(X^{A(\underset{\sim}{i})}) \longrightarrow X \quad .$$

If $\mathcal{E}$ has a natural number object, then the Adjoint Functor Theorem of Paré and Schumacher may be used to construct a free functor for this theory (see [33], Proposition V.2.3.5), which is clearly locally internal; so we obtain an internal algebraic theory in the sense of 7.1. If the category of models of this theory has coequalizers (of which more anon), then we may add an internally-indexed family of equations to it, as we did for finitary theories. A theory which can be described in this way will be said to be presentable; in $\mathcal{S}$, this corresponds to having a rank ([29], Definition 1.5.14).

(b) Let X be an object of $\mathcal{E}$. Then the functor

$$(-)^X\colon \mathcal{E} \longrightarrow \mathcal{E}$$

has a natural (locally internal) monad structure, induced by the morphisms $X \longrightarrow 1$ and $X \xrightarrow{\Delta} X \times X$. This theory is clearly presentable: it has a single X-ary operation and two equations, one unary and one $(X \times X)$-ary. In general, its category of models is not easy to describe; but if X is a subobject of 1, then it is equivalent to $\mathcal{E}/X$.

(c) Let j be a Lawvere-Tierney topology in $\mathcal{E}$. Then the $\mathcal{E}$-topos $sh_j(\mathcal{E})$ of j-sheaves and the quasitopos $sep_j(\mathcal{E})$ of j-separated objects are both locally internal reflective subcategories of $\mathcal{E}$, and so define idempotent algebraic theories in $\mathcal{E}$. If we write J for the subobject of Ω classified by j, and $D(\underset{\sim}{j})$ for the object of $\mathcal{E}/J$ obtained by factoring $1 \xrightarrow{\text{true}} \Omega$ through $J \rightarrowtail \Omega$, then it is not hard to see that $\Sigma_{\underset{\sim}{j}} X^{D(\underset{\sim}{j})}$ is simply the object $\hat{X}$ defined in [17], § 3.3. From the construction of the associated sheaf functor given in [17], it now follows that the theory sh_j is presentable, its operations being indexed by J and having arities $D(\underset{\sim}{j})$. (sep_j is trivially presentable, being a quotient of the initial theory.)

(d) Let $P: \mathcal{E} \longrightarrow \mathcal{E}$ be the covariant power-set functor. It is shown in [1] and [30] that this functor has a monad structure, whose algebras are the internally complete semilattices in $\mathcal{E}$; but P commutes up to isomorphism with logical functors, and so extends to a locally internal monad on $\mathcal{E}$. Similarly, the monadic adjunction

$$\mathcal{E} \rightleftarrows \mathcal{E}^{op}$$

induced by the contravariant power-set functor [32] is locally internal; in this case the algebras may be described as internally complete atomic Heyting algebras in $\mathcal{E}$ (see [10], [30]).

Let $\mathcal{E}$ be a topos. By identifying objects of $\mathcal{E}$ with discrete internal categories, we may identify $\mathcal{E}$ with a full subcategory of $\underset{\sim}{cat}(\mathcal{E})$; and the assignment

$$\underline{A} \longmapsto \mathcal{E}^{\underline{A}}$$

makes $\mathcal{E}$ into an indexed category over $\underset{\sim}{cat}(\mathcal{E})$, in a manner which extends the usual indexing of $\mathcal{E}$ over itself.

<u>7.3 Lemma</u> Let $\mathcal{E}$ be a topos. Then any functor $T: \mathcal{E} \longrightarrow \mathcal{E}$ which is indexed over $\mathcal{E}$ extends (uniquely up to canonical isomorphism) to a functor indexed over $\underset{\sim}{cat}(\mathcal{E})$.

Proof Let $\underline{A}$ be an internal category in $\mathcal{E}$, F an internal diagram on $\underline{A}$. The requirement that $T^{\underline{A}}: \mathcal{E}^{\underline{A}} \longrightarrow \mathcal{E}^{\underline{A}}$ should extend the given indexed endofunctor of $\mathcal{E}$ means that we must have

$$(T^{\underline{A}}(F))(\underset{\sim}{a}) \cong T^{(A_0)}(F(\underset{\sim}{a}))$$

as objects of $\mathcal{E}/A_0$; so it suffices to define an action of $\underline{A}$ on the latter object. But if we regard the action of $\underline{A}$ on $F(\underset{\sim}{a})$ as a morphism

$$\underline{A}(\underset{\sim}{a}_1,\underset{\sim}{a}_2) \longrightarrow F(\underset{\sim}{a}_2)^{F(\underset{\sim}{a}_1)}$$

in $\mathcal{E}/A_0 \times A_0$, we may compose it with the strength of $T^{(A_0 \times A_0)}$

$$F(\underset{\sim}{a}_2)^{F(\underset{\sim}{a}_1)} \longrightarrow TF(\underset{\sim}{a}_2)^{TF(\underset{\sim}{a}_1)}$$

(which exists by Lemma 1.1) to obtain the required action. The remaining details are straightforward. □

7.4 Theorem Let $\mathcal{E}$ be a topos with a natural number object. Let $\mathcal{C}$ denote the category of locally internal endofunctors of $\mathcal{E}$, and let

$$\varphi: \mathcal{E}[U] \longrightarrow \mathcal{C}$$

be the forgetful functor obtained by regarding objects of $\mathcal{E}[U]$ as natural endofunctors of $\mathcal{E}$-toposes, and forgetting their effect on $\mathcal{E}$-toposes not of the form $\mathcal{E}/I$. Then φ is full and faithful and has a right adjoint, denoted $T \longmapsto \check{T}$; and the counit map

$$\check{T}_{\mathcal{E}/I}(X(\underset{\sim}{i})) \longrightarrow T^I X(\underset{\sim}{i})$$

is an isomorphism provided $X(\underset{\sim}{i})$ is a finite cardinal in $\mathcal{E}/I$.

Proof Let T be an object of $\mathcal{E}$. By Lemma 7.3, T induces an endofunctor $T^{(\underline{E}_{fin})}$ of $\mathcal{E}[U]$; let $\check{T}$ be the object $T^{(\underline{E}_{fin})}(U)$. Then if S is any object of $\mathcal{E}[U]$, the uniqueness part of 7.3 implies that

$$(\varphi(S))^{\vee} \cong S_{\mathcal{E}[U]}(U) \cong S \quad ;$$

we take this isomorphism to be the unit of the adjunction. To construct the counit map $\varphi(\check{T}) \longrightarrow T$, consider an object $X(\underset{\sim}{i})$

of $\mathcal{E}/I$, and the morphism

$$X(\underset{\sim}{i})^{[\underset{\sim}{n}]} \longrightarrow T^I X(\underset{\sim}{i})^{T^N[\underset{\sim}{n}]}$$

induced by the strength of $T^{I\times N}$. This transposes to give a map

$$\Sigma_{\underset{\sim}{n}}(T^N[\underset{\sim}{n}] \times X(\underset{\sim}{i})^{[\underset{\sim}{n}]}) \longrightarrow T^I X(\underset{\sim}{i})$$

in $\mathcal{E}/I$, which is easily seen to coequalize the two maps

$$\Sigma_{\underset{\sim}{n}_1,\underset{\sim}{n}_2}(T[\underset{\sim}{n}_1] \times [\underset{\sim}{n}_2]^{[\underset{\sim}{n}_1]} \times X(\underset{\sim}{i})^{[\underset{\sim}{n}_2]}) \rightrightarrows \Sigma_{\underset{\sim}{n}}(T[\underset{\sim}{n}] \times X(\underset{\sim}{i})^{[\underset{\sim}{n}]})$$

whose coequalizer is $\check{T}_{\mathcal{E}/I}(X(\underset{\sim}{i}))$. So we obtain a morphism

$$\varepsilon_{T,I,X}\colon \check{T}_{\mathcal{E}/I}(X(\underset{\sim}{i})) \longrightarrow T^I X(\underset{\sim}{i})$$

in $\mathcal{E}/I$, which is readily checked to be natural in X, I and T, and to define the counit of the required adjunction.

The fact that φ is full and faithful follows from the fact that the unit of the adjunction is an isomorphism.

Finally, suppose $X(\underset{\sim}{i})$ is a finite cardinal in $\mathcal{E}/I$. Then its classifying map

$$\mathcal{E}/I \xrightarrow{\ \tilde{X}\ } \mathcal{E}[U]$$

is induced by an internal functor from the discrete category I to $\underline{E}_{fin}$, so it follows from Lemma 7.3 that the square

$$\begin{array}{ccc} \mathcal{E}[U] & \xrightarrow{T(\underline{E}_{fin})} & \mathcal{E}[U] \\ \downarrow{\scriptstyle \tilde{X}^*} & & \downarrow{\scriptstyle \tilde{X}^*} \\ \mathcal{E}/I & \xrightarrow{\ T^I\ } & \mathcal{E}/I \end{array}$$

commutes up to isomorphism. Hence $\varepsilon_{T,I,X}$ is iso. □

<u>7.5 Corollary</u> Let T be an object of the category $\mathcal{C}$ defined above. Then T is isomorphic to an object in the image of the functor φ iff it "preserves filtered colimits" in the following sense: for any weakly filtered category $\underline{A}$ in $\mathcal{E}$, with colimiting cone $A_0 \twoheadrightarrow L$, the diagram

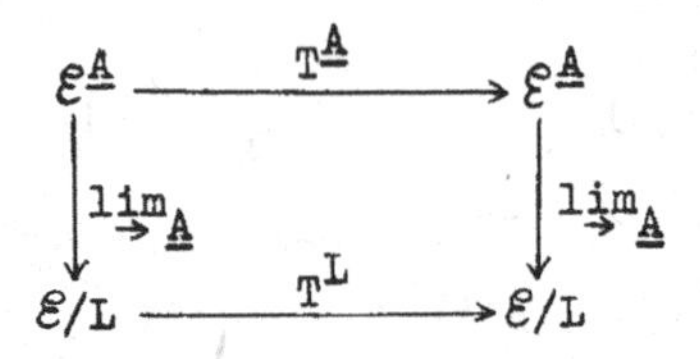

commutes up to isomorphism.

Proof By [17], Proposition 2.55, $\underline{A}$ is weakly filtered iff it is filtered when regarded as an internal category in $\mathcal{E}/L$, and in this case $\varinjlim_{\underline{A}}: \mathcal{E}^{\underline{A}} \longrightarrow \mathcal{E}/L$ is the inverse image of a geometric morphism over $\mathcal{E}$. So if T is in the image of φ, the condition is satisfied. But if the condition is satisfied, consider an object $X(\underset{\sim}{i})$ of $\mathcal{E}/I$. By Lemma 4.4, this corresponds to a flat presheaf $HX(\underset{\sim}{i})$ on $\underline{E}_{fin}$ in $\mathcal{E}/I$, i.e. to a discrete fibration

$$\underline{HX}(\underset{\sim}{i}) \longrightarrow I^*\underline{E}_{fin}$$

where $\underline{HX}(\underset{\sim}{i})$ is filtered. Transposing, we obtain an internal functor

$$\underline{A} \xrightarrow{\ \gamma\ } \underline{E}_{fin}$$

in $\mathcal{E}$, where $\underline{A} = \Sigma_{\underset{\sim}{i}}\underline{HX}(\underset{\sim}{i})$ is weakly filtered with colimit I. But Diaconescu's theorem now tells us that $\tilde{X}^*$ may be factored as the composite

$$\mathcal{E}[U] \xrightarrow{\ \gamma^*\ } \mathcal{E}^{\underline{A}} \xrightarrow{\ \varinjlim_{\underline{A}}\ } \mathcal{E}/I \ ;$$

and so we deduce from the given condition that

$$\begin{array}{ccc} \mathcal{E}[U] & \xrightarrow{\ T^{(\underline{E}_{fin})}\ } & \mathcal{E}[U] \\ \downarrow X^* & & \downarrow \tilde{X}^* \\ \mathcal{E}/I & \xrightarrow{\ T^I\ } & \mathcal{E}/I \end{array}$$

commutes up to isomorphism, i.e.

$$T^I X(\underset{\sim}{i}) \cong \tilde{X}^*(\check{T}) = \check{T}_{\mathcal{E}/I}(X(\underset{\sim}{i})) \ .$$

So $\varepsilon_{T,I,X}$ is an isomorphism. □

Now the functor φ of Theorem 7.4 is clearly monoidal, where $\mathcal{C}$ is given the monoidal structure induced by composition. The functor

$T \longmapsto \check{T}$ is not normally monoidal, but since it is right adjoint to a monoidal functor it comes equipped with a natural transformation

$$\check{S} \otimes \check{T} \longrightarrow (S \otimes T)^{\vee} \quad ,$$

and hence if T has a monoid structure in $\mathcal{C}$ $\check{T}$ has one in $\mathcal{E}[U]$. Thus 7.4 and 7.5 can be "lifted" to the categories of monoids in $\mathcal{E}[U]$ (i.e. finitary algebraic theories) and monoids in $\mathcal{C}$ (i.e. locally internal monads on $\mathcal{E}$). In particular, we have

<u>7.6 Corollary</u> An internal algebraic theory in $\mathcal{E}$ is finitary iff its functor part preserves filtered colimits in the sense defined in Corollary 7.5. □

Clearly, Corollary 7.6 tends to confirm the hypothesis that 7.1 is the correct definition of an internal algebraic theory in a topos. However, there are other properties of internal theories which work well only in a Boolean topos (and therefore not in most of the examples of greatest interest). An important example is the question of whether the free functor for an algebraic theory preserves monomorphisms. The standard proof of this fact in $\mathcal{S}$ ([29], Proposition 1.5.42) can be generalized as follows:

<u>7.7 Proposition</u> Let $\mathcal{E}$ be a Boolean topos, $\mathbb{T}$ an internal algebraic theory in $\mathcal{E}$. Then the functor $T^1 \colon \mathcal{E} \longrightarrow \mathcal{E}$ preserves monomorphisms.

<u>Proof</u> Let $X \xrightarrow{m} Y$ be a monomorphism in $\mathcal{E}$. We consider first two special cases:

(a) Suppose $TX \cong 0$. Then Tm is trivially mono.

(b) Suppose TX has a global element. Then TX is injective in $\mathcal{E}$, and so we can factor the unit map $e_X \colon X \longrightarrow TX$ through m to obtain $f \colon Y \longrightarrow TX$. Then f, being a map from Y to a $\mathbb{T}$-algebra, factors through e_Y to give us a one-sided inverse for Tm. So Tm is mono.

Now let $I \rightarrowtail 1$ be the complement of the image of $TX \longrightarrow 1$. Then since $TX \amalg I \longrightarrow 1$ is epi, the pullback functor $(TX \amalg I)^*$ reflects monos, and so it is sufficient to prove that $(TX)^*(Tm)$ and $I^*(Tm)$ are both mono. But on pulling back to $\mathcal{E}/TX$, the object TX acquires a global element, so case (b) applies; and $I^*(TX) \cong 0$, so case (a) applies. □

<u>7.8 Corollary</u> If $\mathbb{T}$ is finitary, then the word "Boolean" can be omitted from the hypotheses of Proposition 7.7.

<u>Proof</u> We use the Funayama-Freyd-Barr theorem ([17], Proposition 7.54) that for any topos $\mathcal{E}$, there exists a Boolean topos $\mathcal{B}$ and a surjective geometric morphism $p: \mathcal{B} \longrightarrow \mathcal{E}$. Now if m is a monomorphism in $\mathcal{E}$, then $p^*T_{\mathcal{E}}(m) \cong T_{\mathcal{B}}p^*(m)$ is mono in $\mathcal{B}$ by Proposition 7.7; but p^* reflects monos. □

However, the following example shows that 7.7 and 7.8 do not admit a common generalization to arbitrary algebraic theories in arbitrary toposes (and hence suggests that the appeal to the Funayama-Freyd-Barr theorem in Corollary 7.8 was strictly necessary):

<u>7.9 Example</u> Let $\mathcal{E} = \mathcal{S}^{\underline{2}}$ be the <u>Sierpinski topos</u> ([17], Example 4.37(iii)) whose objects are morphisms in $\mathcal{S}$, and whose morphisms are commutative squares. Define a functor $T: \mathcal{E} \longrightarrow \mathcal{E}$ as follows:

$$T(X_0 \xrightarrow{f} X_1) = (X_0 \amalg 1 \amalg N_X \xrightarrow{g} X_1 \amalg 1) \quad ,$$

where 1 is the singleton set $\{*\}$, N_X is the complement of the image of f in X_1, and g is defined by

$$\begin{aligned} g(x) &= f(x) && \text{if } x \in X_0 \\ &= * && \text{otherwise} \quad . \end{aligned}$$

And if $h: X \longrightarrow Y$ is a morphism of $\mathcal{E}$, Th is defined by

$$Th_0(x) = h_0(x) \quad \text{if } x \in X_0 . \qquad Th_1(x) = h_1(x) \quad \text{if } x \in X_1$$
$$= * \quad \text{if } x = * \qquad\qquad = * \quad \text{if } x = *$$
$$= h_1(x) \quad \text{if } x \in N_X \text{ and } h_1(x) \in N_Y$$
$$= * \quad \text{if } x \in N_X \text{ but } h_1(x) \notin N_Y$$

Now T has an obvious monad structure, whose unit is the inclusion map $X \longrightarrow TX$. Moreover, we may extend this monad to a locally internal monad on $\mathcal{E}$: this uses the fact that for any object I of $\mathcal{E}$, an object of $\mathcal{E}/I$ may be considered as a family of diagrams in $\mathcal{S}$ of the form

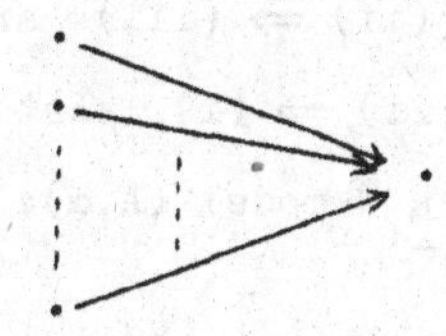

indexed by the set I_1.

However, if V denotes the nontrivial subobject of 1 in $\mathcal{E}$, then T does not preserve the monomorphism $V \rightarrowtail 1$. (In fact every mono preserved by T is either complemented or split.)

As a consequence of 7.7 and 7.8, we note

<u>7.10 Corollary</u> Let $\mathbb{T}$ be an internal algebraic theory in $\mathcal{E}$, and suppose either that $\mathbb{T}$ is finitary or that $\mathcal{E}$ is Boolean. Then the forgetful functor $\mathbb{T}(\mathcal{E}) \longrightarrow \mathcal{E}$ preserves injectives.

<u>Proof</u> Its left adjoint (the free functor) preserves monos, since the forgetful functor reflects them. □

The reader may verify that for the theory $\mathbb{T}$ constructed in Example 7.9, the object $(1 \longrightarrow 2)$ of $\mathcal{E}$, with its unique $\mathbb{T}$-model structure, is injective as an object of $\mathbb{T}(\mathcal{E})$ but not as an object of $\mathcal{E}$.

A similar phenomenon, first observed by P. Freyd, occurs when we consider the concept of consistency for internal algebraic theories. Recall that an algebraic theory $\mathbb{T}$ is said to be <u>consistent</u>

if the free $\mathbb{T}$-model functor is faithful, or equivalently the unit map $e_X: X \longrightarrow TX$ is mono for every X.

<u>7.11 Proposition</u> Let $\mathbb{T}$ be an internal algebraic theory in $\mathcal{E}$, and suppose either that $\mathbb{T}$ is finitary or that $\mathcal{E}$ is Boolean. Then the following conditions are equivalent:

(i) $\mathbb{T}$ is consistent.

(ii) e_2 is mono, where $2 = 1 \amalg 1$.

(iii) There exists a $\mathbb{T}$-model having two disjoint global elements.

<u>Proof</u> The finitary case is reduced to the Boolean case as in 7.8. The implications (i) $\Rightarrow$ (ii) $\Rightarrow$ (iii) are trivial in any case; so it remains to prove (iii) $\Rightarrow$ (i). Let $2 \overset{m}{\rightarrowtail} A$ be a monomorphism from 2 into a $\mathbb{T}$-model (A,α); then for any X we have a monomorphism

$$X \overset{\{\}}{\rightarrowtail} 2^X \overset{m^X}{\rightarrowtail} A^X \quad .$$

But A^X has a $\mathbb{T}$-model structure, obtained by the following deduction:

$$\dfrac{\dfrac{\dfrac{\dfrac{A^X \times X \xrightarrow{\ ev\ } A}{X^*(A^X) \longrightarrow X^*A}}{X^*T^1(A^X) \cong T^X X^*(A^X) \longrightarrow T^X X^*A \cong X^*T^1A \xrightarrow{X^*\alpha} X^*A}}{T^1(A^X) \times X \longrightarrow A}}{T^1(A^X) \longrightarrow A^X} \quad .$$

So X can be mapped monomorphically into a $\mathbb{T}$-model; but this map must factor through e_X, and so the latter must be mono. $\square$

If neither of the hypotheses of Proposition 7.11 apply, we can of course replace 2 by Ω in the statement of (ii). But 2 is not sufficient in general, as can be seen by considering the theory defined (as in Example 7.2(b)) by the functor $(-)^V$, where V is the nontrivial subobject of 1 in the Sierpinski topos. The models of this theory are the objects $(X_0 \xrightarrow{f} X_1)$ for which f is bijective; so the object $(2 \longrightarrow 1)$ is not a subobject of any $\mathbb{T}$-model, although $(2 \longrightarrow 2)$ is.

Further problems arise when we consider the categorical structure of $\mathbb{T}(\mathcal{E})$ for a general $\mathbb{T}$. As we found in the last chapter for many-sorted theories, the arguments involving reflexive coequalizers, which we used in chapter V, will not work even in $\mathcal{S}$ for infinitary theories. However, for (single-sorted) infinitary theories our experience in $\mathcal{S}$ might lead us to hope that we could give an alternative argument using coequalizers of equivalence relations; but even this approach will not work in general.

We shall say that an algebraic theory $\mathbb{T}$ is <u>projectively generated</u> if the functor $T: \mathcal{E} \longrightarrow \mathcal{E}$ preserves coequalizers of equivalence relations.

<u>7.12 Examples</u> (a) Any finitary theory is projectively generated, by Proposition 4.6.

(b) Since the functor $\Sigma_{\underset{\sim}{i}}((-)^{A(\underset{\sim}{i})})$ preserves kernel-pairs, it is easily seen that the free theory of Example 7.2(a) is projectively generated iff this functor preserves epis, i.e. iff $A(\underset{\sim}{i})$ is internally projective in $\mathcal{E}/I$.

(c) Suppose $\mathcal{E}$ satisfies the implicit axiom of choice ([17], Proposition 5.25). Then any coequalizer diagram

$$X \underset{g}{\overset{f}{\rightrightarrows}} Y \xrightarrow{h} Z$$

where (f,g) is the kernel-pair of h is <u>locally</u> part of a split coequalizer diagram; hence it is preserved by any locally internal functor. So every internal algebraic theory in $\mathcal{E}$ is projectively generated.

(d) The single and double power-set theories of Example 7.2(d) are both projectively generated. The proof is very similar to the argument of [32], using the Beck condition for pullback squares of the form

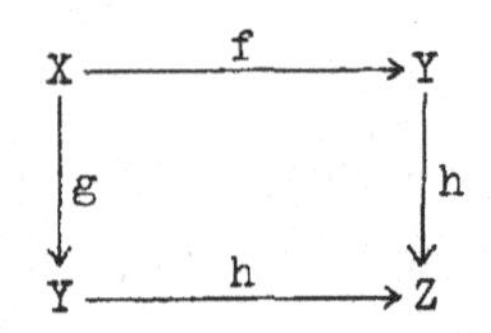

7.13 Proposition Suppose $\mathbb{T}$ is projectively generated. Then $\mathbb{T}(\mathcal{E})$ is an exact category [2] and has finite colimits.

Proof Since T preserves coequalizers of equivalence relations, the forgetful functor $\mathbb{T}(\mathcal{E}) \longrightarrow \mathcal{E}$ creates them, and hence all the properties of an exact category can be lifted from $\mathcal{E}$ to $\mathbb{T}(\mathcal{E})$ as in Corollary 5.7. As usual, the construction of finite colimits can be reduced to that of coequalizers; but if we are given a parallel pair

$$A \overset{f}{\underset{g}{\rightrightarrows}} B$$

in $\mathbb{T}(\mathcal{E})$, we may construct the equivalence relation on B generated by (f,g), by applying the internal intersection operator (in $\mathcal{E}$) to the object of subobjects of $B \times B$ which are simultaneously sub-$\mathbb{T}$-models and equivalence relations, and which contain im(f,g). So we are reduced to constructing coequalizers of equivalence relations; but this is trivial. □

Without the hypothesis of projective generation, $\mathbb{T}(\mathcal{E})$ need not be exact or even regular. To show this, we introduce a class of internal algebraic theories in the Sierpinski topos $\mathcal{S}^{\underline{2}}$: if $\mathbb{T}$ is an algebraic theory in $\mathcal{S}$, we define a theory $\mathbb{T}_f$ ("fibrewise $\mathbb{T}$") in $\mathcal{S}^{\underline{2}}$ by saying that a $\mathbb{T}_f$-model is an object $(X_0 \xrightarrow{f} X_1)$ equipped with a $\mathbb{T}$-model structure on each of the fibres $f^{-1}(x)$, $x \in X_1$. (It is easy to construct a free functor for this theory, and to extend it to a locally internal monad on $\mathcal{S}^{\underline{2}}$.)

Now if $\mathbb{T}$ has no nullary operations, then an object $(X_0 \xrightarrow{f} X_1)$ for which f is mono has a unique $\mathbb{T}_f$-model structure; hence there is a unique way in which we can make $(2 \xrightarrow{\Delta} 2\times 2)$ into a congruence on the constant object $(2 \xrightarrow{1} 2)$. But if $\mathbb{T}$ is the inconsistent theory without nullary operations, then the coequalizer of this equivalence relation in $\mathbb{T}_f(\mathcal{S}^{\underline{2}})$ is $(1 \longrightarrow 1)$, so the equivalence relation is not effective. On the other hand, if we take $\mathbb{T}$ to be the theory of commutative semigroups (without 1), then the coequalizer in $\mathbb{T}_f(\mathcal{S}^{\underline{2}})$ has the form $(3 \longrightarrow 1)$, and the map

$$(2 \longrightarrow 2) \longrightarrow (3 \longrightarrow 1)$$

is not universally regular epi in $\mathbb{T}_f(\mathcal{S}^{\underline{2}})$. So in this case $\mathbb{T}_f(\mathcal{S}^{\underline{2}})$ is not even regular.

The question whether $\mathbb{T}(\mathcal{E})$ always has finite colimits is less easy to answer. We do not know of any counterexamples; and it seems that we should always be able to construct the coequalizer of a pair

$$A \underset{g}{\overset{f}{\rightrightarrows}} B$$

in $\mathbb{T}(\mathcal{E})$ by an application of the Paré-Schumacher Adjoint Functor Theorem ([33], IV.1.5). The solution-set condition should be obtainable from the fact that any morphism $B \longrightarrow C$ coequalizing f and g factors through such a morphism for which the induced map $F(Q) \longrightarrow C$ is a strong epimorphism in $\mathbb{T}(\mathcal{E})$, where Q is the coequalizer of f and g in $\mathcal{E}$ and F is the free $\mathbb{T}$-model functor. Unfortunately, there seems to be no general way of proving that $\mathbb{T}(\mathcal{E})$ is co-well-powered for strong epimorphisms, since these need not be epi in $\mathcal{E}$.

REFERENCES

1. C. Anghel and P. Lecouturier, Généralisation d'un résultat sur le triple de la réunion. Ann. Fac. Sci. de Kinshasa (Zaire), Section Math.-Phys., 1 (1975), 65-94.

2. M. Barr, Exact categories. "Exact categories and categories of sheaves", Springer Lecture Notes in Math. 236 (1971), 1-120.

3. J. Bénabou, Structures algébriques dans les catégories. Cahiers top. et géom. diff. 10 (1968), 1-126.

4. J. Bénabou, Problèmes dans les topos. Univ. Cath. de Louvain, Inst. de Math. Pure et Appliquée, Rapport no. 34 (1973).

5. J. Bénabou, Théories relatives à un corpus. C.R. Acad. Sci. Paris 281 (1975), A831-834.

6. G. Birkhoff, The structure of abstract algebras. Proc. Camb. Philos. Soc. 31 (1935), 433-454.

7. F. Borceux and B.J. Day, Universal algebra in a closed category. Preprint, University of Sydney 1976.

8. R.B. Coates, "Semantics of generalized structures". Ph.D. thesis, King's College, London, 1974.

9. P.M. Cohn, "Universal Algebra". Harper and Row, 1965.

10. M.F. Coste, M. Coste and J. Parent, Algèbres de Heyting dans les topos. Seminaire Bénabou, Université Paris-Nord 1974.

11. R. Diaconescu, Change of base for toposes with generators. J. Pure and Applied Algebra 6 (1975), 191-218.

12. P. Freyd, Aspects of topoi. Bull. Austral. Math. Soc. 7 (1972), 1-76.

13. M. Hakim, "Topos annelés et schémas relatifs". Ergebnisse der Mathematik, Band 64, Springer-Verlag, 1972.

14. P.T. Johnstone, Internal categories and classification theorems. "Model Theory and Topoi", Springer Lecture Notes in Math. 445 (1975), 103-113.

15. P.T. Johnstone, "Some aspects of internal category theory in an elementary topos". Ph.D. thesis, Cambridge University 1974.

16. P.T. Johnstone, Adjoint lifting theorems for categories of algebras. Bull. Lond. Math. Soc. 7 (1975), 294-297.

17. P.T. Johnstone, "Topos Theory". L.M.S. Mathematical Monographs no. 10, Academic Press, 1977.

18. M. Justersen, Ph.D. thesis, Aarhus Universitet 1969.

19. A. Kock, Monads on symmetric monoidal closed categories. Arch. Math. (Basel) 21 (1970), 1-10.

20. A. Kock, Universal projective geometry via topos theory. J. Pure and Applied Algebra 9 (1976), 1-24.

21. A. Kock, P. Lecouturier and C.J. Mikkelsen, Some topos-theoretic concepts of finiteness. "Model Theory and Topoi", Springer Lecture Notes in Math. 445 (1975), 209-283.

22. F.W. Lawvere, "Functorial semantics of algebraic theories". Ph.D. thesis, Columbia University 1963; summarized in Proc. Nat. Acad. Sci. U.S.A. 50 (1963), 869-872.

23. F.W. Lawvere, Variable quantities and variable structures in topoi. "Algebra, Topology and Category Theory: a collection of papers in honor of Samuel Eilenberg", Academic Press 1976, 101-131.

24. D. Lazard, Sur les modules plats. C.R. Acad. Sci. Paris 258 (1964), 6313-6316.

25. B. Lesaffre, "Structures algébriques dans les topos élémentaires". Thèse de 3e cycle, Université Paris VII 1974; summarized in C.R. Acad. Sci. Paris 277 (1973), A663-666.

26. F.E.J. Linton, Some aspects of equational categories. Proc. La Jolla conference on Categorical Algebra, Springer-Verlag 1966, 84-94.

27. F.E.J. Linton, Coequalizers in categories of algebras. "Seminar on Triples and Categorical Homology Theory", Springer Lecture Notes in Math. 80 (1969), 75-90.

28. M. Makkai and G.E. Reyes, Model-theoretic methods in the theory of topoi and related categories. Bull. Acad. Polon. Sci. 24 (1976), 379-392.

29. E.G. Manes, "Algebraic Theories". Graduate Texts in Math. no. 26, Springer-Verlag, 1976.

30. C.J. Mikkelsen, "Lattice-theoretic and logical aspects of elementary topoi". Aarhus Universitet Various Publications Series no. 25, 1976.

31. G. Osius, Logical and set-theoretical tools in elementary topoi. "Model Theory and Topoi", Springer Lecture Notes in Math. 445 (1975), 297-346.

32. R. Paré, Colimits in topoi. Bull. Amer. Math. Soc. 80 (1974), 556-561.

33. R. Paré and D. Schumacher, Abstract families and the Adjoint Functor Theorems. This volume.

34. J. Penon, Catégories localement internes. C.R. Acad. Sci. Paris 278 (1974), A1577-1580.

35. J.D.H. Smith, "Mal'cev Varieties". Springer Lecture Notes in Math. 554 (1976).

36. M. Tierney, Forcing topologies and classifying topoi. "Algebra, Topology and Category Theory: a collection of papers in honor of Samuel Eilenberg", Academic Press 1976, 211-219.

37. H. Wolff, $\underline{V}$-Cat and $\underline{V}$-Graph. J. Pure and Applied Algebra 4 (1974), 123-135.

38. G.C. Wraith, "Algebraic Theories". Aarhus Universitet Lecture Notes Series no. 22 (revised edition 1975).

39. G.C. Wraith, Lectures on elementary topoi. "Model Theory and Topoi", Springer Lecture Notes in Math. 445 (1975), 114-206.

40. R. Wood, $\underline{V}$-indexed categories. This volume.

COEQUALIZERS IN ALGEBRAS FOR AN INTERNAL TYPE

Robert Rosebrugh

§1 INTRODUCTION

An important aspect of the program of studying elementary topoi as categories of variable sets is the consideration of categories which are algebraic over elementary topoi. A question which arises quickly here is that of their cocompleteness (where of course we ask for internal rather than external cocompleteness and this requires working in the setting of indexed categories.) It is a well known result of Linton [Ln] that any category which is monadic over set is cocomplete. That result arises by first reducing the problem to the existence of reflexive coequalizers in the algebraic category and this can also be done in the setting of categories monadic over a topos (at least for an indexed monad). The crucial point in Linton's method is that monads on set preserve epimorphisms since epimorphisms are split. This is far from being the case for indexed monads on a topos.

For example, let $\mathbb{T}$ be the monoid monad on set and $\mathbb{T}_f$ the monad on set^2 which acts fibrewise on objects of set^2 i.e. $\mathbb{T}_f(X \xrightarrow{p} I) = \coprod_{i \in I} T(p^{-1}(i)) \to I$. Applying $\mathbb{T}_f$ to the epimorphism $2 = 2 \twoheadrightarrow 2 \to 1$ shows clearly the sort of problem encountered.

Nevertheless, there are cases in which the standard methods work: if a monad preserves epimorphisms (for example, if it is finitary), then the cocompleteness problem is easily solved [J+W].

Our concern here is with a partial solution to a more difficult case. An internal type or family of arities is given without conditions on the arities or the object indexing them. Paré and Schumacher [P+S] have shown that the algebras for the type are monadic over the base topos provided that the topos has a natural numbers object. We are able to show here that if the base topos is restricted further to be Grothendieck over a topos with the axiom of choice then the algebras for the given type have all coequalizers. As a consequence, although we do not prove it here, the algebras are internally cocomplete. Moreover, in this case, one can impose an internal family of equations and find that the resulting algebras are monadic over the base and cocomplete. For details of these the reader is referred to [Rs] . It should be noted that the result quoted applies, of course, to all topoi which are Grothendieck over set and more (there being topoi with AC which are not Grothendieck), but it does not capture all elementary topoi.

The crux of the method presented here for constructing coequalizers is the construction and use of a "bounded epi-splitting" functor on some topoi. By such a functor we mean a left exact functor which contains the identity functor, not only preserves epimorphisms but splits them and does all this without growing too quickly. This functor serves as a useful replacement for the axiom of choice - it splits epimorphisms and its boundedness allows it to take part in adjoint functor theorem arguments, unlike the epi-splitting functor $\Omega^{()}$ for example.

In §2 we introduce the situation to be studied. The natural factorisation system in the categories of interest is introduced and after studying its properties it is seen that a weak form of cowell-poweredness with respect to the factorisation system is the property needed for existence of coequalizers. Next, it is seen that bounding the endofunctor induced by an internal type with another functor having better properties is sufficient to ensure the existence of coequalizers.

§3 is devoted to the construction and a consideration of the properties of the bounded epi-splitter. This turns out to be an indexed functor which can be defined on any topos which is a functor category over another topos. That it splits epimorphisms depends on the axiom of choice in the base topos.

Finally, §4 utilizes the epi-splitter to construct coequalizers. This is done first for a trivial type i.e. one having only a single arity, on a functor category over a suitable base topos. The next step extends this result to sheaf subcategories of such topoi, that is to all topoi which are Grothendieck over a suitable base. Lastly, it turns out to be a simple extension to consider internal families of arities, that is arbitrary internal types.

The reader will be assumed to be familiar with the Paré and Schumacher theory of indexed categories [P+S], and with the basics of elementary topos theory.

§2 COWELL-POWEREDNESS AND COEQUALIZERS

Let $\underline{S}$ be an elementary topos. A morphism $t:I \to J$ in $\underline{S}$ will be called an *internal type*. We think of t as a J-indexed family of arities. With this in mind, a *t-algebra* or *algebra of type t* should consist of an object X of $\underline{S}$ together with a morphism $(\Delta_J X)^t \to \Delta_J X$ in $\underline{S}/J$. This is equivalent to giving X and a morphism $\coprod_J (\Delta_J X)^t \to X$ in $\underline{S}$. Defining $\Phi X = \coprod_J (\Delta_J X)^t$ gives a functor from $\underline{S}$ to $\underline{S}$ which is $\underline{S}$-indexed, and clearly the category of t-algebras in $\underline{S}$ is the same thing as the category of Φ-algebras i.e. objects X of $\underline{S}$ equipped with a morphism $\Phi X \to X$.

The category of Φ-algebras in $\underline{S}$ is denoted $(\Phi;\underline{S})$, and Paré and Schumacher [P+S] have shown that $(\Phi;\underline{S})$ is well-powered. Indeed this holds for any indexed endofunctor Φ on $\underline{S}$. Moreover, if (A,a) is a Φ-algebra there is a forgetful functor $\underline{\text{Sub}}(A,a) \to \underline{\text{Sub}}(A)$ which has a left adjoint, denoted $\langle - \rangle$. If $A_0 \rightarrowtail A$ is a subobject of A , then $\langle A_0 \rangle \rightarrowtail (A,a)$ is the subalgebra of A generated by A_0 , and has the property that for any subalgebra $(B,b) \rightarrowtail (A,a)$ $A_0 \subseteq B$ if and only if $\langle A_0 \rangle \subseteq (B,b)$. In particular, if A_0 generates (A,a) i.e. if $\langle A_0 \rangle = (A,a)$, then any subalgebra of (A,a) containing A_0 is equal to (A,a) . This construction defines a decomposition of morphisms in $(\Phi;\underline{S})$. Indeed suppose $(A,a) \xrightarrow{f} (B,b)$ is a morphism in $(\Phi;\underline{S})$ with factorisation $A \xrightarrow{e'} B_0 \xrightarrow{m'} B$ in $\underline{S}$, then $(A,a) \xrightarrow{e} \langle B_0 \rangle \xrightarrow{m} (B,b)$ is a decomposition of f in $(\Phi;\underline{S})$. This decomposition is indexed since a morphism in $(\Phi;\underline{S})^I$ can be similarly decomposed and the decompositions are preserved by substitution functors since everything in their construction is. We call this the *image-mono decomposition* of f and e is called the *image* of f . This is the naturally arising epi-mono factorisation system in $(\Phi;\underline{S})$ with which we shall be working throughout the sequel. Images will be denoted $\longrightarrow\!\!\bullet$. It should be noted that an image is the image of itself which is an immediate consequence of:

2.1 LEMMA Images are extremal [Gr,p.127] epimorphisms in $(\Phi;\underline{S})$.

Proof: Suppose $(A,a) \xrightarrow{f} (B,b)$ is an image, then f factors as $A \xrightarrow{e} B_0 \xrightarrow{m} B$ in $\underline{S}$, and B_0 generates (B,b) in itself. Consider

$$\begin{array}{ccccccc} (A,a) & \xrightarrow{e} & B_0 & \xrightarrow{m} & (B,b) & \overset{g}{\underset{h}{\rightrightarrows}} & (C,c) \\ & \searrow^{\ell} & \downarrow{\scriptstyle \ell'} & \nearrow_{k} & & & \\ & & (E,e) & & & & \end{array}$$

where gf = hf in $(\Phi;\underline{S})$, k is the equalizer of g and h in $(\Phi;\underline{S})$ and hence ℓ exists. But now ℓ' exists in $\underline{S}$ since k is the equalizer of f and g in $\underline{S}$ also and e epic in $\underline{S}$ implies gm = hm . Also ℓ' is monic (in $\underline{S}$) since $k\ell' = m$. Thus k is an isomorphism since B_0 generates (B,b) whence g = h and f is an epimorphism.

To see that f is extremal, suppose f = hg with h mono in

$$\begin{array}{ccccccc} (K,k) & \overset{p_0}{\underset{p_1}{\rightrightarrows}} & (A,a) & \xrightarrow{e} & B_0 & \xrightarrow{m} & (B,b) \\ & & & \searrow^{g} & \downarrow{\scriptstyle g'} & \nearrow_{h} & \\ & & & & (C,c) & & \end{array}$$

where (p_0,p_1) is the kernel pair of f . Now since $\underline{S}$ is a topos and kernel pairs in $(\Phi;\underline{S})$ are computed in $\underline{S}$, e is the coequalizer (in $\underline{S}$) of p_0 and p_1 . Also $gp_0 = gp_1$ by cancelling the mono h (which is mono in $\underline{S}$). Hence g' exists (in $\underline{S}$) and hg' = m by uniqueness of maps out of B_0 . As above, since B_0 generates (B,b) , h is an isomorphism.

This argument works as well in $(\Phi;\underline{S})^I$ and that completes the proof. ∎

2.2 COROLLARY Images are strong [Gr,p.127] epimorphisms.

Proof: Since $(\Phi;\underline{S})$ is finitely complete we can apply Prop. 3.6 of Kelley [Kl]. ∎

Recall that a factorisation system in a category is a pair of subcategories $(\underline{E},\underline{M})$ which decompose morphisms, contain isomorphisms in their intersection and satisfy the diagonal fill-in lemma [Ba] . An *indexed factorisation system* is just a pair of stable subcategories of an indexed category with the same properties.

2.3 PROPOSITION The image-mono decompositions of morphisms in $(\Phi;\underline{S})$ form an indexed factorisation system.

Proof: With the observation that monomorphisms in $(\Phi;\underline{S})$ are also monos in $\underline{S}$ and hence form a subcategory, and 2.2, it only remains to show that images compose. Suppose that $(A,a) \xrightarrow{f} (B,b) \xrightarrow{g} (C,c)$ and that (D,d) is the image of gf . Consider

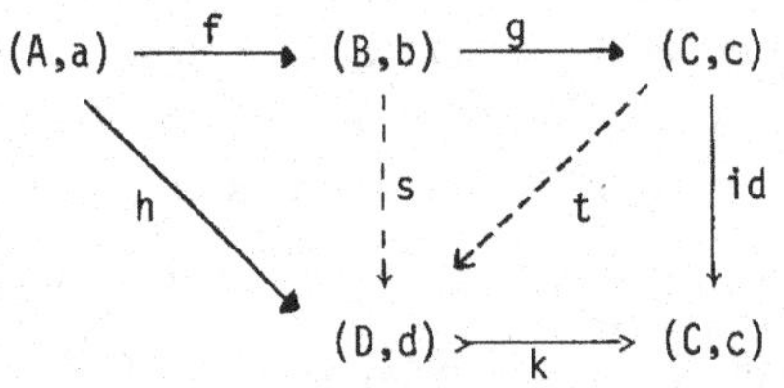

Now s exists since $kh = gf$, k is monic and f is a strong epi. Similarly, t exists since $ks = id \cdot g$, so k is mono and split epi and thus an isomorphism. Hence gf is an image. □

The forgetful functor is denoted $U:(\Phi;\underline{S}) \longrightarrow \underline{S}$. A morphism in $(\Phi;\underline{S})$ is called U-*epi* if its underlying morphism is epimorphic in $\underline{S}$.

2.4. PROPOSITION If Φ preserves epimorphisms, then images are U-epi.

Proof: Suppose $(A,a) \xrightarrow{f} (B,b)$ is a morphism of $(\Phi;\underline{S})$ with U-factorization $f = me$. Consider

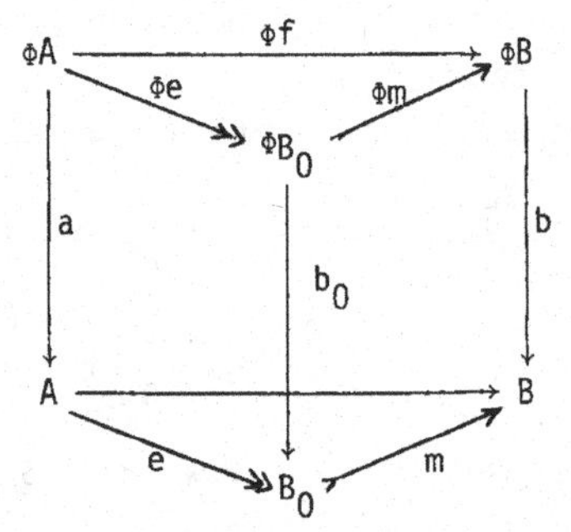

Now b_0 exists by the diagonal fill-in in $\underline{S}$ since Φe is epic. Thus (B_0,b_0) is a subalgebra of (B,b); indeed it is the subalgebra generated by B_0 so is the image of f , but e is epic in $\underline{S}$. □

2.5. PROPOSITION If U has a left adjoint F and $A_0 \overset{j}{\rightarrowtail} A$ generates (A,a) , then the induced homomorphism $(FA_0,a_0) \xrightarrow{\hat{j}} (A,a)$ is an image.

Proof: Consider

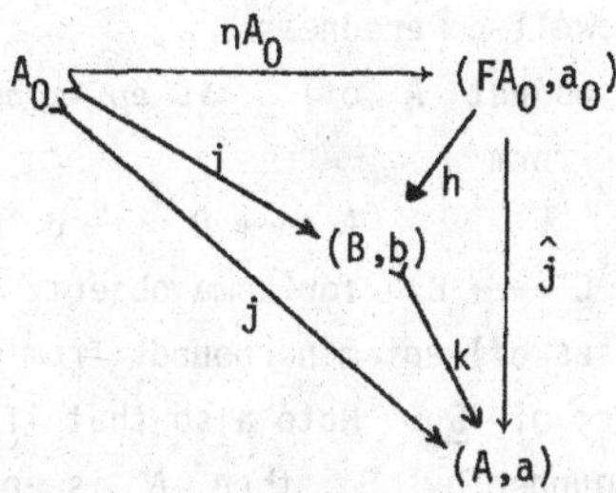

in which kh is the image-mono decomposition of $\hat{j}$, and i is mono since $ki = j$ (as both correspond to $\hat{j}$ by adjointness). Thus (B,b) is a subalgebra of (A,a) containing A_0 which implies that k is an isomorphism, so $\hat{j}$ is am image. □

Propositions 2.4 and 2.5 will be combined below in Proposition 2.9 to obtain a cowell-poweredness condition. Before proceeding to that we define the form of cowell-poweredness which will be useful to us.

2.6. DEFINITION Let $\underline{A}$ be an $\underline{S}$-indexed category with factorization system $(\underline{E},\underline{M})$. $\underline{A}$ is *weakly* $\underline{E}$*-cowell-powered at 1* if for any A in $\underline{A}^1$ there is an object Q of $\underline{S}$ and a morphism $\Delta_Q A \xrightarrow{g} G$ in $\underline{A}^Q$ satisfying: for all $\Delta_J A \xrightarrow{e} B$ in $\underline{A}^J$ with e in $\underline{E}^J$ there exists $J \xrightarrow{a} Q$ with $a^*g = e$.
$\underline{A}$ is *weakly* $\underline{E}$*-cowell-powered* if $\underline{A}/I$ is weakly $\underline{E}$-cowell-powered at 1 for every I.

Intuitively, weak cowell-poweredness means that to each A in $\underline{A}$ there is an object Q of $\underline{S}$ (and a generic morphism) which contains a representation for each $\underline{E}$-quotient of A, but not necessarily uniquely. The generic family of morphisms from A may contain morphisms which are not in $\underline{E}$ and one could always factor the generic morphism to obtain a generic family of morphisms which is in $\underline{E}$, but this is not essential for the sequel. This form of cowell-poweredness is suitable for constructing solution sets. Indeed we immediately have the following application.

2.7. PROPOSITION. Let $\underline{A}$ be complete and have small homs and an indexed factorization system $(\underline{E},\underline{M})$. If $\underline{A}$ is weakly $\underline{E}$-cowell-powered it has stable coequalizers.

Proof: Since $\underline{A}$ is complete the substitution functors a^* for any $J \xrightarrow{a} I$ in $\underline{S}$ have right adjoints and so the a^* preserve any coequalizers which exist. Hence we need only the existence of coequalizers. We can apply the General Adjoint Functor Theorem of Paré and Schumacher [P+S, IV.1.7] at 1 and localize when we observe that a weak quotient object as in 2.6 provides a solution set for the diagonal functor $\underline{A} \xrightarrow{\Delta} \underline{A}^{\cdot\ddagger\cdot}$. Details are left to the reader. □

This has reduced the problem of coequalizers in $(\Phi;\underline{S})$ (which is an internally complete $\underline{S}$-indexed category having small homs and a factorization system) to the question of weak image-cowell-poweredness.

We will say that an object A of $\underline{S}$ is *epi-mono bounded* by an object B if there is a diagram of the form

$$A \rightarrowtail C \twoheadleftarrow B$$

(or equivalently $A \twoheadleftarrow C \rightarrowtail B$) for some object C in $\underline{S}$. As will be seen below, (equivalence classes of) epi-mono bounds from a fixed object B in $\underline{S}$ are representable by an object of $\underline{S}$. Note also that if A is epi-mono bounded by B and B is epi-mono bounded by D, then A is epi-mono bounded by D by pulling back.

An indexed functor $\Phi: \underline{S} \to \underline{S}$ will be called *contained in* $\Psi: \underline{S} \to \underline{S}$ if there is a natural transformation

$$\Phi \overset{t}{\rightarrowtail} \Psi .$$

The remainder of this section is devoted to showing that if Φ is contained in a functor with good properties, then $(\Phi;\underline{S})$ is weakly image cowell-powered. We first need a technical lemma.

2.8. LEMMA. Let $\Phi,\overline{\Phi}: \underline{S} \longrightarrow \underline{S}$ be indexed functors. For an object (A,a) of $(\Phi;\underline{S})$ there is an object of $\underline{S}$ and a generic diagram which represents diagrams of the form

$$(1)\qquad \begin{array}{ccccccccc} \Phi A & \xrightarrow{\Phi fe} & & & \Phi C & & & & \\ \downarrow a & & & & \downarrow c & & & & \\ A & \overset{e}{\twoheadrightarrow} & B & \xrightarrow{f} & C & \rightarrowtail & D & \twoheadleftarrow & \overline{\Phi}B \end{array}$$

in $\underline{S}$.

<u>Proof</u>: First recall that $\underline{S}$ is cowell-powered and let $\Delta_Q A \overset{g}{\twoheadrightarrow} B$ be the generic epimorphism from A. Again since $\underline{S}$ is cowell-powered, there is an object $P \xrightarrow{\alpha} Q$ of $\underline{S}^Q$ and a generic epimorphism $\alpha^*\overline{\Phi}^Q B \twoheadrightarrow D$. Since $\underline{S}$ is well-powered D has a generic subobject, so there is an object $R \xrightarrow{\beta} P$ of $\underline{S}^P$ and a generic monomorphism $C \rightarrowtail \beta^* D$ in $\underline{S}^R$. Since epimorphisms are stable, we have a diagram in $\underline{S}^R$

$$C \rightarrowtail \beta^* D \twoheadleftarrow (\alpha\beta)^*\overline{\Phi}^Q B .$$

The reader will observe that this diagram is the generic epi-mono bound for $\overline{\Phi}^Q B$ and we have shown how to construct the family of objects epi-mono bounded by a fixed object as promised above.

C is the family of objects of $\underline{S}$ which are epi-mono bounded by $\overline{\Phi}$ applied to an object which is a quotient of A. It remains to consider Φ-algebra structures on C and then homomorphisms which factor through a quotient of A.

Since $\underline{S}$ has small homs, there is an object $S \xrightarrow{\gamma} R$ of $\underline{S}^R$ and a generic morphism $\gamma^*\Phi^R C \xrightarrow{h} \gamma^* C$ in $\underline{S}^S$, and an object $T \xrightarrow{\delta} S$ of $\underline{S}^S$ and a generic morphism $(\alpha\beta\gamma\delta)^* B \xrightarrow{k} (\gamma\delta)^* C$. h is the generic algebra structure on C and T is $\mathrm{Hom}^S((\alpha\beta\gamma)^*B,\gamma^*C)$. To complete the construction, let $T_0 \overset{\varepsilon}{\rightarrowtail} T$ be the

equalizer of

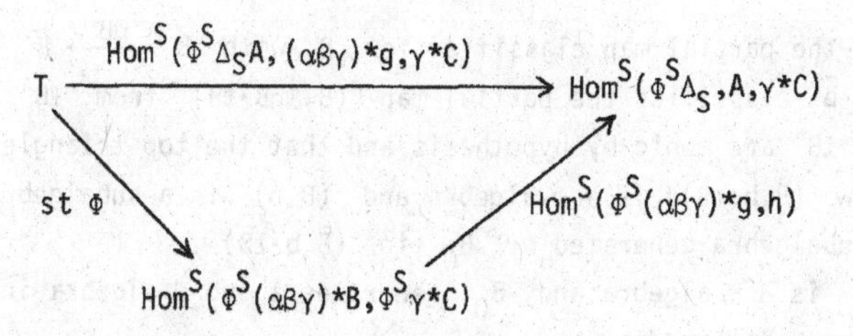

T_0 represents the morphisms from $(\alpha\beta\gamma)*B$ to $\gamma*C$ which are part of a Φ-algebra homomorphism from $\Delta_S(A,a)$ to $(\gamma*C,k)$. Indeed T_0 is the required object of $\underline{S}$, and the generic diagram of type (1) is

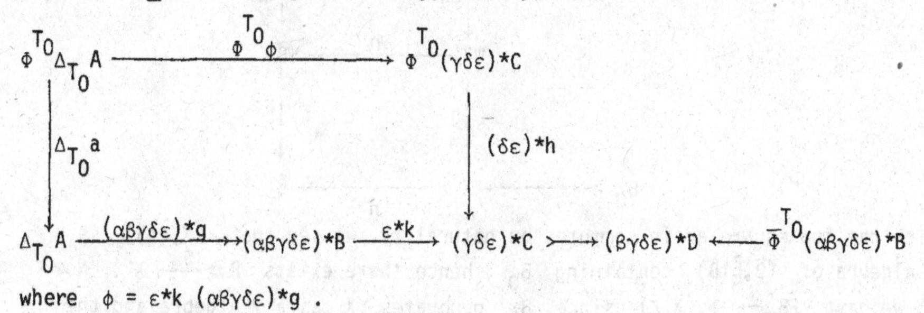

where $\phi = \varepsilon*k\ (\alpha\beta\gamma\delta\varepsilon)*g$. □

Using 2.8 we can now prove

2.9. PROPOSITION Let $\Phi,\Psi\colon \underline{S} \longrightarrow \underline{S}$ be indexed functors with Φ contained in Ψ by $\Phi \overset{t}{\rightarrowtail} \Psi$. If Ψ preserves epis and monos and $U\colon (\Psi{:}S) \longrightarrow \underline{S}$ has a left adjoint F , then $(\Phi;\underline{S})$ is weakly image-cowell-powered.

Proof: For the sake of clarity we work at 1 since everything can be localized. Let $(A,a) \xrightarrow{f} (B,b)$ be an image in $(\Phi;\underline{S})^1$. Let J be the object of $\underline{S}$ representing diagrams of the form (1) as in 2.8 for the functors Φ and $\overline{\Phi} = UF$. Let g be the homomorphism with domain $\Delta_J(A,a)$ given by the commutative square in the generic diagram of the form (1). We will show that the image f above is an instance of (1), so that there is a section $1 \xrightarrow{i} J$ with $i*g = f$. With this we will have shown that J is a weak object of quotients for (A,a) with generic homomorphism g .

To begin, factor f in $\underline{S}$ as $A \twoheadrightarrow^{e} B_0 \overset{m}{\rightarrowtail} B$ so B_0 generates B as a Φ-algebra. Next consider

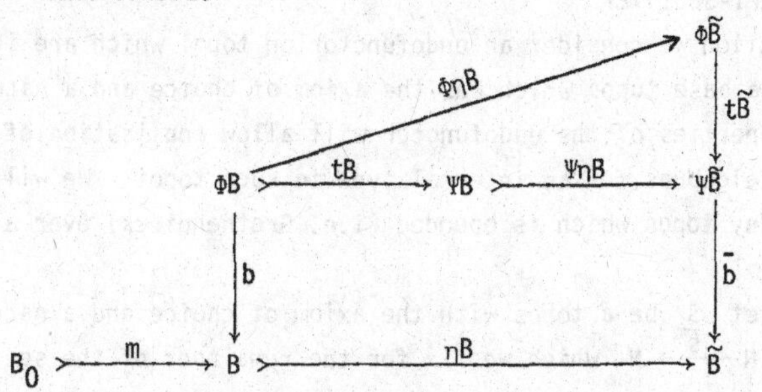

in which $\tilde{B}$ is the partial map classifier for B with $B \xrightarrow{\eta B} \tilde{B}$ the canonical inclusion, and $\bar{b}$ classifies the partial map $(B,\Psi\eta B\cdot tB)$ from $\Psi\tilde{B}$ to B. Note that $\Psi\eta B$ and tB are monic by hypothesis and that the top triangle commutes by naturality. Now $(\tilde{B},\bar{b}\cdot t\tilde{B})$ is a Φ-algebra and (B,b) is a subalgebra. Indeed (B,b) is the subalgebra generated by B_0 in $(\tilde{B},\bar{b}\cdot t\tilde{B})$.

Now $(\tilde{B},\bar{b})$ is a Ψ-algebra and B_0 generates a sub Ψ-algebra in it, (X,ξ) say. From the commutative diagram

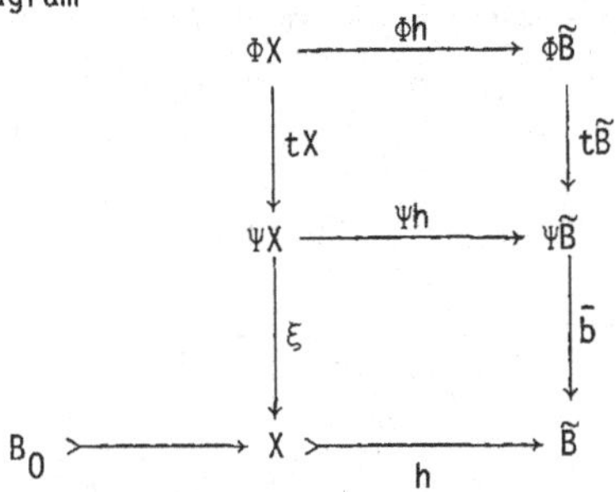

in which the top square again commutes by naturality, we see that $(X,\xi tX)$ is a sub Φ-algebra of $(\tilde{B},\bar{b}t\tilde{B})$ containing B_0, hence there exists $B \xrightarrow{j} X$. Now by 2.5 we have $FB_0 \longrightarrow (X,\xi)$ since B_0 generates X as a Ψ-algebra and then by 2.4 this is a U-epi since Ψ preserves epis. That is $UFB_0 \xrightarrow{e'} X$ in $\underline{S}$.

Using j and e', B is epi-mono bounded by UFB_0 so $A \xrightarrow{e} B_0 \xrightarrow{m} B$ is part of an instance of (1) from 2.8, defining a global section $1 \xrightarrow{x} J$ such that x^* of the generic morphism g is equal to f. Thus $(\Phi;\underline{S})$ is weakly image cowell-powered at 1. Localizing completes the proof. □

Combining this with 2.7 motivates the next section. One can always find a Ψ containing a given Φ which preserves the factorization system. Indeed $\Omega^{(\ )}$ (by direct image) is a functor which contains the identity (using $\{\}$) preserves monos and splits epis. Thus, provided Φ preserves monos (which it does in practice), the functor $\Phi\Omega^{(\ )}$ almost works. The problem is that $\Omega^{(\ )}$ "grows too fast" and it then becomes impossible to take $\Phi\Omega^{(\ )}$ as our Ψ since we need a left adjoint to $U: (\Psi;\underline{S}) \longrightarrow \underline{S}$. What will be done is to find an endofunctor on certain topoi which has the good properties of $\Omega^{(\ )}$ without the bad ones.

§3. A BOUNDED EPI-SPLITTER

In this section we consider an endofunctor on topoi which are internal presheaf categories over a base topos which has the axiom of choice and a natural number object. The properties of the endofunctor will allow application of 2.9 to obtain coequalizers in algebras for an internal type in such topoi. We will be able to extend this to any topos which is bounded (i.e. Grothendieck) over a topos with AC and NNO.

To begin, let $\underline{S}$ be a topos with the axiom of choice and a natural numbers object $1 \xrightarrow{0} N \xrightarrow{s} N$ which we fix for the remainder of the section. Let

$$\mathbb{C}:\ C_2 \overset{\longrightarrow}{\underset{\longrightarrow}{\xrightarrow{e}}} C_1 \overset{\partial_0}{\underset{\partial_1}{\overset{\longrightarrow}{\underset{\longrightarrow}{\xleftarrow{i}}}}} C_0$$

be a category object in $\underline{S}$. Denote the topos of presheaves on $\mathbb{C}$, that is $\underline{S}^{\mathbb{C}^{op}}$, by $\hat{\mathbb{C}}$. Recall that we then have functors

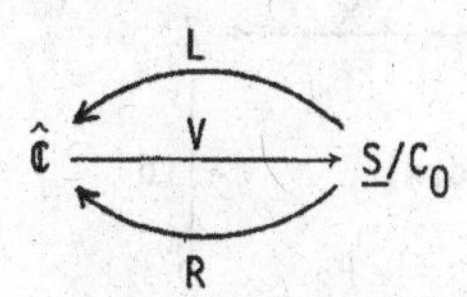

in which V is forgetful, $L \dashv V \dashv R$ and V is both triplable and cotriplable (see e.g. [Dc] or [Jn2]). The functor we shall be using is $\Theta = RV$. We note that it is immediate that Θ has a left adjoint, VL, and recall that the cotriple VR on $\underline{S}/C_0$ is given by $VR = \Pi_{\partial_1} \cdot \partial_0^*$.

3.1. LEMMA 1) $\hat{\mathbb{C}} \overset{t}{\rightarrowtail} \Theta$

2) Θ is left exact and splits epimorphisms. (i.e. the image of an epimorphism under Θ is a split epimorphism.)

Proof: Ad 1): Suppose (A,a) is an object of $\hat{\mathbb{C}}$, that is a coalgebra for VR. Viewed this way we may write a as a costructure morphism i.e. $A \xrightarrow{a} \Pi_{\partial_1} \cdot \partial_0^* A$. Now $\Theta(A,a)$ is the cofree on $V(A,a)$, so the costructure of (A,a) is a cohomomorphism $(A,a) \xrightarrow{t(A,a)} \Theta(A,a)$ and since its underlying morphism, namely a, is a costructure, it is a monomorphism in $\underline{S}$. Then so is $t(A,a)$. Clearly the $t(A,a)$ extend to a natural transformation $\hat{\mathbb{C}} \overset{t}{\rightarrowtail} \Theta$.

Ad 2): V and R being right adjoints, $\Theta = RV$ is surely left exact. Since $\underline{S}$ has AC so does $\underline{S}/C_0$. Now V preserves epimorphisms, so if $(A,a) \overset{e}{\twoheadrightarrow} (B,b)$ in $\hat{\mathbb{C}}$, then $V(A,a) \overset{Ve}{\twoheadrightarrow} V(B,b)$ in $\underline{S}/C_0$ is a split epimorphism and $\Theta e = RVe$ is a split epimorphism in $\hat{\mathbb{C}}$. □

At this point it is probably worthwhile to consider how Θ splits epimorphisms. Suppose $\underline{S} = \underline{set}$ and P is a presheaf on the small category $\mathbb{C}$. It is easily seen that

$$\Theta(P) = \prod_{C \text{ in } C_0} P(C)^{[C,-]} = \prod_{C \to (\ )} P(C)$$

or evaluating at D in C_0

$$\Theta P(D) = \prod_{C \to D} P(C)$$

From this it is clear that to split an epimorphism from P to Q say after applying Θ it is only necessary to split it "pointwise" before applying Θ. In a sense one takes a variable set P and considers all "future" information about P from a given point (if $\mathbb{C}$ is a partial order say) i.e. $P(C)$ for all $C \longrightarrow D$ and makes the "future" information available in the "present" $\Theta P(D)$ which allows splitting to take place. Or again, if $\mathbb{C}$ is the opens of a topological space $\Theta P(D)$ contains information on all refinements of D. To be more specific, suppose $\mathbb{C} = \mathbf{2}$ and $X_0 \xrightarrow{x} X_1$ is an object of $\underline{set}^{\mathbf{2}}$, then $\Theta(x) = X_0 \times X_1 \xrightarrow{p_2} X_1$. Note that the transition x has been lost. It appears in $t(x)$ which is

$$\begin{array}{ccc} X_0 & \xrightarrow{x} & X_1 \\ \downarrow{\scriptstyle (id,x)} & & \downarrow{\scriptstyle id} \\ X_0 \times X_1 & \xrightarrow{p_2} & X_1 \end{array} \qquad \Big\downarrow{\scriptstyle t(x)}$$

Before going on to show that Θ does not "grow too fast", we recall that it is $\hat{\mathbb{C}}$ (and later sheaf subcategories of it) in which we wish to construct coequalizers in algebras for an internal type. Thus we will require Θ to be $\hat{\mathbb{C}}$-indexed.

3.2. LEMMA Θ is $\hat{\mathbb{C}}$-indexed and for all P in $\hat{\mathbb{C}}$, Θ^P satisfies 1) and 2) of 3.1.

Proof: Since $\underline{S}/C_0 \underset{R}{\overset{V}{\rightleftarrows}} \hat{\mathbb{C}}$ is a geometric morphism, V and R are $\hat{\mathbb{C}}$-indexed by [P+S,I.2.4.4] and so is $\Theta = RV$.

To see that Θ^P satisfies 1) and 2) of 3.1 for P in $\hat{\mathbb{C}}$, we consider the definition of Θ^P. Let $Q \xrightarrow{\tau} P$ be an object of $\hat{\mathbb{C}}^P$. $V^P(\tau) = VQ \xrightarrow{V\tau} VP$. If $F \xrightarrow{\sigma} VP$ is an object of $(\underline{S}/C_0)^P$, then $R^P(\sigma)$ is defined by the pullback:

$$\begin{array}{ccc} F' & \longrightarrow & RF \\ \downarrow{\scriptstyle R^P(\sigma)} & \text{P.B.} & \downarrow{\scriptstyle R\sigma} \\ P & \xrightarrow{tP} & RVP \end{array}$$

where tP, as in 3.1, is the front adjunction at P for $V \dashv R$. Thus $\Theta^P(\tau)$ is defined by the pullback

$$\begin{array}{ccc} Q' & \longrightarrow & RVQ \\ \downarrow{\scriptstyle \Theta^P(\tau)} & \text{P.B.} & \downarrow{\scriptstyle RV\tau} \\ P & \xrightarrow{tP} & RVP \end{array}$$

Thus, if $\tau \overset{e}{\twoheadrightarrow} \sigma$ is an epimorphism in $\hat{\mathbb{C}}^P$ and $\underline{S}$ has AC, $RV(e)$ is a split epimorphism. Hence, $\Theta^P(e)$ which is defined by pulling back $RV(e)$ is also split epi. Since finite limits commute with pulling back, and RV is left exact, so is Θ^P. Hence 2) of 3.1 holds for Θ^P.

Finally, to see that $1_{\hat{\mathbb{C}}^P} \overset{t^P}{\rightarrowtail} \Theta^P$, consider

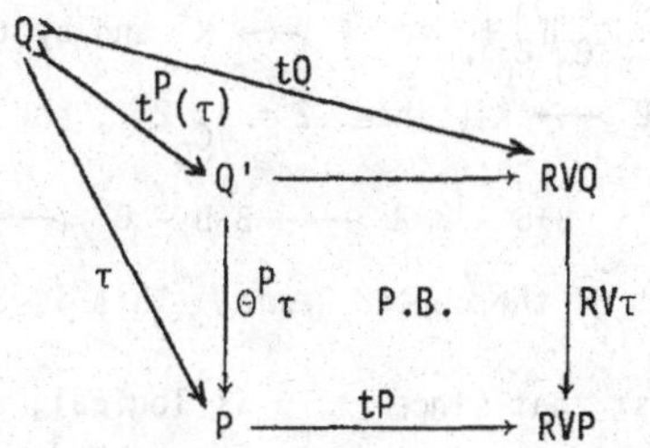

in which the quadrilateral outside commutes by naturality, $t^P(\tau)$ is the unique morphism to the pullback Q' and $t^P(\tau)$ is monic since tQ is. □

It is interesting to note that while $\Theta = RV$ is $\hat{\mathbb{C}}$-indexed, its left adjoint LV may not be indexed. Indeed, when $\hat{\mathbb{C}} = \underline{\text{set}}^{\mathbf{2}}$ we saw above that $\Theta(X_0 \xrightarrow{x} X_1) = X_0 \times X_1 \xrightarrow{p_2} X_1$. On the other hand $L\colon \underline{\text{set}}/2 \to \underline{\text{set}}^{\mathbf{2}}$ is defined by $L(X_0,X_1) = X_0 \xrightarrow{i_1} X_0+X_1$, so that $LV(X_0 \xrightarrow{x} X_1) = X_0 \xrightarrow{i_1} X_0+X_1$. From this it is immediate that LV is not indexed, since an indexed functor on $\underline{\text{set}}^{\mathbf{2}}$ must depend, in the codomain, only on the codomain [see Rs,I.3.7].

The sense in which Θ does not grow too fast is a technical one which allows the application of [P+S,V.2.2], a result which essentially says that if Ψ is a bounded functor (in the technical sense) then $(\Psi;\underline{S}) \xrightarrow{U} \underline{S}$ has a left adjoint.

3.3. DEFINITION An $\underline{S}$-indexed functor $\Psi\colon \underline{S} \to \underline{S}$ is *bounded* if for all X in $\underline{S}$ there exists a B in $\underline{S}$ and monomorphisms

1) $X \rightarrowtail B$
2) $\Psi B \rightarrowtail B$
3) $B+B \rightarrowtail B$

The reader will note that if Φ and Ψ are bounded functors on $\underline{S}$ then it does not follow that $\Phi\Psi$ is a bounded functor. In practice it is often the case that a composite of bounded functors is bounded. We shall meet with several examples of this situation in §4.

Before going on to the lemma which will allow us to show that Θ (and more importantly composites involving it) is bounded, we recall the following result of Paré and Schumacher [P+S,V.2.3.3]:

If K is an object of a topos with natural numbers object N, then $B = \Omega^{K^N}$ has the following properties:

1) there is a monomorphism $K \rightarrowtail B$
2) for any monomorphism $L \rightarrowtail K$ there exists a monomorphism $B^L \rightarrowtail B$.

3.4. LEMMA Let F be in $\underline{S}/C_0$, $K_0 = \amalg_{C_0} F + C_0 + C_1 + 2$ and $K = \Delta_{C_0} K_0$. Then $B = \Omega^{K^N}$ satisfies

1) $F \rightarrowtail B$
2) $VRB \rightarrowtail B$
3) $B+B \rightarrowtail B$

Proof: First $F \rightarrowtail \Delta_{C_0}\Pi_{C_0}F$ so $F \rightarrowtail K$ and by the lemma quoted $F \rightarrowtail B$ which is 1). Also $2 \rightarrowtail K$ (since $2 = \Delta_{C_0}2$), so $2 \rightarrowtail B$ and

$$B+B \simeq B\times 2 \rightarrowtail B\times B \simeq B^2 \rightarrowtail B$$

using the second part of the quoted lemma. This is 3).

For 2) note first that since Δ_{C_0} is logical, $B = \Omega^{(\Delta_{C_0}K_0)^N} = \Delta_{C_0}(\Omega^{K_0^N})$.
Recalling that $VR = \Pi_{\partial_1}\cdot\partial_0^*$, we have

$$\begin{aligned} VRB &= \Pi_{\partial_1}\cdot\partial_0^*\Delta_{C_0}(\Omega^{K_0^N}) \\ &= \Pi_{\partial_1}\cdot\Delta_{C_1}(\Omega^{K_0^N}) \\ &= \Pi_{\partial_1}\cdot\partial_1^*\Delta_{C_0}(\Omega^{K_0^N}) \\ &= \Pi_{\partial_1}\cdot\partial_1^*B \\ &= B^{\partial_1} \end{aligned}$$

However $\partial_1 \rightarrowtail \Delta_{C_0}C_1 \rightarrowtail K$, so by the second part of quoted lemma, $VRB = B^{\partial_1} \rightarrowtail B$ and this is 2). □

We will apply 3.4 to show that composites of functors involving Θ are bounded, but 3.4 shows that Θ is bounded. Indeed, let P be in $\hat{\mathbb{C}}$ and take $F = VP$ in the hypothesis. The bound we find is RB for the B in the conclusion of 3.4. Indeed, $\Theta RB = RVRB \rightarrowtail RB$ using 2) and that R is a right adjoint, and $P \rightarrowtail \Theta P = RVP \rightarrowtail RB$ by 1) of 3.4. That $RB+RB \rightarrowtail RB$ is left to the reader (or see the proof of 4.2).

§4. COEQUALIZERS FOR AN INTERNAL TYPE.

Before considering the general case we restrict our attention to topoi of the form $\hat{\mathbb{C}}$ over a topos $\underline{S}$ with AC and a natural numbers object, and to trivial types $I \longrightarrow 1$. In this case the endofunctor under consideration is $(\ )^I: \hat{\mathbb{C}} \longrightarrow \hat{\mathbb{C}}$. We wish to apply 2.9, so we need a functor containing $\Phi = (\ )^I$ with good properties. Define $\Psi = (\Theta(\ ))^I$.

4.1. LEMMA $\Phi \rightarrowtail \Psi$ and Ψ preserves epis and monos.

Proof: Since $(\ )^I$ is left exact and $\hat{\mathbb{C}} \rightarrowtail \Theta$ we have $\Phi = (\ )^I \rightarrowtail (\Theta(\ ))^I = \Psi$. Ψ preserves monos since Θ and $(\ )^I$ do and epis since Θ splits them. □

All that remains is to find a left adjoint for $(\Psi;\hat{\mathbb{C}}) \xrightarrow{U} \hat{\mathbb{C}}$.

4.2. PROPOSITION U has a left adjoint.

Proof: We apply the theorem of Paré and Schumacher [P+S,V.2.2.2]. First Ψ is left exact and so preserves all pullbacks. It remains to show that Ψ is bounded. Let $X = P+I+2$. Apply 3.4 to $F = VX = VP+VI+V2$ to find B_0 in $\underline{S}/C_0$, where $B_0 = \Omega^{(\Delta_{C_0}K_0)^N}$ and $K_0 = \amalg_{C_0} VX + C_0 + C_1 + 2$. Let $B = RB_0$. Now by 3.4 $VX \rightarrowtail B_0$, so $P \rightarrowtail \Theta P = RVP \rightarrowtail RVX \rightarrowtail RB_0 = B$ since R preserves monos. Thus 1) of 3.3 holds.

Next, $2 \rightarrowtail X$ so $2 \rightarrowtail \Theta 2 = RV2 \rightarrowtail RVX \rightarrowtail B$ (V has a left adjoint so preserves monomorphisms) and $B+B = RB_0+RB_0 \cong RB_0\times 2 \rightarrowtail RB_0\times RB_0 \cong RB_0^2 \cong R(B_0^2) \rightarrowtail R(B_0)$ again using the fact that R is a right adjoint and $B_0^2 \rightarrowtail B_0$ which follows from the construction of B_0 . This is 3) of 3.3.

For 2) of 3.3 observe that $VI \rightarrowtail VX \rightarrowtail \Delta_{C_0}\amalg_{C_0} VX \rightarrowtail \Delta_{C_0}K_0$ so $B_0^{VI} \rightarrowtail B_0$ and $R(B_0^{VI}) \rightarrowtail RB_0$, but $R(B_0^{VI}) \cong (RB_0)^I$. Indeed, for any object T of $\hat{\mathbb{C}}$ we have the following sequence of natural bijections:

$$\frac{T \longrightarrow R(B_0^{VI})}{\dfrac{VT \longrightarrow B_0^{VI}}{\dfrac{VT\times VI \longrightarrow B_0}{\dfrac{V(T\times I) \longrightarrow B_0}{\dfrac{T\times I \longrightarrow RB_0}{T \longrightarrow (RB_0)^I}}}}} \qquad \text{since } V \text{ is left exact}$$

so the isomorphism follows by Yoneda. Thus, $B^I \rightarrowtail B$.

Also $\Theta B = RVRB_0 \rightarrowtail RB_0 = B$ since $VRB_0 \rightarrowtail B_0$ by 3.4, so finally

$$\Psi B = (\Theta B)^I \rightarrowtail B^I \rightarrowtail B$$

since $(\ )^I$ preserves monos.

Thus U has a left adjoint at 1 , and by localization, the result follows. □

4.3. THEOREM. If $\underline{E}$ is a topos which is a functor category over a topos $\underline{S}$ with AC and NNO and I is an object of $\underline{E}$ then $((\ )^I;\underline{E})$ has coequalizers. □

Our next goal is to obtain 4.3 in a topos $\underline{E}$ which is bounded over a topos $\underline{S}$ with AC and a natural numbers object. In this case there is a category object $\mathbb{C}$ in $\underline{S}$ such that $\underline{E}$ is a sheaf subcategory of $\hat{\mathbb{C}}$. To set up the problem we have the following fact in a more general situation:

4.4. PROPOSITION Suppose $\underline{B} \underset{J}{\overset{K}{\rightleftarrows}} \underline{A}$ is a full reflective subcategory and $\underline{B} \xrightarrow{\Phi} \underline{B}$. Then $(\Phi;\underline{B})$ is a full reflective subcategory of $(J\Phi K;\underline{A})$ and in

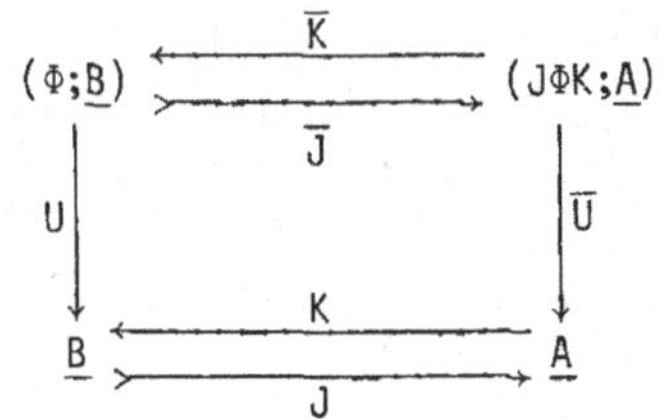

both squares commute where $\bar{J}(B,b) = (JB, Jb\cdot J\Phi\varepsilon B)$ and $\bar{K}(A,a) = (KA, Ka\cdot(\varepsilon\Phi KA)^{-1})$ are the inclusion and reflector respectively.

Proof: Routine, and left to the reader. □

4.5. COROLLARY In the situation above, if $(J\Phi K;\underline{A})$ has coequalizers, so does $(\Phi;\underline{B})$. □

In our situation we have $\underline{E} \underset{i}{\overset{a}{\rightleftarrows}} \hat{\mathbb{C}}$ where $a \dashv i$, and a is the left exact associated sheaf functor. By 4.5 coequalizers exist in $((\ )^I;\underline{E})$ if they exist $(\Phi;\hat{\mathbb{C}})$ where $\Phi X = i(aX)^I = (iaX)^{iI}$. Φ can be contained in a functor Ψ on $\hat{\mathbb{C}}$ in a manner similar to that applied in the case of $\hat{\mathbb{C}}$ above, but this will require that a composite involving ia be bounded. To prepare for that, we have:

4.6. LEMMA If B is an injective object in a topos, then $\tilde{B} \rightarrowtail B\times\Omega$ (where $\tilde{B}$ is the partial map classifier).

Proof: Injectivity of B allows id_B to be extended along $B \overset{\eta B}{\rightarrowtail} \tilde{B}$ by $\tilde{B} \xrightarrow{\overline{id}} B$ say. With the characteristic morphism of ηB , $\tilde{B} \xrightarrow{\chi} \Omega$, we have $\tilde{B} \xrightarrow{(\overline{id},\chi)} B\times\Omega$. This is a split monomorphism. Indeed it is easily verified that

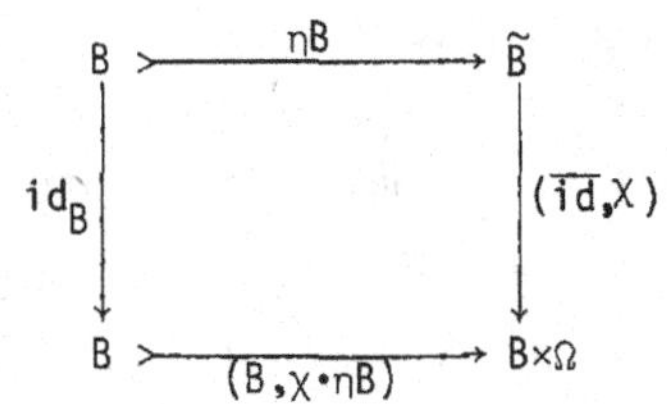

is a pullback. Moreover there is, by the universal property of $\tilde{B}$, a unique $B\times\Omega \xrightarrow{\phi} \tilde{B}$ so that the right hand square below is a pullback. Thus both $id_{\tilde{B}}$ and $\phi\cdot(\overline{id},\chi)$ make the rectangle a pullback: they are equal and $(\overline{id},\chi)$ is a split mono as claimed.

$$\begin{array}{ccccc} \tilde{B} & \xrightarrow{(\overline{id},\chi)} & B\times\Omega & \xrightarrow{\phi} & \tilde{B} \\ \uparrow{\scriptstyle \eta B} & & \uparrow{\scriptstyle (B,\chi\cdot\eta B)} & & \uparrow{\scriptstyle \eta B} \\ B & \xrightarrow{id_B} & B & \xrightarrow{id_B} & B \end{array}$$

□

4.7. LEMMA Suppose $\underline{A} \underset{R}{\overset{L}{\rightleftarrows}} \underline{B}$ and $L \dashv R$. If L preserves monomorphisms and B in $\underline{B}$ is injective then RB is injective. □

The next lemma provides the tools needed to bound our containing functor Ψ.

4.8. LEMMA Let $\underline{S}$ be a topos with AC and NNO and $\mathbb{C}$ a category object in $\underline{S}$. Let P and I be in $\hat{\mathbb{C}}$, then there is a B in $\hat{\mathbb{C}}$ such that there are monomorphisms

1) $P \rightarrowtail B$
2) $B^I \rightarrowtail B$
3) $\Theta B \rightarrowtail B$
4) $\tilde{B} \rightarrowtail B$
5) $B+B \rightarrowtail B$

Proof: Let $X = P+I+\Omega+2$. We apply 3.4 to $F = VX = VP+VI+V\Omega+V2$ (V has a right adjoint), to find B_0 in $\underline{S}/C_0$ satisfying 3.4, and B_0 may be taken to be $\Omega^{(\Delta_{C_0}K_0)^N}$ where $K_0 = \amalg_{C_0} VX + C_0 + C_1 + 2$. Let $B = RB_0$. The reader is now referred to the argument of 4.2 for 1), 2), 3) and 5). For 4) we note that B_0 is injective, hence B is by 4.7 (V preserves monos being a right adjoint). Also $\Omega \rightarrowtail X \rightarrowtail \Theta X = RVX \rightarrowtail B$, so applying 4.6 to get the first monomorphism we have

$$\tilde{B} \rightarrowtail B\times\Omega \rightarrowtail B\times B \rightarrowtail B$$

which is 4) and completes the proof. □

We now define a functor Ψ on $\hat{\mathbb{C}}$ by $\Psi X = (ia\Theta X)^{iI}$.

4.9. LEMMA $\Phi \rightarrowtail \Psi$ and Ψ preserves epis and monos.

Proof: For the first assertion, ia is left exact and by 3.1 we have $ia \rightarrowtail ia\Theta$, and since $(\ )^{iI}$ is left exact we have $\Phi = ia(\ )^{iI} \rightarrowtail ia\Theta(\ )^{iI} = \Psi$. Ψ preserves epis since Θ has already split them and monos since Θ, $(\ )^{iI}$ and ia are all left exact. □

The next observation we make is that $ia = LL$ where L is Johnstone's internalization [Jn1] of the functor which goes by the same name in SGA IV [Gk]. $L{\cdot}L$ is $\hat{\mathbb{C}}$-indexed, so Ψ is. We observe also some features of the construction of L. The functor ia determines a closure operator in $\hat{\mathbb{C}}$. For P in $\hat{\mathbb{C}}$ let $P \rightarrowtail \tilde{P}$ be the inclusion of P in its partial map classifier and $P \rightarrowtail \hat{P}$ the closure of that mono. LP can be expressed as a coequalizer with domain $\hat{P}$ and so $\hat{P} \twoheadrightarrow LP$ in $\hat{\mathbb{C}}$. The existence of this epimorphism is used below.

4.10. LEMMA. Suppose B in $\hat{\mathbb{C}}$ satisfies 2), 3) and 4) of 4.8, then $\Psi B \rightarrowtail B$.

Proof: By the remarks above on construction L we have $\widehat{\Theta B} \twoheadrightarrow L\Theta B$, so $\Theta\widehat{\Theta B} \twoheadrightarrow \Theta L\Theta B$ is a split epi. We have monomorphisms

$$L\Theta B \rightarrowtail \Theta L\Theta B \rightarrowtail \Theta\widehat{\Theta B} \rightarrowtail \Theta\widetilde{\Theta B} \rightarrowtail \Theta\tilde{B}$$

respectively by 3.1, the splitting mentioned, the definition of $\widehat{\Theta B}$ and that Θ preserves monos, and lastly since both Θ and $(\tilde{\ })$ preserve monos and $\Theta B \rightarrowtail B$

by 3) of 4.8. We also have monomorphisms

$$\Theta\tilde{B} \rightarrowtail \Theta B \rightarrowtail B$$

using 4) of 4.8 and that Θ preserves monos, and 3) of 4.8. Altogether, there is a monomorphism $L\Theta B \rightarrowtail B$. Using this and $\widehat{L\Theta B} \twoheadrightarrow LL\Theta B$ so that $\Theta\widehat{L\Theta B} \twoheadrightarrow \Theta LL\Theta B$ is a split epi we get

$$LL\Theta B \rightarrowtail \Theta LL\Theta B \rightarrowtail \Theta\widehat{L\Theta B} \rightarrowtail \Theta\widetilde{L\Theta B} \rightarrowtail \Theta\tilde{B} \rightarrowtail \Theta B \rightarrowtail B .$$

That is $LL\Theta B \rightarrowtail B$. Finally by 2) of 4.8 and since $(\ )^{iI}$ preserves monomorphisms:

$$(LL\Theta B)^{iI} \rightarrowtail B^{iI} \rightarrowtail B$$

that is, there is a monomorphism $\Psi B \rightarrowtail B$ as required. □

This is the last technical lemma needed to prove:

4.11. THEOREM If $\underline{E}$ is a topos which is bounded over a topos $\underline{S}$ with AC and a natural numbers object and I is an object of $\underline{E}$, then $((\ )^{I};\underline{E})$ has coequalizers.

Proof: Using the representation $\underline{E}$ as a sheaf subcategory of $\underline{S}^{\mathbb{C}^{op}}$ and 4.5 it is enough to find coequalizers in $(\Phi;\hat{\mathbb{C}})$ where $\Phi = (ia(\ ))^{iI}$. By 2.9 and 4.9 all that remains is to show that for $\Psi = (ia\Theta(\ ))^{iI}$, $(\Psi;\hat{\mathbb{C}}) \xrightarrow{U} \hat{\mathbb{C}}$ has a left adjoint. This is accomplished by again applying the theorem of Paré and Schumacher [P+S,V.2.2.2]. By 4.8 and 4.10, given X in $\hat{\mathbb{C}}$ there is a B in $\hat{\mathbb{C}}$ with monomorphisms: $X \rightarrowtail B$, $\Psi B \rightarrowtail B$, and $B+B \rightarrowtail B$. Finally Ψ is a composite of left exact functors, so is left exact and preserves all pullbacks. Thus U has a left adjoint at 1 , and to see that U has a left adjoint the above considerations are localized by 3.2 to obtain 4.8 and 4.10 locally. □

It is now not difficult to extend the last result to t-algebras for an arbitrary internal type $I \xrightarrow{t} J$ in $\underline{E}$. In this situation the functor which replaces Φ above is defined on $\hat{\mathbb{C}}$ by $\Phi X = i(\amalg_J(\Delta_J aX)^t)$. Note that for any object J in $\underline{E}$, an iJ indexed family in $\underline{E}$ viewed as a $\hat{\mathbb{C}}$-indexed category is the same thing as a J-indexed family in $\underline{E}$ viewed as a $\underline{E}$-indexed category. Thus, for any $X \xrightarrow{x} J$ in $\underline{E}/J$ we have

$$i(\amalg_J x) = iX = \amalg_{iJ} ix .$$

Similarly, since i is left exact, for any X in $\hat{\mathbb{C}}$

$$i(\Delta_J aX) = i\begin{pmatrix} aX\times J \\ \downarrow p_2 \\ J \end{pmatrix} = \begin{matrix} iaX\times iJ \\ \downarrow p_2 \\ iJ \end{matrix} = \Delta_{iJ}(iaX)$$

as objects of $\hat{\mathbb{C}}/iJ$. Using these facts we have

$$\Phi X = i(\amalg_J(\Delta_J aX)^t) = \amalg_{iJ} i((\Delta_J aX)^t)$$
$$\cong \amalg_{iJ} i(\Delta_J aX)^{it}$$
$$\cong \amalg_{iJ}(\Delta_{iJ} iaX)^{it} .$$

Thus we may replace Ψ above by $\Psi X = \amalg_{iJ}(\Delta_{iJ} ia\otimes X)^{it}$, and Ψ preserves epimorphisms as before, and monomorphisms and pullbacks since all of $\amalg_i J$, Δ_{iJ}, $(\ )^{it}$ and ia preserve them. All that is needed now is to show that Ψ is bounded. To obtain this we observe that all we need is 4.8 with 2) replaced by 2)' $\amalg_J(\Delta_J B)^t \rightarrowtail B$ for then the analogue of LEMMA 4.10 follows with same proof. To obtain 2)' (and 1), 3), 4) and 5) of 4.8) from P in $\hat{\mathbb{C}}$, take $X = P+I+J+\Omega+2$ and apply 3.4 to $F = VX$, find B_0 as before and let $B = RB_0$. Now

$$(\Delta_J B)^t \simeq (\Delta_J RB_0)^t \simeq (R^J \Delta_J B_0)^t \simeq R^J((\Delta_J B_0)^{Vt})$$

using the fact that R is $\hat{\mathbb{C}}$-indexed and V is left exact. But since $Vt \rightarrowtail \Delta_J VX$ and Δ_J is logical, we have $(\Delta_J B_0)^{Vt} \rightarrowtail \Delta_J B_0$ by the lemma of Paré and Schumacher. Thus we have

$$(\Delta_J B)^t \simeq R^J((\Delta_J B_0)^{Vt}) \rightarrowtail R^J(\Delta_J B_0) \cong \Delta_J RB_0 = \Delta_J B .$$

By the definition of B we also have that $J \rightarrowtail B$ so using the fact that $\amalg_J$ preserves monos

$$\amalg_J(\Delta_J B)^t \rightarrowtail \amalg_J(\Delta_J B) \simeq J\times B \rightarrowtail B\times B \rightarrowtail B .$$

This is 2)' which was all that was needed for

4.12. THEOREM Let $\underline{E}$ be a topos which is bounded over a topos $\underline{S}$ with AC and a natural numbers object and $I \xrightarrow{t} J$ an internal type. The category of t-algebras has coequalizers. □

REFERENCES

[Ba] M. Barr, Coequalizers and Free Triples, Math Zeit. 116(1970), 307-322.

[Dc] R. Diaconescu, Change of Base for Toposes with Generators, J. Pure and App. Alg. 6(1975), 191-218.

[Gk] M. Artin, A. Grothendieck, J.L. Verdier, Théorie des topos et cohomologie étale des schemas (SGA IV), Lecture Notes in Math. 269, Springer-Verlag, 1972.

[Gr] P.A. Grillet, Regular Categories, in Exact Categories and Categories of Sheaves, Lecture Notes in Math. 236, Springer-Verlag, 1971.

[Jn1] P. Johnstone, The Associated Sheaf Functor in an Elementary Topos, J. Pure and App. Alg. 4(1974), 231-242.

[Jn2] P. Johnstone, Topos Theory, Academic Press, to appear.

[J+W] P. Johnstone and G. Wraith, Algebraic theories and recursion in elementary topos theory, this volume.

[Kl] G.M. Kelly, Monomorphisms, Epimorphisms and Pullbacks, J. Aus. Math. Soc. 9(1969), 124-142.

[Ln] F.E.J. Linton, Coequalizers in Categories of Algebras, Lecture Notes in Math. 80, Springer-Verlag, 1969.

[P+S] R. Paré and D. Schumacher, Abstract Families and the Adjoint Functor Theorems, this volume.

[Rs] R. Rosebrugh, Abstract Families of Algebras, Ph.D. Thesis, Dalhousie University, 1977.

Vol. 489: J. Bair and R. Fourneau, Etude Géométrique des Espaces Vectoriels. Une Introduction. VII, 185 pages. 1975.

Vol. 490: The Geometry of Metric and Linear Spaces. Proceedings 1974. Edited by L. M. Kelly. X, 244 pages. 1975.

Vol. 491: K. A. Broughan, Invariants for Real-Generated Uniform Topological and Algebraic Categories. X, 197 pages. 1975.

Vol. 492: Infinitary Logic: In Memoriam Carol Karp. Edited by D. W. Kueker. VI, 206 pages. 1975.

Vol. 493: F. W. Kamber and P. Tondeur, Foliated Bundles and Characteristic Classes. XIII, 208 pages. 1975.

Vol. 494: A Cornea and G. Licea. Order and Potential Resolvent Families of Kernels. IV, 154 pages. 1975.

Vol. 495: A. Kerber, Representations of Permutation Groups II. V, 175 pages. 1975.

Vol. 496: L. H. Hodgkin and V. P. Snaith, Topics in K-Theory. Two Independent Contributions. III, 294 pages. 1975.

Vol. 497: Analyse Harmonique sur les Groupes de Lie. Proceedings 1973–75. Edité par P. Eymard et al. VI, 710 pages. 1975.

Vol. 498: Model Theory and Algebra. A Memorial Tribute to Abraham Robinson. Edited by D. H. Saracino and V. B. Weispfenning. X, 463 pages. 1975.

Vol. 499: Logic Conference, Kiel 1974. Proceedings. Edited by G. H. Müller, A. Oberschelp, and K. Potthoff. V, 651 pages 1975.

Vol. 500: Proof Theory Symposion, Kiel 1974. Proceedings. Edited by J. Diller and G. H. Müller. VIII, 383 pages. 1975.

Vol. 501: Spline Functions, Karlsruhe 1975. Proceedings. Edited by K. Böhmer, G. Meinardus, and W. Schempp. VI, 421 pages. 1976.

Vol. 502: János Galambos, Representations of Real Numbers by Infinite Series. VI, 146 pages. 1976.

Vol. 503: Applications of Methods of Functional Analysis to Problems in Mechanics. Proceedings 1975. Edited by P. Germain and B. Nayroles. XIX, 531 pages. 1976.

Vol. 504: S. Lang and H. F. Trotter, Frobenius Distributions in GL_2-Extensions. III, 274 pages. 1976.

Vol. 505: Advances in Complex Function Theory. Proceedings 1973/74. Edited by W. E. Kirwan and L. Zalcman. VIII, 203 pages. 1976.

Vol. 506: Numerical Analysis, Dundee 1975. Proceedings. Edited by G. A. Watson. X, 201 pages. 1976.

Vol. 507: M. C. Reed, Abstract Non-Linear Wave Equations. VI, 128 pages. 1976.

Vol. 508: E. Seneta, Regularly Varying Functions. V, 112 pages. 1976.

Vol. 509: D. E. Blair, Contact Manifolds in Riemannian Geometry. VI, 146 pages. 1976.

Vol. 510: V. Poènaru, Singularités C^∞ en Présence de Symétrie. V, 174 pages. 1976.

Vol. 511: Séminaire de Probabilités X. Proceedings 1974/75. Edité par P. A. Meyer. VI, 593 pages. 1976.

Vol. 512: Spaces of Analytic Functions, Kristiansand, Norway 1975. Proceedings. Edited by O. B. Bekken, B. K. Øksendal, and A. Stray. VIII, 204 pages. 1976.

Vol. 513: R. B. Warfield, Jr. Nilpotent Groups. VIII, 115 pages. 1976.

Vol. 514: Séminaire Bourbaki vol. 1974/75. Exposés 453 – 470. IV, 276 pages. 1976.

Vol. 515: Bäcklund Transformations. Nashville, Tennessee 1974. Proceedings. Edited by R. M. Miura. VIII, 295 pages. 1976.

Vol. 516: M. L. Silverstein, Boundary Theory for Symmetric Markov Processes. XVI, 314 pages. 1976.

Vol. 517: S. Glasner, Proximal Flows. VIII, 153 pages. 1976.

Vol. 518: Séminaire de Théorie du Potentiel, Proceedings Paris 1972–1974. Edité par F. Hirsch et G. Mokobodzki. VI, 275 pages. 1976.

Vol. 519: J. Schmets, Espaces de Fonctions Continues. XII, 150 pages. 1976.

Vol. 520: R. H. Farrell, Techniques of Multivariate Calculation. X, 337 pages. 1976.

Vol. 521: G. Cherlin, Model Theoretic Algebra – Selected Topics. IV, 234 pages. 1976.

Vol. 522: C. O. Bloom and N. D. Kazarinoff, Short Wave Radiation Problems in Inhomogeneous Media: Asymptotic Solutions. V. 104 pages. 1976.

Vol. 523: S. A. Albeverio and R. J. Høegh-Krohn, Mathematical Theory of Feynman Path Integrals. IV, 139 pages. 1976.

Vol. 524: Séminaire Pierre Lelong (Analyse) Année 1974/75. Edité par P. Lelong. V, 222 pages. 1976.

Vol. 525: Structural Stability, the Theory of Catastrophes, and Applications in the Sciences. Proceedings 1975. Edited by P. Hilton. VI, 408 pages. 1976.

Vol. 526: Probability in Banach Spaces. Proceedings 1975. Edited by A. Beck. VI, 290 pages. 1976.

Vol. 527: M. Denker, Ch. Grillenberger, and K. Sigmund, Ergodic Theory on Compact Spaces. IV, 360 pages. 1976.

Vol. 528: J. E. Humphreys, Ordinary and Modular Representations of Chevalley Groups. III, 127 pages. 1976.

Vol. 529: J. Grandell, Doubly Stochastic Poisson Processes. X, 234 pages. 1976.

Vol. 530: S. S. Gelbart, Weil's Representation and the Spectrum of the Metaplectic Group. VII, 140 pages. 1976.

Vol. 531: Y.-C. Wong, The Topology of Uniform Convergence on Order-Bounded Sets. VI, 163 pages. 1976.

Vol. 532: Théorie Ergodique. Proceedings 1973/1974. Edité par J.-P. Conze and M. S. Keane. VIII, 227 pages. 1976.

Vol. 533: F. R. Cohen, T. J. Lada, and J. P. May, The Homology of Iterated Loop Spaces. IX, 490 pages. 1976.

Vol. 534: C. Preston, Random Fields. V, 200 pages. 1976.

Vol. 535: Singularités d'Applications Differentiables. Plans-sur-Bex. 1975. Edité par O. Burlet et F. Ronga. V, 253 pages. 1976.

Vol. 536: W. M. Schmidt, Equations over Finite Fields. An Elementary Approach. IX, 267 pages. 1976.

Vol. 537: Set Theory and Hierarchy Theory. Bierutowice, Poland 1975. A Memorial Tribute to Andrzej Mostowski. Edited by W. Marek, M. Srebrny and A. Zarach. XIII, 345 pages. 1976.

Vol. 538: G. Fischer, Complex Analytic Geometry. VII, 201 pages. 1976.

Vol. 539: A. Badrikian, J. F. C. Kingman et J. Kuelbs, Ecole d'Eté de Probabilités de Saint Flour V-1975. Edité par P.-L. Hennequin. IX, 314 pages. 1976.

Vol. 540: Categorical Topology, Proceedings 1975. Edited by E. Binz and H. Herrlich. XV, 719 pages. 1976.

Vol. 541: Measure Theory, Oberwolfach 1975. Proceedings. Edited by A. Bellow and D. Kölzow. XIV, 430 pages. 1976.

Vol. 542: D. A. Edwards and H. M. Hastings, Čech and Steenrod Homotopy Theories with Applications to Geometric Topology. VII, 296 pages. 1976.

Vol. 543: Nonlinear Operators and the Calculus of Variations, Bruxelles 1975. Edited by J. P. Gossez, E. J. Lami Dozo, J. Mawhin, and L. Waelbroeck, VII, 237 pages. 1976.

Vol. 544: Robert P. Langlands, On the Functional Equations Satisfied by Eisenstein Series. VII, 337 pages. 1976.

Vol. 545: Noncommutative Ring Theory. Kent State 1975. Edited by J. H. Cozzens and F. L. Sandomierski. V, 212 pages. 1976.

Vol. 546: K. Mahler, Lectures on Transcendental Numbers. Edited and Completed by B. Diviš and W. J. Le Veque. XXI, 254 pages. 1976.

Vol. 547: A. Mukherjea and N. A. Tserpes, Measures on Topological Semigroups: Convolution Products and Random Walks. V, 197 pages. 1976.

Vol. 548: D. A. Hejhal, The Selberg Trace Formula for PSL (2,ℝ). Volume I. VI, 516 pages. 1976.

Vol. 549: Brauer Groups, Evanston 1975. Proceedings. Edited by D. Zelinsky. V, 187 pages. 1976.

Vol. 550: Proceedings of the Third Japan – USSR Symposium on Probability Theory. Edited by G. Maruyama and J. V. Prokhorov. VI, 722 pages. 1976.

Vol. 551: Algebraic K-Theory, Evanston 1976. Proceedings. Edited by M. R. Stein. XI, 409 pages. 1976.

Vol. 552: C. G. Gibson, K. Wirthmüller, A. A. du Plessis and E. J. N. Looijenga. Topological Stability of Smooth Mappings. V, 155 pages. 1976.

Vol. 553: M. Petrich, Categories of Algebraic Systems. Vector and Projective Spaces, Semigroups, Rings and Lattices. VIII, 217 pages. 1976.

Vol. 554: J. D. H. Smith, Mal'cev Varieties. VIII, 158 pages. 1976.

Vol. 555: M. Ishida, The Genus Fields of Algebraic Number Fields. VII, 116 pages. 1976.

Vol. 556: Approximation Theory. Bonn 1976. Proceedings. Edited by R. Schaback and K. Scherer. VII, 466 pages. 1976.

Vol. 557: W. Iberkleid and T. Petrie, Smooth S^1 Manifolds. III, 163 pages. 1976.

Vol. 558: B. Weisfeiler, On Construction and Identification of Graphs. XIV, 237 pages. 1976.

Vol. 559: J.-P. Caubet, Le Mouvement Brownien Relativiste. IX, 212 pages. 1976.

Vol. 560: Combinatorial Mathematics, IV, Proceedings 1975. Edited by L. R. A. Casse and W. D. Wallis. VII, 249 pages. 1976.

Vol. 561: Function Theoretic Methods for Partial Differential Equations. Darmstadt 1976. Proceedings. Edited by V. E. Meister, N. Weck and W. L. Wendland. XVIII, 520 pages. 1976.

Vol. 562: R. W. Goodman, Nilpotent Lie Groups: Structure and Applications to Analysis. X, 210 pages. 1976.

Vol. 563: Séminaire de Théorie du Potentiel. Paris, No. 2. Proceedings 1975–1976. Edited by F. Hirsch and G. Mokobodzki. VI, 292 pages. 1976.

Vol. 564: Ordinary and Partial Differential Equations, Dundee 1976. Proceedings. Edited by W. N. Everitt and B. D. Sleeman. XVIII, 551 pages. 1976.

Vol. 565: Turbulence and Navier Stokes Equations. Proceedings 1975. Edited by R. Temam. IX, 194 pages. 1976.

Vol. 566: Empirical Distributions and Processes. Oberwolfach 1976. Proceedings. Edited by P. Gaenssler and P. Révész. VII, 146 pages. 1976.

Vol. 567: Séminaire Bourbaki vol. 1975/76. Exposés 471–488. IV, 303 pages. 1977.

Vol. 568: R. E. Gaines and J. L. Mawhin, Coincidence Degree, and Nonlinear Differential Equations. V, 262 pages. 1977.

Vol. 569: Cohomologie Etale SGA 4½. Séminaire de Géométrie Algébrique du Bois-Marie. Edité par P. Deligne. V, 312 pages. 1977.

Vol. 570: Differential Geometrical Methods in Mathematical Physics, Bonn 1975. Proceedings. Edited by K. Bleuler and A. Reetz. VIII, 576 pages. 1977.

Vol. 571: Constructive Theory of Functions of Several Variables, Oberwolfach 1976. Proceedings. Edited by W. Schempp and K. Zeller. VI, 290 pages. 1977

Vol. 572: Sparse Matrix Techniques, Copenhagen 1976. Edited by V. A. Barker. V, 184 pages. 1977.

Vol. 573: Group Theory, Canberra 1975. Proceedings. Edited by R. A. Bryce, J. Cossey and M. F. Newman. VII, 146 pages. 1977.

Vol. 574: J. Moldestad, Computations in Higher Types. IV, 203 pages. 1977.

Vol. 575: K-Theory and Operator Algebras, Athens, Georgia 1975. Edited by B. B. Morrel and I. M. Singer. VI, 191 pages. 1977.

Vol. 576: V. S. Varadarajan, Harmonic Analysis on Real Reductive Groups. VI, 521 pages. 1977.

Vol. 577: J. P. May, E_∞ Ring Spaces and E_∞ Ring Spectra. IV, 268 pages. 1977.

Vol. 578: Séminaire Pierre Lelong (Analyse) Année 1975/76. Edité par P. Lelong. VI, 327 pages. 1977.

Vol. 579: Combinatoire et Représentation du Groupe Symétrique, Strasbourg 1976. Proceedings 1976. Edité par D. Foata. IV, 339 pages. 1977.

Vol. 580: C. Castaing and M. Valadier, Convex Analysis and Measurable Multifunctions. VIII, 278 pages. 1977.

Vol. 581: Séminaire de Probabilités XI, Université de Strasbourg. Proceedings 1975/1976. Edité par C. Dellacherie, P. A. Meyer et M. Weil. VI, 574 pages. 1977.

Vol. 582: J. M. G. Fell, Induced Representations and Banach *-Algebraic Bundles. IV, 349 pages. 1977.

Vol. 583: W. Hirsch, C. C. Pugh and M. Shub, Invariant Manifolds. IV, 149 pages. 1977.

Vol. 584: C. Brezinski, Accélération de la Convergence en Analyse Numérique. IV, 313 pages. 1977.

Vol. 585: T. A. Springer, Invariant Theory. VI, 112 pages. 1977.

Vol. 586: Séminaire d'Algèbre Paul Dubreil, Paris 1975–1976 (29ème Année). Edited by M. P. Malliavin. VI, 188 pages. 1977.

Vol. 587: Non-Commutative Harmonic Analysis. Proceedings 1976. Edited by J. Carmona and M. Vergne. IV, 240 pages. 1977.

Vol. 588: P. Molino, Théorie des G-Structures: Le Problème d'Equivalence. VI, 163 pages. 1977.

Vol. 589: Cohomologie l-adique et Fonctions L. Séminaire de Géométrie Algébrique du Bois-Marie 1965–66, SGA 5. Edité par L. Illusie. XII, 484 pages. 1977.

Vol. 590: H. Matsumoto, Analyse Harmonique dans les Systèmes de Tits Bornologiques de Type Affine. IV, 219 pages. 1977.

Vol. 591: G. A. Anderson, Surgery with Coefficients. VIII, 157 pages. 1977.

Vol. 592: D. Voigt, Induzierte Darstellungen in der Theorie der endlichen, algebraischen Gruppen. V, 413 Seiten. 1977.

Vol. 593: K. Barbey and H. König, Abstract Analytic Function Theory and Hardy Algebras. VIII, 260 pages. 1977.

Vol. 594: Singular Perturbations and Boundary Layer Theory, Lyon 1976. Edited by C. M. Brauner, B. Gay, and J. Mathieu. VIII, 539 pages. 1977.

Vol. 595: W. Hazod, Stetige Faltungshalbgruppen von Wahrscheinlichkeitsmaßen und erzeugende Distributionen. XIII, 157 Seiten. 1977.

Vol. 596: K. Deimling, Ordinary Differential Equations in Banach Spaces. VI, 137 pages. 1977.

Vol. 597: Geometry and Topology, Rio de Janeiro, July 1976. Proceedings. Edited by J. Palis and M. do Carmo. VI, 866 pages. 1977.

Vol. 598: J. Hoffmann-Jørgensen, T. M. Liggett et J. Neveu, Ecole d'Eté de Probabilités de Saint-Flour VI – 1976. Edité par P.-L. Hennequin. XII, 447 pages. 1977.

Vol. 599: Complex Analysis, Kentucky 1976. Proceedings. Edited by J. D. Buckholtz and T. J. Suffridge. X, 159 pages. 1977.

Vol. 600: W. Stoll, Value Distribution on Parabolic Spaces. VIII, 216 pages. 1977.

Vol. 601: Modular Functions of one Variable V, Bonn 1976. Proceedings. Edited by J.-P. Serre and D. B. Zagier. VI, 294 pages. 1977.

Vol. 602: J. P. Brezin, Harmonic Analysis on Compact Solvmanifolds. VIII, 179 pages. 1977.

Vol. 603: B. Moishezon, Complex Surfaces and Connected Sums of Complex Projective Planes. IV, 234 pages. 1977.

Vol. 604: Banach Spaces of Analytic Functions, Kent, Ohio 1976. Proceedings. Edited by J. Baker, C. Cleaver and Joseph Diestel. VI, 141 pages. 1977.

Vol. 605: Sario et al., Classification Theory of Riemannian Manifolds. XX, 498 pages. 1977.

Vol. 606: Mathematical Aspects of Finite Element Methods. Proceedings 1975. Edited by I. Galligani and E. Magenes. VI, 362 pages. 1977.

Vol. 607: M. Métivier, Reelle und Vektorwertige Quasimartingale und die Theorie der Stochastischen Integration. X, 310 Seiten. 1977.

Vol. 608: Bigard et al., Groupes et Anneaux Réticulés. XIV, 334 pages. 1977.